METAPHYSICS
AND THE
ORIGIN OF SPECIES

SUNY Series in Philosophy and Biology

David Edward Shaner, Editor

METAPHYSICS
AND THE
ORIGIN OF SPECIES

O

MICHAEL T. GHISELIN

STATE UNIVERSITY OF NEW YORK PRESS

Published by
State University of New York Press, Albany

© 1997 State University of New York

For information, address State University of New York Press,
State University Plaza, Albany, N.Y. 12246

Production by M. R. Mulholland
Marketing by Bernadette LaManna

Library of Congress Cataloging-in-Publication Data

Ghiselin, Michael T., 1939-
    Metaphysics and the origin of species / by Michael T. Ghiselin.
        p.   cm. -- (SUNY series in philosophy and biology)
    Includes bibliographical references and index.
    ISBN 0 -7914-3467-2 (hardcover : alk. paper). -- ISBN 0 -7914-3468-0
(pbk. : alk. paper)
        1. Evolution (Biology)--Philosophy.   2. Species--Philosophy.
I. Title.   II. Series.
QH360.5.G48   1997
576.8'01--dc21   96-38957
                                                                      CIP

10 9 8 7 6 5 4 3 2 1

For Ernst Mayr

CONTENTS

PREFACE

When, thirty years ago, I began to draw attention to the fact that biological species are individuals, I sincerely believed that I was merely drawing attention to a silly mistake. In biology 'individual' means one thing, in philosophy something else, and the two ought not to be confused. Imagine my perplexity when I found out that matters are not quite so simple as I had thought. Yes, I was told, the word is indeed equivocal, but the conclusion that species really and truly are individuals could not possibly be correct. Nonetheless it seemed to me that the conclusion followed with such deductive rigor from unassailable premises that I persisted in pressing my case.

Fortunately, David Hull, who had told me in no uncertain terms that I was wrong, soon discovered to his consternation that I had been right after all. He seized upon the opportunity and led a one-man campaign to convince his fellow philosophers not only that species are indeed individuals, but that it makes an enormous difference. Both he and I have, upon numerous occasions, taken the opportunity to develop and extend the "individuality thesis," and it is now recognized as a major contribution to the philosophy of biology, and to evolutionary theory as well.

Nonetheless there has been much dissatisfaction with treating species as individuals. Although it is fairly easy to explain the difference between a class, such as furniture or chair, and an individual, such as the chair upon which I am sitting as I write this book, making that distinction seems hardly adequate. Indeed, it has been justly pointed out that neither Hull nor I, nor, for that matter, anybody else in the history of the world, has adequately explained what it means to be an individual. And the same may be said, with equal justification, of all sorts of other things, including a species. It seems about time.

In rising to the occasion I must beg my reader's indulgence. Such topics are not explicable in isolation, and I find that in explaining one thing, I find myself driven to explain all sorts of other things. The result is an entire system of metaphysics. Admittedly, one might prefer certain alternative systems of metaphysics, but this is the one I came up with, and it seems likely that it or something like it will prove necessary for dealing with the task at hand, for we need something that will allow us to envision all that is knowable in science, and to do so from a unitary and coherent point of view.

I am responding in part to requests for clarification of my own position, seeking to explain many points that to me seem more or less obvious but may not seem so to everybody. Therefore I have felt it justified to emphasize my own point of view, and to argue against some alternatives that have been, or might be, proposed. Efforts on the part of others to do likewise would be most welcome. But the plain fact of the matter is that nobody has yet attempted such a task before. What various authors are thought to believe is largely a matter of guesswork, and the guesses are often wrong. There is always danger that even the most well-intentioned critic will see inconsistencies in an author who presupposes quite different premises, especially when such premises are not made explicit and clear. And given the common habit of treating knowledge as if it were a list of propositions, few are willing to argue the case for a thesis by examining the larger implications.

Those larger implications provide the most compelling evidence that the individuality thesis is true. It helps us to solve a vast range of philosophical problems in biology. Given that such is the manner of reasoning, the components must not be evaluated apart from their context within the whole. The reader who follows the common practice of reading just parts of books had better not waste such time on this one. To be sure, sections can be skimmed, and details passed over. Although one long argument is presented here, it is not a chain of deductions, but something more like a network of explanations. Much of what is said might be considerably revised or reinterpreted without seriously weakening the argument as a whole.

It gives me great pleasure to express my gratitude to a few of the persons who have made this work possible. Above all I wish to thank Ernst Mayr, who sponsored the postdoctoral fellowship at Harvard during the tenure of which the individuality thesis first came to mind. Down the years I have been much sustained by his friendship, which has in no way been diminished by our occasionally seeing things in a somewhat different light. Mayr has done more than anybody in the present century to raise the intellectual stature of systematic biology, and his contribution naturally takes a central place in this work. I should also insist, as I have elsewhere, that although David Hull and I have never coauthored anything, so far as the individuality thesis goes, I have always considered him my closest collaborator.

More generally, it is downright impossible to discuss difficult and contentious issues without disagreeing with all sorts of people. Down the years I have attempted to rise above faction, and it has not always been easy. Nonetheless it is noteworthy how many stimulating and cordial conversations I have enjoyed with colleagues whose views are very difficult to reconcile either with my own or with those of each other.

I am particularly grateful to those who have actually read drafts of the manuscript and provided me with useful advice: Tyrone Cashman, James Fetzer, James Griesemer, Mikael Härlin, David Jamieson, David Kay, Scott Merlino, Jack Wilson, and two anonymous referees.

Many thanks also to the staff of State University of New York Press, especially Clay Morgan and Megeen Mulholland, for expeditiously getting this book through the many stages of production. The Series Editor, David Edward Shaner gave precisely the sort of advice and encouragement that I needed.

Grateful acknowledgment is also due to the following authors and publishers for permission to quote material that is covered by copyright. Ernst Mayr has graciously allowed me to quote a passage from his essay "Darwin and Evolutionary Theory in Biology," in *Evolution and Anthropology: A Centennial Appraisal,* ed. B. J. Meggers (Washington: The Anthropological Society of Washington, 1959). The species definition of Mishler and Brandon, from Brent D. Mishler and Robert N. Brandon, "Individuality, Pluralism, and the Phylogenetic Species Concept," *Biology and Philosophy* 2 (1987), p. 406, is quoted with kind permission from Kluwer Academic Publishers and the authors. Quotations from J. S. L. Gilmour, "Taxonomy and Philosophy," in *The New Systematics,* ed. Julian Huxley (Oxford: Oxford University Press, 1940), pp. 464, 466, and 472, are reprinted by permission of Oxford University Press.

INTRODUCTION

"Origin of man now proved.—Metaphysics must flourish.—he who understands baboon would do more toward Metaphysics than Locke."

—Darwin, Notebook M, August 16, 1838

It happened after a high-table dinner at Saint Catharine's College in Cambridge, over port and madeira in the Senior Combination Room. Sydney Smith, the world's greatest Darwin scholar, was telling about having spent some time in the field with David Lack, and watching that famous ornithologist rappel down a cliff with a telescope strapped to his back. He remarked that "if you are a real ornithologist you don't carry binoculars, you carry a telescope." I turned to him and said, "Sydney, if you are a *real* ornithologist, you carry a shot gun!"

In the same spirit might I suggest that there are not many *real* evolutionists in this world. Merely admitting that evolution has in fact occurred does not make one an evolutionist, any more than believing in birds makes one an ornithologist. To be a real ornithologist, or a real evolutionist, requires a much deeper commitment, a sense of priority, and an enthusiasm that far transcends the ordinary professional loyalties.

Even a real ornithologist, of course, does not always go armed with a shot-gun, and not everybody who carries a shot-gun is an ornithologist, so we need not expect an evolutionist to bear some infallible emblem of his identity. But consider how one might answer some diagnostic questions. When, for example, did the dinosaurs become extinct? A real evolutionary biologist will tell you that they never did become extinct. They are still flying around in the trees, to the delight of mere bird-watchers and real ornithologists alike. Which came first, the chicken or the egg? The egg, of course. Ova and sperm evolved a billion years or so before chickens did. When does human life begin? Never, for it is part of an unbroken series of generations that goes back to Darwin's warm little pond. What does evolution teach us about human nature? It tells us that human nature is a superstition.

To dismiss such answers as hyperbole or persiflage would completely miss the point. Our evolutionist does not abandon his commitments or compromise his fundamental principles simply because his interlocutors might have expected a different answer! The questions are loaded. They presuppose, if not the falsehood, at least the irrelevancy, of a thoroughly evolutionary point of view. A real evolutionist rather starts out with the evolutionary question, and insists upon providing the evolutionary answer.

One may hesitate to use the term 'anti-evolutionist' for someone who simply is not particularly interested in evolution, or whose enthusiasm for it is merely lukewarm. And yet what passes for indifference may actually be antipathy, and sometimes the neglect that seems benign may be downright malevolent. If it is reasonable to call somebody who believes that the world was created in six days an anti-evolutionist, it is hardly unreasonable to apply the same term to somebody who wants others to believe that it was so created—whatever he himself may believe.

All sorts of persons maintain that teaching about evolution, especially evolution by natural selection, is apt to corrupt the morals of the youth. Often enough their motives are political rather than religious. The subject can be ill-taught, and hence removed from public attention, merely by not explaining how evolution makes life intelligible. And getting it out of the curriculum in the universities these days is largely a matter of hiring yet another molecular biologist. Anti-evolutionism is at least partly responsible for the failure of our educational system to train a new generation of systematists at a time when we face a "crisis" in biotic diversity.

Because modern systematics emphasizes the historical aspect of biology, anti-evolutionists have been particularly assiduous in their efforts to suppress it. They may seek to trivialize it by depicting systematics as a sort of mindless cataloging. Or they claim that although phylogenetics has a certain limited interest as an intellectual curiosity, and perhaps was worth attending to in the nineteenth century, what happened in the Paleozoic is purely a matter of speculation. We have no "direct" knowledge of the past, and since we cannot experiment upon it, paleontology is no real exception.

When the history of systematics is depicted from an anti-evolutionary point of view, every effort is made to give the impression that nothing really happened to it when Darwin came along. If Darwin still used the Linnaean hierarchy, and if a few early nineteenth-century authors drew branching diagrams, it is deemed justified to write off the differences as trivial. The notion that no such changes have occurred, even to the present day, has not been a thesis to be defended upon empirical grounds. Rather it is an assumption that is pre-supposed dogmatically, and one that derives from the anti-evolutionary attitudes that ought to have been challenged and rejected. It is about time for serious students of the past to repudiate such Tory history.

The Swedish literary historian Ellegård (reprinted 1990) surveyed the British periodical press of Darwin's day, and documented a whole range of enthusiasm or the lack of it for his views, ranging all the way from biblical literalism to accepting not just evolution, but natural selection as well. Selectionists, especially enthusiastic selectionists like Fritz Müller and Alfred Russel Wallace, represented but a small minority, and real understanding was far less widespread than it might have been. Things at present remain far from satisfactory.

We need hardly dwell upon details that have become so widely known and are so readily accessible in standard histories of evolutionary biology (e.g., Mayr, 1982a, 1991b; Bowler, 1984; Becquemont, 1992). It is notorious that although the general concept of evolution became widely accepted within the decade that followed publication of *The Origin of Species,* natural selection did not. Natural selection was crucial in providing a plausible mechanism, but its very success in that respect helped pave the way for various alternatives. Shortly after Darwin's death in 1882, a reaction set in, not so much against evolution as such, but against natural selection. This situation was exacerbated in the early part of the twentieth century by developments in genetics, which seemed out of line with natural selection. Or so the story goes, though one always wonders whether a reason is more than just an excuse. As what at least appears to have been a consequence of such difficulties, there was a long period during which "Lamarckian" theories, orthogenesis, saltationism, and other alternatives to natural selection were widely advocated.

The emergence of theoretical population genetics, thanks especially to the brilliant work of Chetverikov, Haldane, Fisher, and Wright in the twenties and thirties, set the stage for the emergence of the so-called Evolutionary Synthesis. At any rate it helped. The apparent conflict between genetics and natural selection was seen to be illusory, and there followed a unification of various separate disciplines into what came to be called the Synthetic Theory. The work of Dobzhansky was particularly important because he was a geneticist with a sound grasp of systematics and geographical variation. Mayr (1980) emphasizes the importance of systematists, and his work on animal speciation was one of the most important parts of the Synthesis. Stebbins did much the same for plants, and Simpson gets credit for important paleontological contributions. Other major figures were Schmalhausen, Rensch, and Julian Huxley. In addition to these "architects" as they are called, many others were involved who were far more than just hod-carriers.

I have often entertained, and even publicly expressed, serious doubts as to whether the Synthetic Theory was even a theory, let alone a synthesis. What actually seems to have happened was that the alternatives to natural selection became less and less tenable, and finally lost all credibility, so that the only alternative was natural selection, and that meant a somewhat modernized version of Darwin's original theory. But the sort of unification that is suggested by the term 'synthesis' implies that there were stronger links between the parts than the evidence really seems to justify. This lack of unification has lately been noted by Futuyma (1988) and Mayr (1992a).

The body of knowledge in question did have enough conceptual unity and diversity of content that calling it the Synthetic Theory is, if somewhat hyperbolic, hardly a misnomer. Even so, the pedagogical tradition that has arisen with respect to its historiography seriously distorts matters. The student is told that Darwin's theory was defective because he did not understand genetics, and that genetics came to the rescue. This is hardly what happened. Although genetics was important, it has not played the particular role that has generally been attributed to it. Insofar as the species is concerned, the modern definition that emerged from the Synthesis was often discussed in a genetical context, but the basic idea was that of a reproductive community. Conceptualizing them as genetical populations between which gene flow does not take place may have helped keep the discussion within the mainstream of the discipline, but that does not mean that the Synthesis produced a genetical theory of speciation. Indeed, as Coyne (1992) has emphasized, we still lack anything that really deserves to be called a genetical theory of speciation. No doubt we would be better off if we had one, though how much better off is hard to say.

Darwin was quite successful in creating what remains to this very day our basic evolutionary theory without any knowledge of Mendel's laws or the principles of modern population genetics. Even today, if one wants to understand all sorts of phenomena, one need only ask who out-reproduces whom and when, and that is precisely what Darwin did. The supposed problems with Darwin's theory, such as "blending" inheritance and the inheritance of acquired characteristics, were the consequence of misinformation, not lack of information, about genetics. The problems at hand were soluble without any genetical theory, even though a good one might have helped.

The objections to Darwinism that arose soon after the rediscovery of Mendel's laws were likewise due to misunderstandings on the part of geneticists themselves about genetics, and not due to something having been wrong with natural selection. We must not err in the opposite direction and say that genetics did not help, or that it did nothing more than repudiate its own errors. Genetics made lots of problems easier to solve, and the solutions themselves more compelling. And it provided some very good materials that had not been available before. For instance, if you want a good example of how to do phylogenetics, just consider what has been accomplished using the chromosomes of *Drosophila* and processing the data in exactly the same way that comparative anatomists have (Dobzhansky and Sturtevant, 1938). For a real conceptual novelty, one needs something without clear precedents. There are very few important examples. Genetic drift is one of these. How many more might be listed is hard to say. So without denigrating genetics or denying its legitimate role, we had better seek elsewhere to understand what has been going on.

Instead, let us attack the problem from a very different direction and turn to systematics and the philosophy of classification. The Darwinian revolution may then be seen to have been resisted, not just because it was hard to conceive of how species might get transformed by natural selection, but because it was so difficult to conceive of change in general. And rather than focus upon Darwin's lack of a proper theory of heredity, let us emphasize his lack of a proper species concept. For him, species had not yet taken on the sort of theoretical importance that they have in modern speciation theory. The fundamental units of classification had not yet become the fundamental units of evolutionary theory. Darwin was indeed the great reformer of taxonomy, but this was because he advocated strictly genealogical classification, and not because of anything he added to our concept of species beyond the fact that they give rise to genealogical lineages.

The Synthesis made a positive advance, insofar as it developed a new conception of the species, but it retrogressed insofar as it waffled and compromised with respect to the higher categories. Continued resistance, both to genealogical classification and to the new conception of species, is symptomatic of failure to attain the sort of unity that ought to be provided by good philosophy. Although Darwin's original theory was neither perfect nor complete, it was comprehensive and unitary: it envisioned the entire cosmos from a single point of view. And it was more than just consistent. It was coherent, in the sense that its various elements were interconnected so that they lent one another mutual support.

Darwin's theory was one, not many, and the very lack of such unity is what makes the term 'Synthetic Theory' somewhat misleading. One might even go so far as call it the "Syncretic" rather than the "Synthetic" theory. The term 'syncretism' means a flagrant and illogical disregard for basic principles, and was rightly applied by Hennig (1966) to what was then common practice in systematics above the species level. Organisms were classified according to an arbitrary and capricious mixture of genealogical relationships on the one hand, and something called "similarity" on the other. Consequently systems of classification were uninformative and downright misleading to anybody who wanted to use them for scientific research. At the species level, the term 'pluralism' is used instead of 'syncretism,' and it is used in an honorific sense, rather than a pejorative one, even by some professed followers of Hennig, for a similar laxity of standards with respect to principles. The effect is the same. So many different kinds of things, both real and imaginary, can be called "species" that figuring out what, if anything, is meant when this term is used can be pure guesswork.

The differences between the old ways of thinking and the new are not always apparent. When one reads old books, or for that matter new books on an unfamiliar topic, it is easy to deceive oneself into thinking that one has understood what the author meant when one really hasn't. We are more than just tempted to use words in the sense to which we have been accustomed. We do not always ask whether the author presupposes exactly what we presuppose. Therefore when we go back to writings of Darwin's predecessors and contemporaries, we may fail to appreciate the difference between his view of things and those of his contemporaries. If we see, for example, a branching diagram, we might wrongly interpret it as a phylogenetic tree, and get the false impression that the author believed in evolution. The word 'evolution' itself is tricky, for all too often it has meant "ontogeny" rather than "phylogeny." Upon the basis of such considerations we can cross a lot of names off our list of early "evolutionists" and so-called "precursors" of Darwin—and some later figures as well.

But there is a profound, albeit not always obtrusive, difference between what Darwin and such quasi-evolutionists had in mind. For Darwin, change was real. It was a fundamental aspect of reality as he envisioned it. It was not something that could be glossed over or dismissed as superficial or trivial. His predecessors had indeed moved in the direction of imposing an historical interpretation upon their traditional perspectives, but it was hardly a matter of thinking in evolutionary terms as Darwin understood them, let alone as they are understood by a few of us today. They were still trying to make sense out of Darwin's accomplishment by forcing it to conform to the very assumptions that Darwin had deprived of their legitimacy. And those assumptions have continued to be presupposed in efforts at coming to grips with Darwin's accomplishments, especially from a philosophical point of view. We have, for example, a vast literature on the "form-function" controversy. The usual approach is to contrast the Platonist Geoffroy with the Aristotelian Cuvier, and ask who was "right." How often are we given Heraclitus and Darwin as the answer?

For Darwin, as for Heraclitus, change was the fundamental reality. It was not something superficial, or illusory, or something to be explained away. Ancient Greek philosophers treated change as a very serious problem indeed, and many features of their systems were designed to cope with it (Popper, 1945; revised edition 1966). Although we have learned a great deal in the last two thousand years, the problem is still with us. And so are its mostly unsatisfactory solutions.

Plato dealt with the problem of change by positing a timeless and unchanging ideal world that was populated by eternal objects. The particular things that we encounter through daily experience are but imperfect copies of "ideas" or "forms" that supposedly exist in the ideal, or as it came to be called, "archetypal" world. The object of knowledge was not particular things, which were deemed inferior, but the ones in the archetypal world. Because the archetypal world was to some extent conceptualized as a world of ideas, and because it had a sacred character, the attainment of knowledge was pursued as a kind of divine mind-reading. Plato was a follower of Pythagoras, and therefore it was perfectly reasonable for him to take mathematics as the ideal for knowledge in general. If we look upon the inhabitants of the archetypal world as things like "the triangle" and "the circle," then the timeless and immutable character of reality thus conceived makes a lot of sense. After all, the triangle will always be the triangle, not the square.

In a famous passage in his dialog *Meno,* Plato has Socrates interrogate a slave boy and lead him to prove a theorem in geometry. This Plato takes as compelling evidence that learning is really a kind of memory. Here we have a fine example of somebody seeming to have changed, but the change is, if not downright illusory, at least superficial. There is a profound difference between remembering something that one already knows and the active process of making what might be called a discovery, even if it be not an original one. In his dialog *Timaeus,* Plato presents a sort of creation myth in which the earth is compared to a divine animal, which was created. Plato's own views on such matters are of little concern in the present context, so if the foregoing account strikes the erudite as a bit oversimplified it doesn't really matter. The point is merely that systematic biology has been profoundly influenced by Platonism. If one believes that organisms are somehow "modeled" upon the inhabitants of some timeless and ideal world, then one hardly qualifies as an evolutionist, even if one admits that different forms have been manifest at different times.

Plato's forms and things like them are often called "essences," and the term 'essentialism' (often called "typology") is a general rubric for the doctrines, thought-habits, and other things that go along with the acceptance of essences. Rather than attempt to give a precise definition to this term, a task that seems downright hopeless given the vagueness and diversity of usage that we might have to deal with, it seems better to develop the topic of essentialism as we proceed, and come back to it from time to time as the arguments unfold.

As is well known, Aristotle did not endorse the notion of an archetypal world. Although he did invoke "essences" that functioned much like Plato's Ideas, he did not believe that they exist apart from particular things. In Scholastic jargon, he treated them as immanent rather than transcendent. They remained, however, the basic object of theoretical, or we might even say scientific, knowledge. To understand a thing was to know its essence. The essences, however, retained their timeless and unchanging character. Individual men might change, but for the class "man" to change would not even have made any sense. Species could not originate. Furthermore, and in some ways this is even more important, Aristotle did not believe that the universe itself originated: it is eternal.

Aristotle's universe did allow for a certain amount of change, but only within distinct limits. Much of the work of providing for a world that was basically static was accomplished by means of cycles. For example, what we might call a "steady-state" universe was accomplished by circular motion. The stars might move around a bit, but they always come back to the same place. In the *Meteorologia* Aristotle shows how the water moves from land to sea, to air and back to the land again; the erosion and deposition of sediments is treated from the same point of view. So too are the life cycles of animals. The same thing keeps recurring with perhaps minor variations upon the same basic theme.

The kind of geology that Darwin practiced was profoundly affected by revisionist Aristotelianism. It was more or less explicit in the writings of early uniformitarians. Lyell, whom Darwin esteemed as a major role model, advocated a "steady state" model for geology (Lyell, 1830, 1832). So too did Darwin at times, with his elevation and subsidence of the continents (Darwin, 1842, 1846). Lyell long denied that there is any evidence for progressive change in the fossil record, and his stubborn resistance to evolution in general and natural selection in particular fits the same pattern. So although one might think that research upon historical geology would be strongly conducive to evolutionary thinking, that impetus was far from sufficient.

When pre-Darwinian biologists were entertaining the possibility of something like evolution, they were often misled by the sort of change that is undergone by a developing embryo. They had various ways of conceiving how and why embryos develop, ones that were largely incorrect, and the changes they had in mind were bound to be superficial and more apparent than real. To understand their mind-set, just think about how we now understand the development of an embryo. It would not happen at all were it not for the developmental "program" that exists in the egg and the zygote and provides a set of instructions that guide embryogenesis along stringently restricted lines, and which, albeit modified somewhat by mutation and recombination, is not much altered from generation to generation.

Now, if we attempt to explain the last few billion years of phylogenetics upon the assumption that the ancestral prokaryote contained the "blueprints" of all the subsequent radiations of organic beings that have ever taken place, we run into some theoretical difficulties. We could simplify matters just a trifle by denying cladogenesis (branching), and running every species as a single, autonomous lineage all the way back to the *Urschleim*. Nonetheless, the "blueprints" themselves would not change, and that is what makes such thinking just quasi-evolutionary. Furthermore one has to account for the initial existence of each lineage, and to endow the successive forms that are produced with the capacity to deal adaptively with future circumstances, down to the last minute adjustment between flowers and their pollinators. Given an omnipotent Creator, that was perfectly acceptable to such luminaries as Richard Owen (1849:86), in whose words "the Divine Mind which planned the archetype foreknew all its modifications." The necessity of invoking supernatural agency here is all too obvious, but we do need to emphasize once more that the reality of change has been denied. Such "preformationist orthogenesis" is fundamentally different from what we moderns call "evolution."

One alternative is "epigeneticist orthogenesis," which has its roots in another misconception about embryology. Rather than think of God as a Divine Architect stuffing blueprints into primordial ova, think of Him as a Divine Legislator ordaining Laws of Nature that determine the properties of organic matter. One would rather compare the developing embryo to a growing crystal, with its form determined by physical forces. From here it was a minor bit of extrapolation to make the laws of nature do a little bit more work and have them determine what goes on upon a geological scale as well. No matter if nobody had the foggiest notions what such laws were supposed to be, or in what particular manner they were supposed to operate! Thomas Henry Huxley, whose career was virtually one long series of battles with Owen, initially presupposed that some kind of physico-chemical laws would explain organic diversity. That evidently explains a lot about Huxley, not the least the surprise he experienced when Darwin came up with the solution, and very likely his preference for evolution by leaps or saltations. He never did come out and advocate an epigeneticist orthogenesis, but we have plenty of examples of biologists who did, including Owen later in his career.

We will have more to say about epigeneticist orthogenesis later on, for to this very day it remains one of the most popular forms of anti-evolutionary thinking. For the moment, however, we need to emphasize, once again, that it treats change as if it were not real. The various "species" of crystals differ from the species of evolutionary biology in a most fundamental manner. There is nothing historical about them. To be sure, any given mineral crystal that you can pick up and hold in your hand has a history, a location, a beginning, and an end. But there is nothing fundamentally different about the crystals of calcite that formed in the Cambrian from those that are being formed today. The laws of nature that determine their structure have not changed in the least. Calcite is calcite, it always has been, and it always will be, for ever and ever, everywhere. A Cambrian mollusk may have incorporated calcite in its shell, and so too may its descendants at the present day, though some of them might have shells of calcite, others of aragonite, yet others no shell at all. Although it makes perfectly good sense to say that the mollusks in question have speciated, and otherwise evolved, it would be utter nonsense to say that one mineral "species" has proliferated or evolved into another.

For some of us moderns it may be difficult to imagine how an advocate of some epigeneticist version of orthogenesis would be able to make sense out of the fossil record and out of the data of comparative biology in general. But that is mainly because we reason from different assumptions and because certain alternatives do not occur to us. If one does not assume that the various groups of organisms are related to one another by descent, but instead that they come into being by special creation, by spontaneous generation, by the fortuitous concourse of atoms, or by natural or supernatural means that are unknown and perhaps unknowable, then the problem has a ready solution. Just as different kinds of crystals precipitate under different conditions, so too might different kinds of organisms come into being in analogous fashion. Given local differences in temperature, or a gradual cooling of the earth through time, it stands to reason that different kinds of organisms would predominate at different places and at different times. But this would not be evolution. Rather it would be the sort of change that happens when winter sets in and what falls is no longer rain, but snow.

Now, if one wishes to propound an epigeneticist version of orthogenesis, in which the present inhabitants of the globe are literally the descendants of the earlier ones, things obviously become a bit more difficult. One has to think of the changes that are undergone within lineages as analogous to the stages through which a crystal passes as it grows. The analogy becomes all too forced, especially if one tries to allow for the splitting and proliferation of lineages. But the result is only quasi-evolutionary again, for inevitably the changes that take place are superficial, at least in the sense that nothing really new has come into existence. The laws of nature that are responsible for the organisms having the properties that they do remain unchanged.

So the mere fact that someone believes that things seem to be different now from what they were in the past hardly justifies calling him an evolutionist. And it seems to me that the application of that epithet to all sorts of people has been anachronistic and grossly misleading. It would be interesting to reconsider such "forerunners" of Darwin as De Maillet and Buffon from this point of view, but doing that here would divert attention from the tasks at hand. Nonetheless it deserves emphasis that virtually everybody seems to take it for granted that persons like Lamarck and Herbert Spencer were evolutionists, and of course, in a certain sense, they were. But the very failure to realize how different their views were from those of Darwin can only serve to illustrate how fundamental were the differences. Darwin was contemptuous of both Lamarck and Spencer, and for the best of reasons, as can be seen from the correspondence and the unexpurgated autobiography (Barlow, 1959).

Darwin was much vexed by Lyell's persistent incapacity to distinguish Lamarck's theory from his own, as is clear from their correspondence with respect to *The Origin of Species*. Lyell found it far easier grasp Lamarck's theory than Darwin's largely because of Darwin's fundamentally different outlook upon change. For Lyell and for Lamarck the mutability of species meant that they were not "real"—not the sort of fundamental units that numbers are (see Coleman, 1962). Lamarck conceived of life originating spontaneously, as a consequence of physico-chemical laws, then heading in a particular direction, namely, toward Man. His position is a bit complicated by the fact that he allowed for a certain amount of deviation from the straight path by virtue of the fact that animals diversify upon an ecological basis. Nonetheless, his was definitely a version of epigeneticist orthogenesis. Lamarck seems to have allowed for a certain amount of branching, but his branching diagrams were not real genealogies and did not function as such in his research. In at least one important respect, however, Lamarck's views agreed with Darwin's. The organism was conceived of as playing an active role in the evolutionary process. Behavioral evolutionary ecology owes a great deal to Darwin, and, in one sense, through Darwin to Lamarck. But that is not what is generally thought to have made Lamarck, or, for that matter, Erasmus Darwin, an "evolutionist."

Herbert Spencer is also widely perceived as an important evolutionist, especially by historians who write from an anti-evolutionary point of view (Greene, 1977). Of course he was very influential, but the very fact that such interpretations are forced upon Spencer's views merely serves to underscore how badly understood have been Darwin's. Having read all of Spencer's collected essays and his entire *Synthetic Philosophy* from beginning to end, I get more than a mere impression that he was just another advocate of orthogenesis upon a cosmic scale (Ghiselin, 1974a). He treated human society and the world biota alike as the analogue of a developing embryo. And if the various groups of plants and animals diversified, they did so in a manner that is analogous to the specialization of tissues within the body of an organism. In his sociopolitical writings the message was clear: we should not interfere with the course of nature during the genesis of either embryos or societies. In his earlier writings, Spencer (e.g., 1850) had God take a more active role in arranging things, whereas later more of the burden was laid upon the laws of nature. The theology fades into the background, and some contingency was admitted, but the "evolution" remained orthogenetic.

Darwin, as should be much better known, did indeed believe that there is an important relationship between embryology and evolution in the modern sense of that term. He thought that the properties of the developing embryo impose all sorts of constraints upon the course of evolutionary history. Variation, the "raw material" of evolution, resulted from modifications in developmental processes. Because the developmental processes could be modified in some ways but not in others, there was an order to the evolutionary process over and above that which results from selection alone. Darwin by no means believed that selection is "random" in the sense that all mutations are equally probable. All this is clearly spelled out, and in great detail, in his book *The Variation of*

Animals and Plants under Domestication (Darwin, 1868, second edition 1875). The details need not concern us here, for what really matters is what differentiates Darwin from his predecessors. Namely, he believed that what causes the organisms to develop as they do is something that itself evolves. In modern parlance, again, some persons would say that the "genetic program" evolves. Neither preformationist nor epigeneticist orthogenesis would allow for this. If the developmental processes change, then we have to attribute the properties of the organisms to historical contingencies, and not to laws of nature as envisioned by the advocates of orthogenesis.

Having explained some basic ideas about evolution, let us now consider how they relate to philosophy. The distinction between "real" and "apparent" change, or between "real" and "superficial" change may be said to be a "metaphysical" one. Discussing it at some length has made it possible to avoid having to define the term 'metaphysics' in abstract terms. It seemed easier to present a little essay on a philosophical topic, and let that stand as an example, and a contextual definition, of the term 'metaphysics'. Metaphysics deals with such topics as the "what," "how," and "why" of reality. The example of the reality of change suggests that metaphysics is something of great importance to evolutionary biologists, probably much more important than most of them realize. Even for those readers who have a strong background in philosophy, this exercise may not have been altogether superfluous. It provides an introduction to the kind of issue that will be treated in later chapters, and may help to avoid "metaphysics" being taken in a sense that was not intended. (Emphatically not in the sense of "occult metaphysics" that fills the shelves of what are called "metaphysical bookstores.")

Metaphysics is usually considered a branch of philosophy, equal in autonomy—perhaps even in dignity—to ethics, aesthetics, epistemology, and logic. Here, however, it will be treated as one of the natural sciences. Indeed, it will be treated as the most fundamental among them, and in some ways the most important. To take this step, and thereby in a sense to redefine metaphysics, is neither more nor less radical than my earlier efforts to redefine that supremely fundamental category of systematic biology: the species. Indeed, it would seem to be little more than carrying an investigation to its logical conclusion. If a systematic biologist is justified in revising a genus of mollusks, why not the entire Kingdom Animalia? Why stop with the animals? Why not revise everything? At the very least, the exercise might prove instructive, especially in telling us about classification in general.

Some readers may perhaps respond by saying that classification is not particularly interesting. It is common knowledge and intuitively obvious that classification schemes are merely matters of convenience and do not tell us anything really important. So, for example, my suggestion that metaphysics is a natural science is like deciding to shelve my books by size, rather than by author or subject. It should be obvious that such an argument begs the question: it presupposes that which is at issue. If we consider the larger issues of classification in a scientific spirit, we come to quite different conclusions.

I stress this point if for no other reason that in the past some of my philosopher critics have taken it for granted that I am even less enterprising and innovative than they themselves. As they see it, the way to do what is called philosophy of science is to learn what professors of philosophy say, and use that supposed wisdom to cast light upon what goes on in the laboratory. But one does not have to read much philosophical literature to learn that philosophers have been wrong in the past, and that they disagree among themselves at present. Why not treat philosophical ideas as hypotheses? Why not develop their implications and test them by seeing how well they allow us to deal with the data of scientific experience? If one does that, one is behaving like a natural scientist. Indeed it is not always easy to draw a clear line between philosophy and science. The two are inseparable, and if they are different, the distinction is not a matter of academic convention.

My investigations on species illustrate what I have in mind. As a young comparative anatomist trying to develop a better phylogenetic tree of a subclass of gastropods, I became concerned about the methodological controversies that had lately begun to rage about me. The knowability of the topic of my doctoral dissertation itself was subject to skeptical attack. Partly in order to defend myself, I ransacked the philosophical literature, and did not give up simply because so much of it was of very little use. Among other things I learned how to deal with equivocations, and one of these that struck my attention was the two senses of the word 'individual'. In biology, 'individual' is usually synonymous with 'organism', as it is in everyday life. In metaphysics and logic, it has a more general sense, namely a particular thing, including not only an organism like Fido or me, but a chair, the Milky Way, and all sorts of other things. The chair upon which I sit as I write this book is obviously an individual, and it is a member of the class of chairs. I cannot even imagine what it would mean to sit on "chair" in the abstract, nor have I ever heard of anybody having done so. The distinction between classes and the individuals which are said to be their "members" is not, therefore, particularly difficult. Indeed, the particular chair that I sit on is so familiar a part of daily life, that we all take it for granted. But since one cannot sit upon the chair in the abstract, such general terms as "chair" have seemed quite a puzzle. Therefore a lot of attention has been paid to the so-called "problem of universals" whereas the particular things have been neglected.

The literature devoted to the "species problem" includes a lot of discussion about the so-called "reality" of species. According to one very popular philosophical notion, nominalism, individuals are "real" but classes are not. This makes a certain amount of sense: a nominalist would say that this chair is real, whereas 'chair' in general or in the abstract is not. Nominalists generally go somewhat further than that, and might even say that chairs share nothing at all except the name 'chair'—hence the etymology. One does not have to be a nominalist to admit that "chair" and the one on which I sit do not have the same ontological status. In other words, one does not have to go so far as they do and deny that "chair" in general has some kind of "reality"—whatever that is supposed to mean.

In the controversies that surround the species problem, one position that has been taken is the so-called nominalistic species concept. According to this view, species are classes, classes are not real, therefore species are not real. Consequently, perhaps, they are mere conventions and have no role to play in biological thinking. The traditional response to this kind of nominalism was to deny nominalism itself, and take a "realistic" view of species. Species are classes, and they are real, so there was no problem. But the nominalistic argument made a certain amount of sense: if species are classes, how could they evolve or become extinct? It would be like sitting on chair in the abstract. So I turned the problem on its head. Species are not classes; they are individuals. The truth of the solution seemed to me self-evident, one that followed from the definitions of 'species' and of 'individual'.

At this point the impatient reader might call for a precise and simple definition of the term 'individual' and perhaps of 'class' and 'species' as well. I can only answer that one cannot have it both ways. To be sure, I might take a hint from Aristotle and say that a class is a "such" and an individual a "that" or from Dobzhansky and say that species are the largest Mendelian populations. But such definitions do no more than serve as guideposts. If you really want to know what a word means in scientific discourse, such terse little formulae as one encounters in glossaries not only do not suffice, they can be downright misleading. This follows from the very nature of definitions themselves, a point which will be developed in later chapters.

I pointed out that species are individuals in a paper that I submitted to *Systematic Zoology* in 1965 and was published the next year (Ghiselin, 1966b). It was sent for review to David Hull, the leading philosopher of taxonomy, and he hit the roof. A long series of letters followed. He remained unconvinced, but I left the manuscript as it was. I did, however, try to explain matters more fully in my book *The Triumph of the Darwinian Method* (Ghiselin, 1969d:53–54):

> It should thus be clear that metaphysical preconceptions profoundly influence the course of scientific investigation. The effects of divergent philosophical points of view are readily seen in attitudes toward species. Aristotelian definition leaves no room for changes in properties. In a sense the species is the set of properties which distinguish between the individuals of different groups. If species change, they do not exist,

for things that change cannot be defined and hence cannot exist. A Platonist, by way of contrast, is interested only in the ideal organism, and ignores the individual differences which are so crucial to an understanding of changes in species by natural selection. If one is a radical nominalist, species cannot exist in principle, and any study of change must treat only individuals. It is possible to accept species as real and still embrace a kind of nominalism, if one looks upon species as individuals. Buffon (1707–1788), for example, would seem to have entertained the notion that a species is a group of interbreeding organisms. This point of view has certain analogies with the biological species definition of the modern biologist: "Species are groups of actually or potentially interbreeding natural populations, which are reproductively isolated from other such groups." A species is thus a particular, or an 'individual'—not a biological individual, but a social one. It is not a strictly nominal class— that is, it is not an abstraction or mere group of similar things—because the biological individuals stand in relation to the species as parts to a whole. To attain this divergence in attitude required more than simple affirmation. It was necessary to conceive of biological groupings in terms of social interaction and not merely in terms of taxonomic characters.

The new manner of thinking about groups of organisms entailed the concept of a population as an integrated system, existing at a level above that of the biological individual. A population may be defined as a group of things which interact with one another. A group of gaseous molecules in a single vessel or the populace of a country form units of such nature. Families, the House of Lords, a hive of bees, or a shoal of mussels—in short, social entities—all constitute populations. The class of red books, thanatocoenoses, or all hermaphrodites do not. The best test of whether or not a group forms a population is to ask whether or not it is possible to affect one member of the group by acting on another. Removing a worker from a hive of bees, for example, should influence the number of eggs the queen may lay, and therefore both worker and queen are parts of the same population. The concept in use is often exceedingly abstract. If one is to conceive of a population at all, one must use a logic in which the relationships are not treated in terms of black and white. The emphasis must be placed on dynamic equilibria and on processes. The interaction between the parts may be intermittent, dispositional, or even only potential. Entities are components of the same population to the degree that their interaction is probable. That the definition of 'population' allows for no distinct boundary between populations and nonpopulations creates difficulties for traditional logic which can only be surmounted with some effort.

Hull still remained unconvinced, even going so far as to take a public stand against the individuality of species (Hull, 1969). However, his own investigations soon led him to realize that species function as individuals in evolutionary theory. He changed his mind as a result of reading a book by J. J. C. Smart (1963), who claimed that biology is unscientific, because there are no laws for such entities as *Homo sapiens*. But Hull realized that the reason why there are no laws for *Homo sapiens* is the same reason why there are no laws for the Solar System or the Milky Way: these entities are all individuals, and there are no laws for individuals in any science. Hull spent years getting the point across to philosophers and biologists alike, and made the individuality of species the nucleus of a very productive research program. Perhaps the most important outcome of this program was a book in which he treated evolving concepts as individuals in the history of systematic biology (Hull, 1988b).

Except for Hull, nobody seems to have paid any real attention to my efforts to explain the individuality of species before the publication of my paper entitled *A radical solution to the species problem* (Ghiselin, 1974c). This is generally considered the starting point for subsequent discussion on such matters, even though, as Mayr (1987b; response in Ghiselin 1987e) has pointed out, there were many precursors, both real and imagined. The reaction was a mixture of entrenched and determined resistance in some quarters, a bandwagon effect in others. In general, heresy has evolved into consensus. However, there has been much discussion, especially of the broader issues, and people continue to explore alternatives. That is what has led me to write this book.

It began as a little essay responding to the commentaries that had been appearing in the literature. The manuscript grew all out of proportion. I found that there was no way that I could explain anything without explaining just about everything. For one thing, nobody seems to have given a really satisfactory explanation of what it means to be an individual. For another, many implications of the individuality thesis remain unexplored, or at least inadequately appreciated. Darwin's book *On the Origin of Species* said very little about the origin of species, but it said a lot about all sorts of other things. This one says a lot about species as well as about metaphysics. The title alludes to those of earlier works by Dobzhansky and Mayr, which seems fitting insofar as it aims at a metaphysical synthesis that continues the tradition of its justly illustrious predecessors.

Some persons may object to my using the term 'metaphysics' in the title, and for much the same reason that they may react negatively when they encounter 'individual' in an ontological sense. Perhaps they are unfamiliar with anything but "occult" metaphysics, such as astrology and kindred mysticism. Or someone told them that scientists are not supposed to be involved in such things. But it seems better to raise the world's consciousness of truth, rather than to perpetuate misunderstanding. After all, in 1859 the very idea that species might originate, especially through some natural process, seemed downright paradoxical.

A glance at the Contents may suggest the structure of this work and see how it constitutes an integrated whole. We have already begun to define 'metaphysics' and otherwise to explain what the book is all about. The chapter that follows explains a lot of basic metaphysical concepts, and Chapters Three and Four explain what an individual is and is not. Chapter Five deals with the various kinds of definitions. At last, in Chapter Six we are in a position to define such terms as 'hierarchy' and 'species' and to explicate the biological species concept, then proceed, in Chapter Seven, to examine its alternatives. In Chapter Eight we rebut all sorts of objections to the thesis that species are individuals.

Having explained the individuality thesis and shown how it applies to species, puts us in a position to further justify it by showing that it has all sorts of important, and largely unanticipated consequences. The rest of the book explores the broader implications, beginning, in Chapter Nine, with language and other cultural entities. Then we see, in Chapter Ten, how some controversial issues in evolutionary theory might be clarified. Chapters Eleven through Thirteen address some long-standing issues in philosophy of systematics, such as what is meant by such terms as 'character' and 'homology'.

The individuality thesis has led to important new insights about the role of laws of nature and that of history in biology and in other sciences. These implications are addressed in Chapters Fourteen and Fifteen, then illustrated by means of a discussion on embryology in Chapter Sixteen. Further implications, having to do with macroevolution and the fossil record, are sketched out in Chapter Seventeen. In the final chapter, it is argued that the historical narrative aspect of evolutionary biology is far more important than has generally been recognized.

An Appendix encapsulates many of the more important points. It can be used to get an overview of the work. With the help of the index, or by just scanning for words in **boldface,** it can also serve as a kind of glossary.

Such a look at the work as a whole should make it abundantly clear that its subject matter extends far beyond the topic of what a species is. Classification, contrary to what its etymology suggests, is not just the making of classes, or the product of that process. Rather it is the organization of knowledge. Therefore the revolution that is going on in systematic biology has profound implications for our understanding of everything that is known and is knowable, especially of knowledge itself.

2

BEYOND LANGUAGE

"To study Metaphysic[s], as they have always been studied appears to me to be like puzzling at Astronomy without Mechanics."

—Darwin, Notebook N, p. 5, 2 October, 1838

Although things are changing, 'metaphysics' has tended to be a term of opprobrium. One reason has been the popularity of what is called "positivism." Auguste Comte (1798–1857), one of the founders of sociology, portrayed intellectual evolution as a progressive affair, with three stages: the theological, the metaphysical, and the positive. Positivism denies the legitimacy of metaphysics as well as that of theology, and rejects as meaningless everything but matters of scientific theory and fact. In the present century the Logical Positivist school attempted to get rid of metaphysics by creating a so-called philosophy of science based upon logic and language analysis. One result was the creation of some of the most preposterous systems of metaphysics in the whole history of philosophy. Another was the tradition of casting discussions of metaphysical topics in logical terminology. There is no way in which one can divorce science from metaphysics altogether: to deny metaphysics is itself a metaphysical thesis. One can no more have science without metaphysics than a drink without a beverage: the only choice is that between good metaphysics and bad metaphysics, good science and bad science.

The notion that metaphysical problems are purely linguistic matters to be handled simply by cleaning up the language continues to enjoy considerable popularity. Nobody in his right mind would deny that the analysis of terms and concepts often leads to progress in philosophy. In the present work we repeatedly examine the equivocal use of terms that seem on the face of it quite unproblematic. But the notion that such discourse is purely linguistic—that arguments are purely semantic, that all one needs is a good dictionary—completely misses the point. We need to get beyond the language and come to grips with what the discourse is all about, which in science is the entities that populate the universe and what really goes on in world of objective reality. So we need to take stock of the fundamental units according to which things known and knowable might be classified. At a lower level we need to consider what it is to be an organism or a species. The distinction between class and individual is one of the most profound we might make, but there are a host of others.

To drive home the point that such distinctions are not just matters of language or scientific fact as commonly understood, let us begin this analysis with a discussion of various kinds of possibility. Logicians treat possibility as one of several so-called modalities of propositions. Carnap (1956:175) lists six of these: necessary, non-necessary, possible, impossible, contingent, and non-contingent. These modalities are clearly related, and are often treated together under the rubric of "modal logic."

If something is necessarily the case, it could not possibly be otherwise. If something is the case as a matter of contingent fact, it could possibly be otherwise. When we speak of something as being possible, we usually are interested in states of affairs that could be realized, but as a matter of (contingent) fact are not. Now, if something is in fact the case, then it is possible, however trivially, that it may be the case. If I have in fact written a book, then obviously it was possible for me to do so. Or at least we would use 'possible' that way in ordinary discourse. Some philosophers have defined the possible as something that might be but is not; this is a purely definitional matter that need not be discussed further here. We are interested at present in various kinds of possibility, in this case: 1. physical possibility; 2. logical possibility; and 3. metaphysical, or ontological possibility. Let us consider them in that order.

To say that something is *physically* possible means that it is consistent with the laws of nature. We will have more to say about the ontological status of laws of nature later on. Suffice it to say for the meanwhile that both in our daily lives and in the conduct of scientific inquiry we rely upon the assumption that nature is uniform and that its uniformities are necessarily true irrespective of time and place; when things happen, they must happen according to some regular pattern no matter when and where they happen. Anything out of line with such regularity we would consider physically impossible. Conversely, that which occurs in conformity with such regularity is physically possible. Because such laws are necessary in that sense, we never experience a situation in which they do not obtain, although of course we experience situations in which they do not apply.

A miracle is often defined as something that happens contrary to the laws of nature, supposedly as a consequence of some supernatural act. Because science operates upon the assumption that nature is strictly uniform, with no real exceptions to its laws, it is generally accepted that miracles are beyond the grasp of scientific knowledge, irrespective of whether or not they occur. For present purposes we need not go into the problem of whether laws of nature are necessary in some other, perhaps stronger sense, for example that perhaps no universe could possibly have different laws. We may likewise avoid considering the fascinating possibility that laws of nature might change from time to time or vary from place to place.

We can easily imagine what it would be like for something that is physically impossible to happen. Magicians make their living in part by giving the appearance of doing just that. Or just run a motion picture backward and watch paratroopers rising from the ground into an airplane, or somebody uncracking an egg. We can also, without straining too much, imagine a world in which there' were laws of nature, but ones different from those with which we are familiar. For example I can imagine what it would be like if water did not expand upon freezing: the oceans would consist of solid ice to their very depths. Our imagination may not be an infallible guide here, and we certainly would not wish to rule out the possibility of something that we cannot conceive of actually happening. The point is only that the physically impossible is often imaginatively conceivable.

Logical possibility means conformity with the rules of valid inference. When we say that something is logically impossible, we mean that it is contrary to the most fundamental of those rules, the principle of contradiction. Something cannot be unequivocally A and not-A at the same time. The truths of pure logic in general, including mathematics in particular, are true because they follow from definitions. If $2 + 2 = 4$ and $2 + 4 = 6$, it follows deductively that $2 + 2 + 2 = 6$. We might redefine the symbols. Thus for example if $2 + 2 = 5$ and $2 + 5 = 9$, then $2 + 2 + 2 = 9$. The choice of symbols is purely conventional. A logical impossibility would be for example: $2 + 2 + 2 = 6$ and $2 + 2 + 2 = 9$ and $9 \neq 6$. Anything that contradicts the definitions of terms must count as a logical impossibility. A round square. A married bachelor. To know that something is logically impossible, we need know nothing more than the definitions and how to follow the rules. In this it differs from physical impossibility, because we need to know what the laws of nature actually are to decide what is and is not possible. Also we find it hard to imagine what it would mean for something that is logically impossible actually to be the case.

All this is of course very elementary, and familiar to anybody who knows a little philosophy. It only becomes interesting when we use it as a basis for discussion of what is *metaphysically* or *ontologically* possible, and for considering how metaphysical possibility differs from physical and logical possibility, if indeed there is any real difference. Often I walk from my home to the nearest store, which is about a mile away. Usually it takes me something like twenty minutes. Although I can imagine what it would be like to run there in a minute, this is physically impossible, though there is nothing self-contradictory, and therefore logically impossible, about that. I have in fact estimated the distance between those two points, and could easily measure it. But could I *weigh* that distance? Of course not. I can weigh myself, but not the distance between any two points. I cannot weigh my walk itself either. Or eat it. I cannot even imagine what it would be like to do such things. Why not? One answer is that in speaking of distances, myself, and walks, we refer to different ontological categories.

The categorical structure of the world (Grossmann, 1983) is such that some things are possible—metaphysically possible—whereas other things are not. A metaphysical impossibility is not just unimaginable, or self-contradictory, but downright unintelligible. Even a round square makes a certain amount of sense, for roundness does apply to some geometrical figures. To render the notion of eating a walk intelligible, we have to assume that someone has made a so-called category mistake, and perhaps really meant that he munched on an ice-cream cone while walking.

It seems to me that such category mistakes are much more than just linguistic infelicities, like breaking with convention and asking "who" one gave something to, rather than "whom" one gave it to. But before we entertain that possibility, let us examine the natural history and the taxonomy of categories. (For a good review by a philosopher see M. Thompson, 1967.)

As one might expect, the term 'category' is equivocal. Often it is used as a general term for any kind of group whatsoever. We "categorize" people and other things. The word is frequently used in that sense in the technical literature on what is called the psychology of categorization. This sense is so broad and vague that it is not much use, and it sometimes leads to confusion with respect to different kinds of groups that would better be kept conceptually distinct. Another sense of 'category' is that of taxonomic category, which is a level in a hierarchical system of classification. In the familiar Linnaean hierarchy used by systematic biologists, groups of organisms are given what is called "categorical rank." They are assigned to a level, which is by definition a category. The groups themselves are called taxa. Listed below are some examples of categories, followed by taxa, which are their members:

Class: Mammalia, Aves . . .
Order: Primates, Carnivora . . .
Family: Hominidae
Genus: *Homo*
Species: *Homo sapiens*

In explaining such matters, it generally helps to provide examples of other hierarchical systems with both categories and taxa. For example:

Nation: Canada, Australia . . .
Province: British Columbia, Ontario . . .
City: Victoria

Later we will examine such hierarchical systems in detail. For the moment we should mention that in older biological literature, the term 'category' was used to designate not just categories such as the genus, but taxa such as *Homo*. The equivocal usage led to serious lack of communication, and a failure to recognize the distinction is a major source of error to this very day. When people talk about defining species, it is still not obvious whether they are concerned with something analogous to the city on the one hand or to Victoria, B.C., on the other.

Returning to the ontological categories, we should observe from the outset that they have been classified in more than one way. Indeed, a primary concern here is to see if we can develop a better classification. In his book entitled *Categories,* Aristotle provides the first and one of the most influential classifications, though a systematic biologist might be inclined to say that it is not a classification, but a mere list.

Aristotle's categories are as follows: 1. substance; 2. quantity; 3. quality; 4. relation; 5. place; 6. time; 7. posture; 8. state; 9. action; and 10. affection. Although the names may have a familiar ring, it seems a good idea to explain them and give some examples before we proceed with this discussion.

A *substance* is a concrete particular thing or class of such things. For an anthropocentric philosopher like Aristotle, the best example would be a man, and we can use, as he did, a Greek named Callias as an example. Any number of other material objects might do just as well, such as one of his shoes. In our days we tend to identify substance with matter.

Quantity answers the question "how much?" Callias might be six feet tall. Quantitative differences have to do with "more or less." For Aristotle, there were no degrees of quantity, for example, Callias being six feet tall would be a statement of quantity, but being taller than Socrates would be a relation. In modern physics we have come to disbelieve in absolute size. We often associate quantity with measurement, and with exactitude, and in science we consider quantification an important goal.

Quality refers to certain kinds of attributes of substances. For example, Callias might be generous. In Aristotle's system, however, certain transient features were not considered qualities. For example, Callias might be embarrassed, and this would be an "affection" rather than a quality. This distinction would seem to follow from the notion that some properties are "essential" whereas others are not. Aristotle and a lot of other philosophers found it very difficult to cope with change, and as we shall see, this difficulty has had many important consequences. Be this as it may, the qualities of things are familiar parts of scientific discourse. The chemistry curriculum includes both qualitative and quantitative analysis.

Relation compares at least two things, as when we say that Callias is taller than Socrates, or that he stands between Plato and Alcibiades. Relations obviously must have terms: somebody may be just plain fat, but he is not "fatter than" without reference to some other organism. I used the example of "fat" rather than "tall" here because in the case of what looks like a simple property, there may actually be a concealed relation.

Place tells us where the substance is located, much as *time* tells us when it is there. Callias is in the Agora at noon. Later we will consider what might be called "other kinds of place" such as one's place in a society or in an ecosystem.

Posture or attitude would mean erect, recumbent, or whatever. Aristotle himself slights this category, evidently because he did not consider it problematic or particularly important.

State means in what condition a thing is, such as warm or shod. *Action* and *affection* might be illustrated by Callias being warmed up by the sun. The sun acts upon him, and he is affected.

We have suggested that Aristotle provides what may look more like a mere list than a classification. A few words about this distinction would seem to be in order. Material has to be "arranged" somehow before we are able to deal with it. An arrangement allows us to get at the items that interest us, but a classification also tells us something about those items themselves. A shopping list does not have to be a classification. It may consist only of memoranda put together as a class of things not to forget. Making it alphabetical would not turn it into a classification, though it would make it easier to locate the items on it. If one grouped the items according to which store was apt to sell them, that would be a classification. In systematics it is understood that "keys" allowing one to identify specimens are just "finding aids" and not classifications. It has been proposed that the term 'classification' be restricted to hierarchical systems of classes, whereas 'systematization' should be used for arrangements of individuals (Griffiths, 1974; O'Hara, 1993). The point is very well taken, but the proposal seems a bit immodest. Granting the individuality of things like nations and the branches of genealogical trees, we would even have to stop calling the Linnaean hierarchy, with its genera and species, a classification. It seems to me that "classification" in its most general sense is the process of organizing knowledge, and the product is therefore virtually coextensive with knowledge itself. So it seems reasonable to call the products of that process classifications, even though we make some finer distinctions for special purposes.

Be this as it may, Aristotle does at least appear to have provided more than just a list, for we may detect an implicit rationale behind the arrangement. He begins with what seems to him most important, namely substance. He also puts related terms in close proximity to each other: quantity and quality, place and time, action and affection. In modern biological classification, the use of "standard sequences" is quite common (Mayr and Bock, 1994). Unfortunately, putting things into a single line often distorts the evolutionary situation, and the rationale for putting one item next to another may not be evident to the uninitiated.

Aristotle does not seem to have been aiming at a complete or exhaustive system of categories. Rather he sought to find a place for them in his philosophy, which was heavily dependent on logic, especially upon his theory of predication. Substance was the basic category, and the others were predicable of substance. One could say that Callias is a man, or a substance. Or one could predicate a term, falling under another category, of a substance such as a man: Callias is hot. He pointed out that although what falls under other categories can be predicated of substances, substances cannot be predicated of other categories; for example, it is nonsensical to say that tall or angry is Callias. And the members of some categories have contraries whereas others do not: the state hot has the contrary cold, but Callias has no contrary. The only opposite of Callias is the contradictory: not Callias. The details are rather intriguing in and of themselves, but they are of little immediate interest for our present endeavor, which is to see if we can come up with a natural system of categories, something that probably never even occurred to Aristotle. But before we do that, it seems as good a place as any to mention some other efforts to come up with systems of categories.

After Aristotle, the Stoics advocated four basic categories. These were substratum, quality, state, and relation. Substratum and relation were approximately the same as the Aristotelian substance and relation. Qualities were essential attributes, whereas states were accidental ones. The essential attributes were the ones that really made things what they are, whereas the accidents would be things that might be otherwise. So far as I am aware, this arrangement has not had much influence. Mediaeval philosophers who followed Aristotle generally assumed that the list of ten categories is exhaustive, something that Aristotle himself did not maintain. They were, however, much concerned about questions that had been raised by Porphyry in his commentary on the *Categories,* namely whether general terms, or universals, exist; so the disputes about nominalism were closely related to the problems that surround categories.

Kant tried a quite different approach. His categories involved a classification of judgments, rather than a classification of things and what might be said about them. He came up with a system that has four kinds of categories, each with three subheadings. Thus for any statement we get:

Quantity
    Universal
    Particular
    Singular
Quality
    Affirmative
    Negative
    Infinite

Relation
  Categorical
  Hypothetical
  Disjunctive
Modality
  Problematic
  Assertoric
  Apodictic

And this adds up to twelve categories of the pure understanding. These categories, in keeping with Kant's metaphysics, are applicable to phenomena (appearances), not to things in themselves. According to Kant, it is only the phenomena that are knowable. Moreover, they are not what we would consider ontological categories, but epistemological ones. Therefore the Kantian categories are of little interest, except insofar as they provide us with an example of a classification, rather than just a list.

Among twentieth-century philosophers who have worked extensively on categories, Ryle (1949, 1953) concluded that there is no natural hierarchy of categories, and in effect despaired of what has been attempted in the present chapter. Strawson (1959), in his book on *Individuals,* made an important distinction between "descriptive metaphysics" on the one hand, and "revisionary metaphysics" on the other. Descriptive metaphysics is based upon the conceptions and distinctions that we find in ordinary language, whereas revisionist metaphysics attempts to try something that departs radically from that scheme. Aristotle and Kant would be descriptive in this sense; Descartes, Leibniz, Berkeley, and Hegel would be revisionary. Aristotle's scheme does have the advantage that it makes sense of language as we use it in every-day life, and this is true of his physics as well as his ontology. But we know that the *Physics,* with its notions of absolute motion and natural place, left much to be desired. Science has to start somewhere, and everyday ordinary language is perfectly acceptable, but only as a place to start.

Just as zoologists and botanists need no excuse for revising what began as the *Systema Naturae* of Linnaeus, we have the best of reasons for attempting to revise the categories. As a preliminary step, however, we may observe that each and every category contains both classes and individuals. Or does it? Many authors have claimed otherwise. In some systems, for example that of Grossmann (1983), the only individuals are those which fall under the category of substance. For him there are no individuals under the other categories. And yet the above examples that I have given of the Aristotelian categories suggests that there are. Just as I have instanced Callias, a particular organism, so I have instanced where he was, and what particular qualities he had, such as his weight. It seems to me that a given standing in relation is just as much an individual as that which stands in that relation.

The issue of whether there are individual attributes has often hinged upon discussions of color. Sometimes a particular shade of red, for example, has been said to be an individual. Intuitively this seems wrong, if only because classes of material bodies might consist of members that are identical in that given shade of color. So one is tempted to deny that there are individual rednesses. But this would be a mistake. What we are talking about is the "this-redness-here" of "this-thing-here" when we consider redness at the individual level. To reach the level of the individual redness, one has to go a step further than had been proposed. Color is more general than red, quality more general than color.

Some philosophers, for example Carnap (1956:32) and Hattiangadi (1987), have treated particular numbers as individuals. But it seems to me that they are mistaken, and in parting company with them I should stress that doing so is by no means a trivial step. In every case that I have examined, the instances are something other than the numbers themselves. If, for example, we speak of the bilateral symmetry of an animal's body, we are not speaking of two, but of a duality. If we speak of a second cervical vertebra, the individual here is the secondness of that particular vertebra, a twoness, not two. Numbers are more general, and of course more abstract, than anything that may be instantiated of them. The individuals here are individual quantities, or individual meristic attributes, and suchlike. One might try to make numbers the "individuals" in some nonontological universe of discourse, such as a world of pure mathematics and pure logic. But to do so would be to evade the ontological issues through a linguistic ploy that conflates individuality with particularity.

Now let us see how the Aristotelian list might be transformed into a classification. Recollect that he gives us:
Substance
Quantity
Quality
Relation
Place
Time
Posture
State or condition
Action
Affection

To begin, let us dissect out the last two, namely action and affection, and group them together under a larger heading, which may be called "process." Action and affection have to do with the active and passive aspects of change. The things that act and are acted upon are referred to as "agents" and "patients"; a physician is an "agent" who affects his patient. However, a thing can be affected without an agent's really acting upon it. For example, if I am the second tallest man in town, and the tallest man in town is run over by a truck and killed, the consequence is that I become "promoted" to the status of the tallest man in town without the agent (truck) acting upon me. This is sometimes called a "Cambridge change" because they were first discussed by philosophers at that university. So our meta-category of "process" might include two categories, action and affection, but one might want to subdivide the latter; alternatively there might be three categories of equal rank. The individual changes that go on are often spoken of as events, yet not all goings on, or processes, actually result in change; consider terms like 'lurking' or 'enduring.'

One might want to put substance together with process in a single taxon, on the basis of things and what they do being so closely interconnected from the point of view of causality. Let us leave them separate, however, to emphasize the importance of the distinctions between them, about which more will be said later.

Next I propose that we remove from Aristotle's list the "when and where" of such goings on, and put it in its own metacategory, using "place" in a much broader sense than he did, so as to include socioeconomic places such as jobs and ecological niches. This leaves us with a sort of garbage-can taxon for the rest, but all of these are what we generally call "properties"—or attributes or characters (as taxonomic jargon puts it). Although definitely in need of revision, we may leave the properties as they were for the moment. We get:

Substance
Process
   Action
   Affection
Place
Property
   Quantity
   Quality
   Relation
   Posture
   State or condition

This arrangement distinguishes four fundamental kinds of entities that play very different roles. Basically, we have what changes, the changing itself, where it might be when undergoing such changes, and that in terms of which it changes. Anything that exists, insofar as scientific knowledge tells us, must have a material substratum, although the word 'material' may be too narrow; energy is problematic here, and although I do not believe in immaterial substances like souls, they are an interesting possibility. There seem to be no "disembodied" processes. However, a process is not that which engages in, or undergoes, the process in question. I walk, but a walk does not. Nor is the place where I walk the same thing as me or my walking. Places are peculiar in that they may be quite empty. On the other hand every individual substance, process, or event that we know of has one property or another. A "bare particular," in the extreme sense of a piece of matter or an activity without any properties whatsoever, is something that is hard to imagine and seems to be a metaphysical impossibility (Baker, 1967). Even places are here or there.

The arrangement just suggested may or may not seem much of an improvement over Aristotle's, and everyone should feel free to propose alternatives. Be this as it may, it does bring out what would appear to be a major defect in Aristotle's metaphysics. Like other Greek philosophers, he had a very difficult time dealing with change, as has already been pointed out. Plato treated the material universe, with its changeable and perishable material objects, as inferior to a transcendent realm of immutable and eternal Ideas, or essences. Aristotle did not go so far. He believed in essences but maintained that they do not exist apart from individual things. Nonetheless, the various doctrines that invoke essences—essentialism or typology—emphasize stasis at the expense of change. Later we will consider essences and essentialisms in greater detail. For the moment the crucial point is that Aristotle's treatment of the categories de-emphasizes processes, activities, and change. If one's goal is a metaphysics that can adequately cope with evolution, it leaves a great deal to be desired.

Although the "ordinary language" approach to metaphysics can at best serve as a first approximation, it is nonetheless instructive to note that substances are generally designated by the part of speech called a noun, or a "substantive." The categories grouped above as attributes are mainly named by means of adjectives, though there are some exceptions. For Aristotle this all makes a great deal of sense, since his treatment of the categories was linked up with his theory of predication. States, actions, and affections can be discussed in "attributive" terms, as when we say that something is warm, or warmed up. But what has happened to the verbs? And what about those nouns that designate, not substances, but the activities of substances—what the substances do? What about adverbs?

Verbs are very useful. They lend a sense of concreteness to our utterances that is hard to obtain otherwise. One way to improve one's writing is to go through the manuscript and see what happens when one crosses out nouns, especially abstract nouns that end in *ation,* and replaces them with verbs. The same maneuver sometimes helps in metaphysical research. If one can recast the sentences in terms of events and processes, one may see things rather differently. Or one searches for items in one's experience that might be looked at from the point of view of process but usually are not.

A walk, for instance, is obviously an instance of walking; less obviously it has parts, such as a going forth and a coming back. The walk I took yesterday afternoon was not a substance. It was an activity or a process. Although I walked, the walk and I can readily be distinguished. My walk lasted one hour, I did not. We might try one way or another to equate the doer of the action with the action itself, the walk with the walker walking. In the case of walks, the action and the actor seem inseparable. But consider a dance. The dance is not the dancers. We can imagine a dance in which one dancer replaced another to the point that there were no participants at the end who had been present at the beginning, but it would still be the same dance.

One might wish to argue that processes are somehow derivative of the entities that engage in them, and that therefore they are of lesser importance in what may be called a scale of ontological priority. Thus things would be ontologically prior to what they do. But this decision seems arbitrary. Yes, to be sure, there cannot be an action without an actor. A thing devoid of attributes would make no sense either. Consider what the universe would be like if there were no processes. It would be perfectly static. Nothing would happen.

In a "process metaphysics" the order of priority would be reversed. The fundamental reality would be the processes, and the substances or things would be like froth on the ocean, or perhaps their very existence would be denied. Efforts to develop a process metaphysics have an ancient and honorable history going back at least as far as Heraclitus. Such efforts have not been altogether successful. At the very least, however, it may be said that the history of science gives us plenty of examples of substances that turn out to be process. The best example is heat, which some people tried to identify with a hypothetical substance called caloric. The kinetic theory of heat treats it as a kind of motion.

Biology is the study of life, and living would seem to be a process. So too with evolution. Living evolves. If one's goal is to understand the living world, then one had better take processes very seriously indeed. This is not to say that one ought to get rid of things and attributes altogether. When life evolves, things change their attributes. Nonetheless an evolutionary biologist would be ill-advised to treat things as more important than what they do, and might well entertain the possibility of going one step further along the road to a fullblown process metaphysics.

Even if we do not go that far, biology would certainly be better off if process were more clearly conceptualized. In elementary textbooks the students are often provided with lists of ecologically important "interactions" such as predation, competition, mutualism, commensalism, and parasitism. Here we have some interesting failures to distinguish between processes and certain other categories. Let us consider an analogy first. The act of begetting a child may reasonably be considered an interaction between two partners; their marriage, however, is definitely not an interaction: it is a relationship that is perhaps intended to facilitate such interactions. When a predator pursues a prey and the prey attempts to escape, there is again a kind of interaction. And although mutualists, commensals, and parasites can and do interact, such terms as mutualism designate the relationship and not those interactions. Let us hope that the students are not as badly confused as the authors of the textbooks.

Now, what about competition? The term is notoriously difficult to define, a point that I have addressed elsewhere (Ghiselin, 1974a), and would rather not belabor again here. I would, however, urge that much of what biologists say about it seems quite muddled, that economists have a better idea of what it means, and that the confusion seems to arise from a lack of ontological clarity. I would prefer to call competition a *situation* or a standing-in-relation under which certain kinds of interactions occur. But these competitive interactions might better be called competitive "interaffections" because the competitors do not have to "act upon" each other. To explain, let us consider a competitive situation in academia: a course in which the teacher grades on a curve. The scarcity of grades creates a circumstance such that how one student performs *affects* the welfare of the others, and they behave accordingly. A noncompetitive situation would arise if the teacher graded by means of some standardized test so that all could do equally well or equally badly. Grades not being in short supply, the performance of any one student in the course would be wholly indifferent to all the others. A noncompetitive situation might lead to what is easier to conceive of as a true interaction, namely cooperation, in which the students actually try to help one another.

Evidently people find it easier to conceive of straight-forward occurrences in which one entity, the agent, affects another entity, the patient, by acting directly upon it, than to conceive of more subtle goings-on in which an agent affects the patient without actually acting upon it. Predation obviously involves a predator acting upon something in the course of affecting it and is easy to picture in the mind's eye. And competition tends to be misconstrued as if it likewise had to involve one participant acting directly upon another. Now, although boxers may do just that when they compete by slugging it out, the participants in a foot race by no means "interact" by means of direct physical contact. So when we find Gould and Calloway (1980) asserting that brachiopods and bivalves have not competed but rather have been as "ships passing in the night," it seems appropriate to point out that there could hardly be a better way for the captain of a merchant ship to drive his competitors out of business, than by overtaking and then passing them, be it by day or be it by night (Ghiselin, 1987d).

Natural selection, a process that can only occur in competitive situations, is especially subject to metaphysical confusion with respect to agency. In particular, we have the notion of whether natural selection "acts upon" the genotype, the phenotype, or both. There are some perfectly legitimate considerations here, such as the point that a gene, in order to be selected, has to be "phenotypically expressed." This in addition to some silly talk about how it is hard to conceive of natural selection acting upon the inside of an organism.

The very idea of natural selection "acting upon" anything is a category mistake. Natural selection is not an agent, and therefore cannot "act upon" anything. In explaining this point it helps to use the analogy of "undressing" (Ghiselin, 1981). When somebody undresses before taking a bath, he (the agent) acts upon his clothing and affects himself by becoming naked. But what is it upon which undressing acts? Nothing, of course. What is affected by the process is quite another matter.

To my knowledge there is no good precedent for treating place as a particularly important category. Nonetheless it makes a great deal of sense in terms of the individuality thesis. If individuals are spatio-temporally restricted, then the where and when of things is very important. Physics-minded philosophers have tended to treat place as simply a matter of position in space and time, or space-time. Logicians are apt to treat it as just a minor subheading under relation, or to dismiss it as just one example of an extrinsic property. It may indeed be true that place is somehow derivative, but much reflection upon biological and social places makes me rather skeptical of such efforts. In particular I have in mind things like social roles, offices, and ecological niches. A lot of sophistical talk about "a hole is a negative thing" notwithstanding, the concept of an empty place seems to play a very important role in certain kinds of theories. And that which occupies the place is definitely not the same thing as the place itself, as is clear from the fact that organisms, dances, and instantiations of properties move from one place to another (cf. Dretske, 1967).

I suppose one can treat emptiness as a kind of "disposition" of some whole to accommodate additional parts that might be incorporated into it. But this view immediately raises the general difficulty that students of logic and metaphysics have experienced with dispositional properties. They want to deal with what is definitely the case, not with what would be or might be the case, or might have been the case but is not. In technical language, they do not approve of "counterfactuals" or "contrary-to-fact conditionals," which seem to imply the reality of alternative worlds. A lot of others share this sentiment, witness those who dislike having species be defined as groups of "potentially interbreeding" populations. On the other hand, dispositional properties are exceedingly important in the behavior and ecology of all sorts of organisms. Much of our time is spent deciding whether something might or might not poison us, or want to let us in the door. The mere fact that a certain possible state of affairs was not in fact realized may be quite irrelevant in the conduct of our daily lives, for example when we punish somebody for driving when inebriated even when the accident that might have happened did not.

Admitting that in elevating the rank of place we may go too far, let us briefly consider what a "place metaphysics" might be like. We might try to imagine a metaphysics in which the places are ontologically prior to their occupants. The appropriate attitude is suggested by situations in which what really matters is the office rather than the official, and the same person can without contradiction be treated with deference and respect as head of state, yet be held in contempt for his failures as a politician. But this is not strong enough; a real place metaphysician would have to argue that the office is somehow the fundamental reality, and that the office-holder is epiphenomenal. It is hard to imagine somebody going to the extreme of thinking that places exist, but their occupants are nothing more than figments of our imaginations.

Be this as it may, it seems obvious that our thinking about places in biology has been seriously muddled at times. This is particularly obvious with respect to the niche, a term that has been variously defined in ecology, wherein there is a fundamental distinction between the Eltonian and the Hutchinsonian niches (see Griesemer, 1992). As is well known, the Eltonian concept really goes back to Darwin, and the term was at least popularized by Grinnell (1904). Darwin referred to a "place in the economy of nature" that might become seized upon or occupied by an evolving species. This conception was an economic one, and fundamental to his thinking on competition and the "divergence of character"—or, as we might prefer to put it, adaptive radiation. It was also crucial to the work of Elton (1927) and Gause (1934) who applied essentially the same niche concept in their work on diversity and competitive exclusion. In their ecological theories, niches were definitely places, albeit economic rather than spatial or temporal ones. As with places generally, there was no problem with respect to their being full or empty, occupied or not. On the other hand, when Hutchinson tried to develop a new approach to ecology, he used the traditional term in a vastly different sense, for a class of ecologically significant properties possessed by the organisms of a particular species (see Hutchinson, 1965). So instead of a place it now meant a complex property of an occupant of that place. An occupant of course cannot be "empty" any more than a place can metabolize. Hence some confused talk. Colwell (1992), in an excellent discussion on niche concepts, cannot see much use for the Eltonian niche, albeit granting its limited utility for those who are interested in such phenomena as "unsaturated" environments. But this seems to me precisely the point: if one is an evolutionary biologist rather than a synchronic ecologist, that is precisely what one should be interested in. The Hutchinsonian niche is typological and ahistorical, and it muddles the categories of historical and functional biology.

Elton's explication of the niche concept was cast in terms of some economic metaphors and might better have been grounded in economic identities. He equated an organism's niche with its "profession," but this is a confusing simile that does not really bring out the underlying order of things. A niche is an economic place, a sort of professional position or business opportunity that may or may not be filled, whereas a profession is an economic capacity that fits one to occupy such a position or to take advantage of such an opportunity, something which again may or may not be achieved. So an empty niche would be like a job opening in the History Department at Harvard, left vacant by a professor who retired and ceased to occupy it or practice his profession, and sought out by aspiring young professionals who would like to, but might not yet, be practicing the profession for which they were trained.

The problems we encounter with dispositions and empty places are by no means unrelated. For an empty niche or an unfilled position at a university would imply that an economic whole is disposed to provide an opportunity to additional parts. So, far from treating such "emptinesses" as something trivial or negative, we ought to view them as potentially valuable resources, like "vacancies" when we are in search of housing. And when we think of natural selection, we ought to conceive of it, not merely as culling out the unfit, but as taking advantage of such opportunities.

Simpson's notion of an "adaptive zone" was a great failure, as he himself admitted, though he could not explain why (Simpson, 1944, 1953). He had the idea that, in Australia and North America respectively, a marsupial wolf would occupy the same adaptive zone as a placental wolf, and there was indeed a vague notion of ecological equivalence suggestive of a class of niches. The problem can now be understood in the light of the individuality thesis. It arose from trying to combine the historical units of taxonomy with the functional classes of ecology, so that there was an incoherent mixture of homology and analogy, or of individuals and classes. The ecological correspondences are there, but they have to be ecological equivalents, such as cursorial predators.

The cavalier move suggested above, of dumping the remainder of Aristotle's categories into a garbage-can taxon called "property," by no means provides us with a satisfactory classification. Aristotle himself, we may remember, did not intend his list to be exhaustive. He also discussed various subdivisions of these categories. It may not be possible to come up with a good system, and we are always vexed with the possibility that we are merely making linguistic groups that have little if anything to do with ontology.

Be this as it may, there do seem to be some important ontological distinctions to be made, and logic can be of some use here. Perhaps the most daunting challenge is how to deal with the intrinsic versus the relational aspect. An intrinsic property is one that something may be said to possess, in and of itself, such as maleness. A relational property, on the other hand, is one that applies to a thing if and only if it "stands in relation" to some other thing, such as brotherhood. A lot of properties that may seem intrinsic may turn out actually to be relational. Solubility, for example, has been treated as a relational property, since paraffin, for example, is soluble in xylene but not in water; it is not "just plain" soluble, and to say that it is "soluble in," without providing a term that tells us in what it is soluble, is to speak nonsense. On the other hand we may suspect that such properties as solubility are the indirect consequence of the molecular structure of the material in question, and that, therefore, the relationality is of secondary ontologial significance.

More importantly, there seems to be a fundamental distinction between relations among the parts of individuals, on the one hand, and relations between members of classes, on the other. Modern physics has come to treat things from a relativistic point of view. There may or may not turn out to be something that ultimately answers to the concept of absolute size. Nonetheless the length or volume of an object is measured in terms of a common meterstick for all the items in the universe. Things are larger or smaller than anything anywhere at any time. On the other hand if one is a sibling, one has to be a sibling of some particular part of a particular genealogical nexus.

And of course things interact with other things as parts of some larger whole—be they parts of organisms, of species, of biotas, or whatever. The causal interactions that interest scientists do indeed include the gravitational attraction of every body in the universe, but for biological interactions at least, the important ones occur among bodies in relatively much closer proximity—often, but by no means always, in direct, physical contact.

To see why this point really matters, even though it may seem all too obvious, we need only reflect upon how often things are taken out of context. Overlooking context underlies much naive reductionism. In genetics, we sometimes find genes conceived of as corpuscular entities, and such phenomena as position effects happily ignored. Likewise systematics has been plagued by the notion that there must be "unit characters," which bear comparison with the "logical atoms" of a long since dated metaphysics. In efforts to understand language and thinking, context seems to have been the last place where people have looked, when it should have been the first. Is the meaning or, to borrow a term from what verges upon occult metaphysics, the information in a sentence to be sought in the words? In the sentence itself? A sentence that will be taken as a joke by one person might be taken as an insult by another, and without real contradiction. What a sentence means depends upon the individual situation.

Even from a practical point of view this is no trivial point. It is very easy to program a computer to go through a manuscript in one language and have it substitute equivalent words in another, but exceedingly difficult not to have the results turn out downright ludicrous. Anyone who has tried to use an optical character recognition program knows how much easier it is for a human being, who knows what the text is all about, to get things right, than it is for a machine that has no real knowledge at all. Professional Darwin scholars working in the Cambridge University Library and trying to decipher Darwin's handwriting in order to make sense out of the marginalia have to read the text, forming and rejecting hypotheses like paleontologists struggling to reconstruct the entire organism from fragments of a skeleton.

Thus the individuality thesis suggests that if we are going to make any deep metaphysical cut within a natural system of the properties, perhaps we ought not to treat relation as a monolithic ontological category. Nor does it seem clear that we can start with a bunch of "simple" properties of a given kind and group them in larger and larger assemblages. One approach, which has often been invoked without really attempting to develop it, is to say that everything follows from the intrinsic properties of substances. Accordingly redness would be somehow derivative of the atomic, molecular, and other structure of the red object, with appropriate qualifications, such as the redness being manifest only when the object is exposed to a certain kind of light. If there are indeed "simples" among the properties, it is rather hard to enumerate them.

It is not hard to come up with examples of important kinds of properties, such as qualitative and quantitative, dispositional, relational, intrinsic, extrinsic . . . one could go on and on. But such kinds seem more a matter of logic than of ontology, even though they are not without ontological interest. If for no other reason, we need not attempt an ontological classification of the properties, but at least for the time being treat them on an ad hoc basis.

As to the question of whether one might propound a "property metaphysics," the answer is that these have been exceedingly popular among philosophers and have had no dearth of enthusiasts among scientists. It is very easy to say that all we know is sense impressions, which are indicative of the properties of things, and that we know nothing of things in themselves, to get a good start. It is likewise easy to abandon the notion of a material, substantial substratum and to dismiss things as metaphysical delusions. Given such a position, the things *are* the properties. A more restrained approach is to admit the existence of substances, but to make the properties ontologically prior to them. In general, however, the commonsense approach is to treat substances as prior to the properties, and everything else as hardly worthy of attention. Such a substance-attribute metaphysics seems to be presupposed by a large number of taxonomists, and the distortions that result are a major theme in this work.

3

What an Individual Is

"Natural Selection cannot effect what is not good for the individual, including in this term a social community."

—Darwin, letter to Wallace, 6 April, 1868

As was belabored in the previous chapter, Aristotle's metaphysics might be characterized as more of a thing, or substance-property, metaphysics than as a process metaphysics, given the relative importance, or priority, that he ascribed to those two categories. Likewise it can be described as giving priority to classes, rather than to individuals, and this is particularly clear with respect to their role in knowledge. Know the universals, and the particulars are a mere chore. Throughout history the universals have been considered the problematic entities. Hence there is a vast literature on the "problem of universals" (Armstrong, 1978), but almost nothing, comparatively speaking, on what is not even referred to as the "problem of individuals." Most of that indeed is concerned with what are called the criteria of individuation: how we decide, for example, that you are you rather than somebody else (Wiggins, 1980). Here, as already pointed out, we reverse the priority, treating the individuals as what we need to understand.

We can apply the terms 'class' and 'individual' to all bona fide ontological categories. Everything that falls under such an ontological category is either a class or an individual. We shall in due time examine the possible alternatives, which are firstly that something might be neither a class nor an individual but some third kind of entity, and secondly that it might be both a class and an individual. For the moment let us say only that I shall reject these alternatives. We should also note that certain entities that are not bona fide ontological categories might well be recognized.

In the usual biological sense, 'individual' is a synonym for 'organism' but the ontological term is a much, much broader one. Although all organisms are individuals in the ontological sense, not all individuals in the ontological sense are individuals in the usual biological sense. We have suggested all sorts of things lacking the defining properties of 'organism' that might be given as examples of an ontological individual. A chair is a piece of matter that an organism might sit on, and the world is full of such things. Or consider a part of such an individual: one of a person's legs, or one of the legs of a chair. A part of an organism can be an individual, including not just each and every organ, but each and every cell, each and every molecule, and each and every atom. It is problematic whether going further and further toward ever smaller components one would ultimately end up with "simples" that have no parts.

Likewise we can say that larger things can be individuals. An individual society would be a good example. If you do not like a society as an example of an individual, try the Earth, the Solar System, the Milky Way, and the Universe. Whatever an individual is, an individual is not by definition a nongroup. Every material object that has ever been observed by a human being is understood by modern physical science to be made up of parts. It is hard to think of examples of individuals that are definitely not, in one sense or another, groups. But saying what an individual is not, although it leads us toward an understanding through elimination, can be tedious. So let us rather discuss the legitimate criteria of individuality and defer the illegitimate ones to the next chapter.

In the first place, it is possible for a class, but not for an individual, to have instances. This gives us the principle of noninstantiability, which, if properly applied, would seem to be infallible. Gracia (1988) treats it as the sole criterion of individuality. I will give some others which he might argue are derivative, but it is hard to say what is fundamental and what is derivative in this situation. When we say that something is an instance of something else, we provide one kind of example. I am an instance of "organism"—more briefly I am an organism. But what would be an instance of me? Surely a part of me would not qualify as an instance of me. We would not say that my right hand is an I, and my left hand is yet another I, for that would populate the world with at least two of us. Part of a person, such as some blood drawn for medical diagnosis, may provide useful information about the whole, but it is a sample, not an example, of that person.

One metaphysical ploy, by no means endorsed here, would be for me to say that the world is populated by an infinite number of me (plural), rather like a succession of frames in a motion picture. Each of these would be an instance of me. If we make such a claim, we are treating me as if I were a class of such entities. Bertrand Russell tried to argue along such lines, and the results were rather odd. One response we might give to such a claim is that the so-called stills are purely fictive and do not correspond to what actually goes on in the real world. Another possibility would be that, yes, there are quanta of being that exist as stills, but they are nonetheless not instances. Rather each is a part of a larger whole which is the film itself, which is likewise an individual, and the stills are no more instances than my thumbs are instances of me.

The instantiability test also works very well with social groups. Because of its historical interest I quote a passage to this effect from John Stuart Mill's *Logic*:

> 'The 76th regiment of foot in the British army,' which is a collective name is not a general but an individual name; for though it can be predicated of many soldiers taken jointly, it cannot be predicated of them severally. We may say, Jones is a soldier, but we cannot say, Jones is the 76th regiment and Thompson is the 76th regiment, and Smith is the 76th regiment. We can only say, Jones, and Thompson, and Smith, and Brown, and so forth (enumerating all the soldiers), are the 76th regiment. (Mill, 1872:17.)

The instantiability test has to take account of the relationship between individuals that stand to each other as whole to part. When we say that an individual "falls under" or "is an element of" a class, we invoke class membership. This must not be confused with "membership" in a club or with being one of the "members" of the human body. Clubs and bodies are individuals, and the relation here is that of whole to part, not class membership. When we say that Jones is a soldier, we are saying that he is a member (in the logical sense) of the class of soldiers. When we say that the 76th regiment is a regiment, we are saying that it is a member (in the logical sense) of the class of regiments. Jones is also a member of the class of servicemen, and also of the class of organisms. The relationship between the classes of organisms, servicemen, and soldiers is called "class inclusion"—the larger or more inclusive class includes the smaller, and the smaller likewise includes the still smaller one. This relationship is transitive: that 1. organisms includes servicemen, and 2. servicemen includes soldiers implies that 3. organisms includes soldiers. Translating into what may be more familiar terms, all soldiers are servicemen and all servicemen are organisms; therefore all soldiers are organisms. This of course is a valid syllogism.

There is, however, no relationship of inclusion between the class of soldiers and the class of regiments. Nor is the relation between a soldier and the class of regiments transitive, i.e., the 76th regiment is a regiment, and Jones is an element of the 76th regiment does not imply that Jones is a regiment. However, Jones is a part of the 76th regiment, and the 76th regiment is part of the British army. Therefore Jones is a part of the British army. The relation is a transitive one. Consequently one can reason with whole-part relationships much as one can with class inclusion relationships. This is one reason why biological classification systems based upon whole-part relationships have often been misinterpreted as depending upon class inclusion, and the classificatory units themselves as classes rather than as individuals.

Here is as good a place as any, perhaps, to explain the difference between individuality and particularity. "Serviceman" includes "soldier," and the former is the more general, the latter the more particular, term. We might also say that the British Army is a more general term than the 76th Regiment of British Foot. But we would not say that the 76th Regiment of British Foot is more individual than the British Army. In that sense, all individuals are particulars, but not all particulars are individuals. As we will explain later on, only "concrete" particulars are individuals by definition.

In order to avoid such confusion it would help to reorganize our language and adopt a terminology that makes it clear and unequivocal whether we are dealing with the logic of classes or of individuals. Unfortunately the efforts of logicians along such lines have not produced the sort of logic (mereology) that would seem useful for our purposes (Leonard & Goodman, 1940; Clarke, 1981). We need something that treats individuals as more than just the sums of their parts. The following guidelines are suggested. One might want to express a very abstract relationship that is more general than class membership, class inclusion, or whole-part. In this case we would merely say that the "elements" "fall under" that of which they are elements. Thus my liver would be an element of me, and of liver, and liver would be an element of organ. Then to make things less abstract, we distinguish between the different relationships along lines already suggested. We restrict membership to class membership, so that my liver is a member of liver. Likewise we restrict inclusion to class inclusion: organs includes livers. Let us call the relationship between wholes and their parts "incorporation": my body incorporates my liver, or my liver is incorporated in my body. This may take some getting used to, but it will have definite advantages. For example, as we pass from lower to higher hierarchical levels, irrespective of whether it is a class-inclusion hierarchy or a whole-part (incorporative) hierarchy, we make statements that are sometimes said to be more general. But something very different is involved. Organ is more general than liver, and when a liver is infected, then by definition an organ has to be infected. But when a liver gets infected, an entire organism gets sick. There is causal connection between a whole and its parts that does not exist between a class and its elements.

Due to an unfortunate choice of words (for which I admit some personal responsibility) philosophers of taxonomy have come to speak of taxa such as species and subspecies as more or less inclusive. ("Species are the most inclusive units such that. . . .") Little mischief seems to have resulted thus far, but to recast such definitions in mereological terms would at least stress the distinction between inclusion and incorporation: "Species are the most incorporative units, such that. . . ."

One might wish to argue, as has Salthe (1985), that not all of the relationships between wholes and their parts are transitive. For example, his liver is part of Jones, Jones is part of the 76th regiment. Is Jones's liver part of the 76th regiment? At least some people intuit that such is not the case, but are their intuitions reliable? To get around this kind of anomaly, one solution is to invoke the notion of a logically proper part. Only Jones the organismal whole, and not his components, would count as a part of a regiment. Such a move seems unnecessary as well as misleading. It is only in some kinds of circumstances that such a problem crops up. It seems not to be a problem when we deal with molecules, organisms, and machines. All the matter in an organism or in a machine is conceived of as a part. We have a little more difficulty perhaps in conceiving of an atom in an automobile as a component in quite the same sense as we do a wheel. The reason would seem to be that "component" suggests a functional subunit, and a given atom might be looked upon as dispensable. We should mention that such apparent intransitivities in componential hierarchies have attracted some attention from students of psycholinguistics (Cruse, 1979; Hampton, 1982).

When we shift from Jones to his organs, we are obviously shifting from a military context to an anatomical one. When, in ordinary discourse, we talk about military organization, we are concerned only with the parts that are particularly interesting. For all practical purposes the hierarchy stops at the organism, which is, after all, the entity that enlists, obeys orders, and so forth. So in such a context we tend to read "participant" rather than "part" in the usual sense. The military and the anatomical language do not recognize the relationship, even though a wounded soldier and his physician might be painfully aware of it. Ontologically, however, it seems quite unobjectionable to speak of Jones's liver as part of the British Army. Otherwise stated, we have here a failure of ordinary language to reflect the actual structure of the world, and our intuition fails us. We are puzzled, not by properties of the things that we have experienced, but rather by the properties of a hierarchical classification system that we have created to cope with certain aspects of our experience. A part of a classification scheme and a part of an organism are not the same thing.

Lumping the last two criteria together as aspects of a single one, let us pass to a second, namely that individuals are spatio-temporally restricted, whereas classes are spatio-temporally unrestricted. In other words, an individual occupies a definite position in space and time. It has a beginning and an end. Once it ceases to exist it is gone forever. In a biological context this means that an organism never comes back into existence once it is dead, and a species never comes back into existence once it has become extinct. And although it might move from one place to another, there has to be a continuity across space as well as through time. Such continuity is obviously crucial to our conception of individuality (Rorty, 1961; Ayers, 1974). Perhaps there are cases in which something can get from one place to another, or can get across gaps in the fossil record, without traversing the intermediate space or time. But strictly speaking even the authors of science fiction stories do not go that far.

An individual such as a chair obviously exists at just one place at any given time. The class of chairs is another matter. It has no location apart from the location of its members. If we destroyed all the chairs in the world and then made new ones, we would not say that chair had ceased to exist and been replaced by something else. Classes need not have members. Now, of course certain classes of objects do in fact have members that are to be found only in a certain restricted area or at a given time. We don't encounter chairs on the moon. And we can define a class in a way that restricts the members spatially and temporally by means of defining properties, for example: everything in North America between the size of an orange and a cantaloupe at 10:00 A.M. on December 2, 1933. Does this mean that classes, like individuals, can be spatio-temporally restricted? I think not, for what is spatio-temporally restricted in such entities is the members, and such a claim would seem at the very least to be equivocal. Consider, for example, an extensionally-defined class (set) consisting of the chair, North America, and Paris, France. Would we say that "it" is spatio-temporally restricted? No, for only some of them are. The idea of a spatio-temporally restricted class would seem therefore to be a category mistake. Spatio-temporal restriction, even if one rejects such arguments, is at least a necessary condition for being an individual. This criterion again would seem to admit of no exceptions. Sometimes we have little else, if anything, to go on when deciding that something is an individual.

A third criterion is that all individuals, without exception, are concrete rather than abstract, and this concreteness, unlike the abstractness of classes, does not admit of degrees. Individuals are absolutely and without qualification concrete. But I hasten to point out that such terms as 'abstract' and 'concrete' have been used in various ways, so here it seems a good idea to explain what I mean in considerable detail.

According to some authors, such as John Stuart Mill, abstractness is a property of attributes. Thus a cat would be concrete, but its affectionateness would be abstract. The idea here would seem to be that we "abstract" certain properties from the concrete objects in our experience. A nominalist would say that the properties, which are universal, are nothing more than abstractions from those things. Mill contrasted his own view with that of Locke, who related abstractness to generality. Locke's is somewhat more like what I have in mind. If we take a class with included subclasses, for example, domesticated organisms, pets, and housecats, we proceed from the more general to the more particular until at last we come to an individual kitten, such as Puff. The more inclusive classes have more members, but the individual members of those classes have fewer traits in common. If we examine a hierarchy that is made up of wholes and parts, with the higher levels more incorporative than the lower ones, the things ranked at higher levels are in a certain sense more general, and those ranked at lower levels more particular. Puff is more general than one of her claws. And yet although I can pet the individual cat, I cannot pet pet in general, nor, *a fortiori*, can I pet domesticated organism. I can indeed be scratched by an individual kitten with any or all of its individual claws. So that is the sense in which I say that all individuals, without exception, are absolutely and without qualification concrete.

Other senses of concreteness are perfectly legitimate, but they are not what makes the difference between individuals and classes. We might, for example, want to consider a part without thinking of the whole: perhaps to think of Puff's digestive system abstracted, or apart from, the rest of her body. Or we might want to apply some "abstract" class term such as 'appendage' to the one that bore the claw that scratched me. There is a tricky, technical sense in which one would say that a class consisting of all of Puff except for her claw that scratched me is an abstraction, and this point will be discussed in due course. In common parlance 'abstract' is sometimes loosely used to mean something like "abstruse." And by 'concrete' one might want to mean "tangible" or something akin to that. I can pick up a kitten and caress the creature, but inaccessible objects like the sun (which at least does shine upon us) are no less concrete.

One reason why it may seem counter-intuitive to claim that species are individuals is that we do not normally conceptualize species as having the sort of concreteness that we associate with kittens. But that is a problem with our thinking, not something that inheres in the species themselves. Likewise some people might not like to apply the notion of concreteness to some of the ontological categories. But once we admit that a particular scratching, a particular purring, an instance of furriness or perversity, and so forth counts as an individual, then the application of concreteness to all of them is necessary if one is going to be consistent.

One might argue that "concreteness" thus defined merely plays upon the whole-part criterion as theme and variation. Indeed, one might ask if the two are altogether equivalent. I maintain that all classes are to some degree abstract, though they can be more or less so, whereas all individuals are concrete, and that their concreteness does not admit of degrees. To speak of "concrete universals," as have Hegel and his followers, makes no sense, at least in the system of metaphysics being expounded here. Either some word is being used in a different sense or, as seems likely, individuals are being mistaken for classes. However, the possibility remains to be considered of whether some concrete things are something other than individuals. If this be so, then there must be some alternative. Perhaps there is some third category or metacategory under which might fall an entity that is concrete but is neither a class nor an individual.

Let us see if we can find some entities that qualify for such an ontological status. To do so is instructive because it introduces some of the other criteria of individuality that have been proposed. We need to consider things that do not occur as anything like what might be called a "bolus"—a lump of more or less definite magnitude. It helps, furthermore, to clarify what we mean when we try to say that an individual is "a single thing." Common language is instructive on this account. A lot of substantives are "mass nouns"; they do not imply a particular amount of stuff. Good examples are "blood" and "water." We shed blood, not a blood, or drink water, not a water. On the other hand we can drink beer or a beer. Liquids move around easily and do not seem to occur as what might be called "natural units." Solids seem to be a bit different. We nonetheless speak of both "rock" and "rocks." Organisms seem still less likely to need mass-noun terms. I never heard anybody say that he saw some organism, but then again we do eat some beef or some chicken.

In the case of water, we quite often encounter it in more or less definite bodies: droplets, lakes, and oceans. We also recognize that water consists of molecules. In none of these cases do we encounter insurmountable difficulties in identifying what might be called individuals. When I put ice in my drink, it is in the form of individual ice cubes, or at least definite fragments. And the drink itself is a definite amount, with a restricted position in space and time. Invariably what we drink or eat can be interpreted as an instance of some class of something being ingested. Likewise, when I say that we are both "covered with sweat," I mean that I am covered with this particular sweat, and you are covered with that particular sweat, not that either or both of us is covered with sweat in general or in the abstract. Thus I reject the suggestion (Laycock, 1978; Hirsch, 1982) that mass nouns are "abstract" terms rather than ordinary sortals. Rather the terms are noncommittal about what the individuals are. On the other hand, the term 'sortal', which evidently goes back to Locke, is significant in the present context (Jack Wilson, MS). Apparently Thomas Henry Huxley (1852) had sortal terms in mind when he defined 'individual' as "a single thing of a given kind." For reasons that will be developed in due course, I consider such a definition of 'individual' too restrictive.

In the case of mixtures (see Sharvy, 1983) we get some further difficulties. For example it is hard to tell how far one can subdivide a glass of milk and still have milk, not a molecule of casein or whatever. But none of this seems fatal to the general proposition: even if definite individuals are hard to make out, mass nouns refer to entities with individuality, whether or not one has difficulty in singling out discrete individuals. There are indeed serious problems for both the logician and the metaphysician here. They illustrate the point that our ordinary language by no means always mirrors the underlying reality that we seek to understand.

Similar reasoning applies when we switch from everyday life to science. And this provides a convenient opportunity to introduce a fourth criterion of individuality. Namely, in scientific theories, classes and individuals function in very different ways, and mass nouns do not present exceptions. Later we shall discuss in considerable detail what are laws of nature in general, and what is their role in biology in particular. For the present we may be satisfied with the point that there are no laws for individuals as such, only for classes of individuals. Laws of nature are spatio-temporally unrestricted, and refer only to classes of individuals. Thus, although there are laws about celestial bodies in general, there is no law of nature for Mars or the Milky Way. Of course laws of nature apply to such individuals; they are true, and true of physical necessity, of every individual to which they apply. In testing an hypothesized law of nature, a physicist might observe an individual planet to see if it behaves as the hypothesis predicts. If the prediction is not born out by experience, then the hypothesis has to be rejected. In dealing with things referred to by mass nouns, scientists of all disciplines would seem to behave, at least when they are doing their jobs properly, in a manner that is not fundamentally different from what they do with respect to ordinary nouns. Water is a class, and any observation upon water that is used to test an hypothesis about water has to be made upon a spatio-temporally restricted body, or sample, thereof.

One might wish to argue that what is true of classes is necessarily true, and in a certain trivial sense that is correct. Water simply must freeze or thaw if certain conditions are met. This is what makes the laws about such "natural kinds" as water and the various elements so useful under a wide range of circumstances. But a lot of classes are not natural kinds, but rather what are sometimes spoken of as artificial kinds—groups that we make rather than discover. Mill's term "natural kind" is unfortunate because the natural groups of systematic biology are individuals, and hence by definition they are not natural kinds. Some years ago (Ghiselin, 1981a) I suggested an example of a grotesquely artificial kind: everything in the universe, the name of which in English begins with the letter g—including garbage, Ghiselin, and geometry. The members of this class all share one property, and they share it of logical necessity. The class is defined in terms of that property, so they simply must have it. But there is nothing physically or logically necessary about any of those things having those particular names. That is a purely contingent matter. It could have been otherwise, and in fact the English language has changed a lot during the last millennium or so, and we may expect it to continue such evolution.

So we might say that what is necessarily true of natural kinds is true of physical necessity, whereas what is necessarily true of artificial kinds is true of logical necessity, and what is true of individuals is true only as a matter of contingent fact, and therefore not necessary at all. Let us recast this, however, to stress the definitional aspect, which is more a matter of logic than of metaphysics. To do so makes it convenient for us to introduce a fifth characteristic of individuals, namely that they have no *defining properties*. We have already alluded to this matter above, and we will address it in more detail later on. When we define the name of a class by means of defining properties, we state the conditions necessary and sufficient for that name to apply to any given instance. As mentioned earlier, a bachelor is an unmarried male, and if someone is married, he cannot possibly be a bachelor. "Unmarried" is a defining property that is necessary for the name to apply, though by itself it is not sufficient, since some unmarried persons are female. There are other ways to define class names, but this is definitely one of them.

With respect to proper names, which by definition are the names of individuals, there are no such defining properties. This is by no means obvious to everybody, but consider what such a property might be. At a party somebody might mention a name of a person whom you could not identify. You ask who he is, and get told that he is the bearded man conversing with the hostess. The property of beardedness is not defining, for he would be the same person were he to shave off his beard, and it seems most unlikely that he was bearded as a child. And even if he is bearded now, this is simply a matter of contingent fact, for he might have shaved his beard off earlier that day. He would not have been, by definition, somebody else. His conversing with the hostess is not defining for the same sort of reason. Some philosophers have attempted to treat such descriptions as if they were definitions (saying, for example that Sir Walter Scott is by definition the author of *Waverly*), but such notions are rejected here for reasons that will be explained later on.

Now, in defining that person's name, somebody might have gotten along without any use of properties at all, simply by pointing at him. Such a definition is called an ostensive definition, from a root that means "to show", as in "ostentation." Mentioning the fact that the person whose name is being defined has a beard makes it easier to single him out, but it does not make the definition any the less ostensive. Any other character might have been used, and would have, had he happened to have been beardless at the time. At a "christening" a name is bestowed upon a child, and this act often serves as a metaphor for, or an example of, ostensive definition of proper names.

Ostensive definition is rather problematic. Class names are often defined ostensively by giving examples. However, in such cases the thing shown is an instance of the class and neither a part of the entity named nor its entirety. Furthermore, defining properties are presupposed, and the person for whom the word is defined infers what these properties are supposed to be. The inference may or may not be correct. With respect to the ostensive definition of the names of individuals, there are all sorts of other problems. Not the least of these is whether perhaps some property really is implicit when such names are defined.

This brings us to a fifth and last criterion of individuality, which answers the question of whether there is such an implicit property with an emphatic no! An individual is not an individual by virtue of its membership in some class, except in the trivial sense that all individuals are individuals. Another way to put this is to say that being an individual is ontologically prior to being the member of any class. This position is very strong, and in taking it I side with Kant and Bertrand Russell against Aristotle and many others.

The notion that individuals are somehow ontologically dependent upon classes relates to that of an "essence." Essences are a topic that crops up repeatedly in the present work, and an appropriate point has been reached to say a bit more about them. Sometimes an essence is nothing more than a class of defining properties, but in many metaphysical systems they play a far more grandiose role. I only mention them at this point to draw attention to the connection. If one believes that individuals have essences, and that an essence is what makes a thing what it is, then loss of some "essential" property would seem to imply the very termination of existence. In order to do justice to this and related issues, essences will have to be considered in detail, and this job will be deferred until a more appropriate time.

In ordinary language it does seem that we treat the names of individuals as if the individual's existence were contingent upon its having some property that is defining of a class. An organism, for example, is often treated as a sortal. Thus we may feel that when somebody dies he ceases to be anybody and therefore no longer exists. But not everybody is so inclined (see Ayers, 1981). We would perhaps say that somebody who is dead still exists, but only in the form of his remains. When an organism becomes a corpse, it is still the same organism, albeit a dead one. Then again, to some people a dead organism does seem to be a contradiction in terms, though we might agree that this is largely a conventional matter. Most people would no doubt agree that when the remains have broken down and dispersed altogether, it is hard to say that the individual exists at all, apart from the persistence of some immortal soul that would complicate the argument if such in fact exist.

However, the very conception of an organism does suggest a certain kind of activity. If our organism concept were more that of a process than of a material body, and we identified the organism with that activity and nothing else, then a case for living creatures ceasing to exist when they die would be much more compelling. It would be like a dance after all the participants had gone home and climbed into bed. But what would we say if the dance turned into a brawl? If we conceive of an organism as both the material body and what it does, then we are hard pressed to say when it ceases to exist. The way in which we commonly conceive of organisms would again seem far from adequate for dealing with the underlying reality.

When somebody passes from childhood to adolescence, we may or may not consider it a good thing. But we would hardly assert that the child has ceased to exist. Rather, an individual person has ceased to be a member of one class and become a member of another class. And this is how we treat "phase sortals" in general (Jack Wilson, MS). If any individual has ceased to exist, it is an individual attribute or a particular condition of sexual infancy—a child-hood. Individuals need to be envisioned in the context of the temporal dimension, in other words diachronically rather than just synchronically, and not as if they were different things at different times.

We might generalize a bit and say that all changes are of such character. Change does not suffice to make an individual cease to exist in an unequivocal sense. Consider what happens to a unicellular organism when it divides mitotically. It splits in two. Where there was one organism, there now are two. What became of the original cell? Did it cease to exist? No, it has multiplied; one might say that the cell has ceased to exist as such, but it has not been subtracted from the universe. In multiplying, it has converted itself into an individual that we call a clone, the parts of which are individual cells. The clone is a lineage. The parts are autonomous organisms. Things would be somewhat different had we to do with a zygote undergoing the first cell-cleavage in embryogenesis. The unicellular organism would transform itself into the rudiment of a multicellular organism—an individual, which, unlike a clone, would be "cohesive"—and of course still an individual.

Breaking with the common, but by no means universal, practice of treating things as depending for their existence upon the sortal character of some class to which they belong creates some need for clarification. Individuals, and only individuals, can come into existence, go out of existence, and otherwise undergo what we ordinarily would call change. But the issue of what constitutes beginning to exist or ceasing to exist in that broad and unequivocal sense has to be considered from a different point of view than just what properties might be predicated of an entity. We need to consider the question, not from the point of view of what exists, but of what is possible. Although individuals can change indefinitely, they can do so only within the limitations of what is metaphysically possible. Of course they can only do what is logically and physically possible too, but that is not particularly interesting. It is physically possible for an individual chair to be transformed into an individual stool, or perhaps an individual table. Is it then the same individual? A traditional analysis would say that the individual chair has ceased to exist, and a second individual, which happens to be an individual stool, has come into being. Alternatively, and traditionally again, one might maintain that an individual piece of furniture has continued to exist all along, but was transformed from a chair into a stool. One might just as well have painted it a different color.

But such logical maneuvers are most unsatisfying to the ontologist. Rather let us consider what kinds of changes might possibly occur, such that an individual can, by some stretch of the imagination, be considered the same thing. In a fairy tale, changing a frog into a handsome young prince is purely a matter of routine, and with the growth of biotechnology we begin to wonder how far one might go in that direction. At any rate, we can imagine such goings-on, so it does seem a metaphysical possibility. By gradually modifying the prince, we might change him into any number of things, perhaps an ugly old king, then a large bar of soap. But there are certain changes that would make no metaphysical sense at all. How, for instance, could the prince turn into his own brother? In this one sense, at least, it is metaphysically impossible for one individual to become another individual. In simpler terms, this cannot become that. Monozygotic twins are no exception.

Just as it is metaphysically possible for a frog to become a prince, and both metaphysically and physically possible for a prince to become a king, it is also both metaphysically and physically possible for a dance to turn into a fight and a fight into a funeral. But it is hard to imagine what it would mean, were somebody to claim that a prince has turned into a dance. The reason is quite straightforward. Individuals can change only within the boundaries of their own category. A substance can change into another kind of substance—a prince into a king and then into a corpse. And an activity or process can change into another kind of activity or process—a dance into a fight or whatever. As to the suggestion that according to modern physics matter can be changed into energy, we may simply reject the notion that there has been a real leap from one ontological category to another.

Summing up, we have enumerated six criteria by virtue of which individuals may be recognized and individuality may be defined: 1. non-instantiability, 2. spatio-temporal restriction, 3. concreteness, 4. not functioning in laws, 5. lack of defining properties, and 6. ontological autonomy. Of these, non-instantiability is perhaps of greatest interest to the logician, concreteness to the ontologist. Quite a number of items that might have been added to this list have been deliberately left off for reasons that will be explained in the following chapter.

4

WHAT AN INDIVIDUAL IS NOT

"It is quite curious how such words as 'zoological parentage' [and] 'descent from different types' are used metaphorically. In future ages such language will be a wonder."

—Darwin, manuscript comment on a paper by Milne-Edwards, 1844 (Darwin Archives, Cambridge, Volume DAR 72)

Turning things around, whatever an individual may be, not all individuals are organisms. But all organisms are individuals, so any argument against treating a species (or anything else) as an individual can be rebutted by showing that it would apply just as well to an organism. Often in the present work we thus argue by counter-example. However, the converse is not true. The mere fact that all organisms have a given feature by no means implies that all other individuals must have it. The moon does not grow by intussuception. Nonetheless quite a number of misconceptions of the individuality thesis have resulted from such misguided extrapolation. The reason is not far to seek. Not only does ordinary language conflate two senses of 'individual', but for thousands of years it has been traditional to consider a person as the paradigm example of an individual. A beginner in such an exercise may have no solid guideline about which properties of the more particular concept to attribute to the more general one as well. A much better approach, which has been adopted here so far as possible, is to start with the broader concept and then focus in on the more particular. Having explained what is meant by an individual, it is now appropriate to consider such misconceptions, as well as a few related ontological problems.

Organisms are "cohesive" in the sense of being tightly held together by sinews, connective tissue, and all sorts of other things. Species are held together by sex, and if species are individuals, one might wonder whether all individuals are held together by something of the sort. The notion that cohesiveness is a necessary condition for individuality has been maintained by such able thinkers on metaphysical subjects as Mishler and Brandon (1987), Ereshefsky (1988), Grene (1989) and Mary Williams (1992). They attribute the idea to

David Hull, but when I telephoned Hull and asked him, he told me that he was by no means of that opinion. Nor do I believe that he ever has been of that opinion, though what he said in his early writings might be construed that way. Nor have I ever been of that opinion, though my conviction has grown down the years. If something is cohesive, then it is capable of doing things or participating in processes, and is like the chair that I can actually sit on, rather than like chair in general. It is concrete and has a definite position in space and time. Cohesiveness is therefore sufficient, but not necessary, for individuality. But what of noncohesive entities, such as lineages? Is something an individual if all that the elements share is a common ancestor? What else could it be? There are four logical possibilities, namely: they are classes, they are individuals, they are neither, or they are both.

If one believes that species are classes, then it would be difficult to conceive of higher taxa such as genera and families as anything but classes. If one believes that species are individuals, one can still treat the higher taxa as classes. For if it were true that individuals must be cohesive, as just suggested, and if higher taxa are not cohesive, perhaps one would see no other alternative. And indeed such a manner of reasoning helps to explain the very common sentiment that supra-specific taxa are "less real" than species are. There is another interesting possibility. If species arose by spontaneous generation or by special creation, and therefore shared no common ancestry with one another, the higher taxa would indeed be classes. And those taxa that do not include the common ancestral stock, especially ones that were erected to accommodate organisms that are similar as a consequence of evolutionary convergence, are widely interpreted as classes and dismissed as "artificial." Slugs are a good example. They have repeatedly evolved from different groups of snails.

If, as I maintain, "individual" and "class" are contraries, entities have to be one or the other, or perhaps neither, but they cannot be both. But not everybody thinks that they are contraries. Van Valen (1976) is the only person of whom I am aware who has explicitly argued that species are both classes and individuals. He refers to "individualistic classes" for what I would be more happy to call "classistic individuals" if that were not a dreadful neologism. Van Valen receives high marks for ingenuity, but in order to make a real case for such a thesis one would have to develop a suitable ontology to go with it. To be sure, every material body known to us is a group as well as a single thing, and this seems to be the source of much perplexity. Yet I know of no object in the entire universe of which it may be truly and unequivocally said that it both is and is not spatio-temporally restricted, that it both is and is not concrete, that it both does and does not have instances, that it both can and cannot participate in processes. Colless (1972), one of the few authors to discuss my views on individuality before my paper of 1974, tried to treat species as both classes and individuals, but at the same time he treated everything, including organisms, as classes of phenomena. Such ploys really don't solve the problem.

Another logical possibility is to say that an entity could be something other than a class or an individual. This one is much more popular than saying that it could be both, and on the face of it at least seems more or less plausible. Nelson (1985), Platnick (1985), and others have asserted that such is indeed the case, but without telling us what the third alternative might be. One can hardly be expected to answer what is not an argument, but rather an assertion unsupported by evidence. Löther (1972, 1991) proposes that lineages are neither classes nor individuals, but something *sui generis*. A professional philosopher of biology who began to deny that species are classes at about the same time that I did, Löther preferred to call species "material systems" rather than "individuals." Unfortunately he has merely stated his position, so we are in no position to criticize it one way or another.

Among professional philosophers, perhaps the most adamant opponent of the individuality of species has been Philip Kitcher (1984b, 1987, 1989). According to Kitcher, species are neither classes nor individuals, but "sets." His views have been criticized upon various grounds by his fellow philosophers (Sober, 1984). As Mayr (1987b) points out, it is rather difficult to rebut Kitcher's claims, because he does not explain what it means to be a set. Max Black (1971), in a well-known essay, shows that mathematicians and philosophers alike have been notoriously vague and downright muddled by what they mean by "set," so Kitcher is in distinguished company. Until he explains himself we can only guess what he is trying to say. Bradley Wilson (1995) interprets him to mean that species are artificial classes, but rejects that view. His own solution is to shift the emphasis from populations to lineages, which of course are individuals too.

The basic, elementary argument against the notion that species are sets rather than individuals is that sets are indeed classes. But rather than being intensionally defined classes, they are extensionally defined classes. One merely lists the things that are, by definition, the members of the set. Thus, the moon, Kitcher's right ear, and the waste basket next to me "is" or "are" a set. The moon, Kitcher's right ear, and my telephone "is" or "are" another set. A species or other population is composed of a "set" of organisms at any given instant. As time passes, some of these organisms reproduce, adding new ones to the population, and old ones are subtracted through mortality. But when this happens, what we started out with does not have the same extension as what we ended up with. Therefore it is not the same set. An organism is not a set of molecules, for we are always anabolizing and catabolizing, building tissues up and breaking them down again. The water in our bodies is rapidly replaced as we drink and excrete it.

The argument of course gets a bit more complicated than that, and logicians might hope to develop a kind of set-theoretical treatment of species that would not have such a problem. However, it seems to me that such efforts are a complete waste of time so far as metaphysics goes. The term 'set' as used in such discourse does not assert anything about the ontological status of species or of anything else. When I say that a species is an individual or that it is not a class, I am asserting something about metaphysics, placing it within an ontological classification scheme. But when Kitcher says that a species "is" a set, he is asserting nothing of the sort, if indeed he is enunciating a meaningful statement at all. Set theory is indifferent to ontological status. When one says that an entity "is" a set, one is merely enunciating the rules for how it shall be handled from a logical point of view. One speaks in the imperative mode, not the indicative. It is a proposal to talk about it in a given way, not a proposition about it that can meaningfully be said to be true or false. There is no question that one can formulate sentences in which species are interpreted as sets—perhaps even as something other than sets defined merely by the enumeration of their members. However, the same thing can be done with organisms and anything else in the universe that has parts, including Kitcher. To say that species and organisms are sets is like saying that we can talk about them in English or Chinese.

Wiley (1980) accepted the individuality of species, but proposed that the sort of lineages that share only common ancestry be accorded a special ontological status: "historical entity." Thus, as I have suggested in earlier publications (Ghiselin, 1981a), we get two different ontological classification systems:
Wiley:
> Individuals
> Classes
> Historical entities
Ghiselin:
> Classes
> Individuals
>> Cohesive individuals
>> Historical individuals

In other words, the distinction between lineages, on the one hand, and those composite wholes that are held together by more than just common ancestry, on the other, can be acknowledged either with or without denying that the former are individuals. (Sometimes the term 'historical entity' is used in a broad sense to mean "individual" whether cohesive or not, but the basic arrangement is the same.)

What ought one to do under such circumstances? In biological systematics, it is perfectly acceptable to have what is called a trichotomy, with three items of equal rank. For instance, Linnaeus (1758) in his *Systema Naturae* divided the tellurian creation into three kingdoms: Animalia, Plantae, and Lapides. His was not a *Systema Vitae*, and discussions of biological classification that appear in textbooks to my knowledge never even suggest that one might place all of life on Earth into a single empire, perhaps called *Vita*. But the advantages to recognizing the existence of a single taxon for the common ancestor of tellurian organisms and all of its descendants are obvious. Yes, we might argue for a better system of kingdoms, and many have. But we do not consider plants as something *sui generis*—as much like stones as animals—merely because they lack some of the properties that occur in animals.

Those who would deny individuality to lineages do so primarily on the grounds that these lack cohesiveness. Granted, but consider what they do have in common with cohesive things. They are indeed spatio-temporally restricted, each of them having a beginning and an end. They are not instantiable, though perhaps one might speak of them as if they were. They function as individuals in scientific theories, and for the same reason that cohesive ones do. They are concrete in the sense discussed above. This is the basic argument, which should suffice for the present, though further evidence will be provided by means of examples. In later sections of this book we will examine the properties of hierarchically-structured entities that incorporate some historical units. Among these there are the objects of linguistics, such as languages with their language families.

At this stage, however, a more appropriate consideration is what we make of cohesiveness in those organisms and other common objects that may actually lose their cohesion for a time but regain it later on. A situation in which an organism breaks up into component parts that never get back together again is familiar even to the lay person. Propagation by cuttings, budding, and the fission of some animals such as starfish are good examples. There are fewer examples of organisms that break up, then fuse back together, but slime-molds are an example. These fungi, which form lineages produced by asexual reproduction, forage independently on organic materials. Later in their life cycle they come together and form a single mass, complete with reproductive organs that give rise to spores. Sometimes they are called "social amoebae," and the term aptly compares them to societies that are united only from time to time. Sponges too can be taken apart and made to reaggregate in the laboratory, though the result is not, to my knowledge at least, treated as if it were the reconstitution of the same organism.

Nonetheless, our conception of individuality is sorely tried, and perhaps expanded, when we contemplate objects that can be taken apart and be put back together again. Things get even more frustrating when we consider what happens when parts are substituted. There is a substantial philosophical literature on such topics, and for fear of giving the wrong impression I hesitate somewhat to discuss it. People might think that I have something to add to philosophical discourse about mereological essentialism and the like. No, it is simply a matter of providing an exercise that helps us to appreciate the problems of individuality, some of which may be insoluble in principle.

Many readers will no doubt be familiar with the famous "Ship of Theseus" puzzle of Hobbes. A ship has its parts replaced, one by one. The original parts are discarded, then somebody takes the parts that had been discarded and puts them together into a second ship. Is the first ship, the second ship, neither, or both the individual ship that was initially present? Identifying the ship with a continuously cohering entity might lead someone to prefer the first of these alternatives, as does Hirsh (1982:69), but those who tend to identify an individual with its components might prefer the second. To identify the original ship with neither or with both of the later ones would be more draconian solutions. The latter of these, which amounts to turning the original ship into a something like a clone, comes naturally to a biologist like me, but seems not to have been entertained in the philosophical literature.

It seems to me that more draconian solutions, which acknowledge the inherent difficulty of applying the concept of an individual to certain cases, are probably necessary. However, it is well to point out that the ship of Theseus problem, as formulated by Hobbes and by modern philosophers such as Hirsch (1982) and Simons (1987), tends to be simplistic and to suffer from misguided efforts to apply the traditional metaphysical categories. Thus, following Locke, we might say that some identify the ship with its matter, some with its form; but from the point of view of physiological anatomy a ship is a ship because of what it does—its function, which is not to be confused with its final cause, but rather means its capacity, at least, to transport objects from one place to another across water. Thus conceived, the continuous existence of a ship in its functional state would imply that the one that was reconstituted out of discarded parts is a new and distinct individual, even though these were the components of the original one.

Having introduced the problem by means of that traditional ship, let me discuss two somewhat different individuals: a pistol and a committee. Firearms are often disassembled, perhaps for cleaning or repairing them, or for safety. A disassembled pistol is less hazardous to children precisely because it lacks cohesiveness and interconnectedness of parts and in that sense resembles a mere lineage. But when somebody breaks the pistol down and then puts it back together again, we do not ordinarily say that it has ceased to exist and that another one has been constructed. We can imagine such a view to have some legal applications. Suppose that somebody claimed under oath in court that he did not own a firearm. Accused of perjury, he might plead innocent on the grounds that all he owned was a bunch of unassembled components. Whether or not we reject his argument as sophistical, it does seem that mere loss of cohesiveness is not the sort of change that would cast a thing into a radically different metaphysical category. Cohesiveness would seem to be sufficient, but not necessary, to maintain individuality.

Even so, a pistol or anything else continues to constitute a particular item in the inventory of its possessor's belongings in spite of having been disassembled. For that reason if no other, it does not make much sense to say that it ceases to exist when disassembled and then comes back into existence when reassembled, as has been argued by Burke (1980). On the contrary, we had better conclude that we have yet another example of how that notoriously equivocal term 'exist' misleads philosophers. It is problematic whether we should say that the intentions and dispositions of the owner provide a kind of "external cohesiveness" upon such items.

Whether one owns a pistol when some functionally indispensable parts are missing is another issue. At least one owns an individual something, even when it lacks a firing pin. But consider what happens when we swap parts among a group of pistols. One might take half a dozen of them, disassemble them entirely, and put half a dozen back together again with randomly selected parts. What has happened to the original pistols? One solution would be to say that they have lost their cohesiveness but continued to exist as historical individuals, with their parts randomly distributed among half a dozen cohesive individuals, which are not the same cohesive individuals as we started out with. We might also say that the pistols, by virtue of their parts having been commingled and reconstituted, now form an individual of higher order. It is a population of pistols. And if we look at modern means of arms manufacture, we may note that one reason why the parts are interchangeable is that all are deliberately constructed according to the same plan, specifications, or blueprint. In the case of manufactured items, they are generally produced with numbers that identify the lot, and serial numbers that identify the individual functional unit, such as a pistol. The serial number may be stamped on just the "main" part, though it might be stamped on several, or all of them.

Let us now try to make matters not better, but worse, by proceeding to the second example that was promised: a committee. Because committees are social units, we rather expect them to be more evanescent than their component organisms. When a committee is actually meeting, it seems more or less tangible, a functioning whole with definite modes of interaction among the parts. Some of these interactions are crudely physical. Indeed, the "members" might even come to blows. When the committee is not meeting, one might want to say that it has passed out of existence. And yet it has not ceased to exist as a functional unit. If the committee was empowered to vote itself out of existence, and had in fact done so, the committee members might behave very differently, and so might the larger organization of which the committee is a part. Consider, for example, the possibility that it is a grievance committee. Persons who felt aggrieved would behave very differently if the Grievance Committee were not in session, on the one hand, and if it had ceased to exist, on the other. We might want to say that a committee has a certain kind of social cohesiveness. It is a very interesting kind of cohesiveness, for it illustrates the point that a whole can exist without the kind of physical continuity that we observe among the parts of a chair.

With committees, furthermore, we are perhaps less apt to equate the whole with a mere enumeration of the components, that is, to treat the thing as if it were an extensionally-defined set. If John resigns from the Membership Committee and is replaced by Barbara, it is still the same committee. If the Membership Committee consisted only of Barbara, it would still be that individual committee. There is nothing paradoxical about a committee of one. Indeed, the Membership Committee might be thought of as continuing to exist even if John, Barbara, and all the other persons on it were to resign. In this case we might say that the Membership Committee still existed, but none of the positions on it were occupied.

A committee might be thought of as the occupants of social places ("offices") within the committee, of which the committee itself occupies a social place of a higher order. Indeed, one might treat the committee as a whole that incorporates those offices. On this basis one might reasonably elaborate a kind of "place metaphysics" in which the office was prior to its occupant.

As Marcus has urged, committees are not the "set" of office-holders (Marcus, 1974). Exactly the same persons might constitute two different committees. But they would not be the same committee, for they would not have the same capacities and functions. Suppose, for instance, that the Grievance Committee were to meet, consisting, by chance, of persons who were also the components of the Rules Committee. Now suppose that a motion were laid upon the table, and it was pointed out that only the Rules Committee had jurisdiction, and no action could be taken. Of course one possibility might be to call a quick meeting of the Rules Committee, but maybe committee meetings were not allowed without two weeks notice. On the other hand it might be perfectly possible for two committees, consisting of exactly the same persons, to meet in joint session. This provides an instructive example of two individuals (say the Rules Committee and the Grievance Committee just considered) occupying exactly the same physical space at exactly the same time. And it flies in the face of what is so often said about individuals. If an individual were simply a given blob of matter, then it would stand to reason that no two individuals can occupy exactly the same space at exactly the same time. Social wholes are by no means thus constrained.

I hasten to point out, however, that this counter-example is not so devastating as one might think. The two committees are not the same part of the same social whole, and in joint session each person occupies two different places in the society. Why should we say that an individual is to be identified with its component matter, rather than with a certain functional role that gets discharged within a functional whole (any more than we must identify it with the matter, rather than with the activity, as in the case of a dance)? This is not to deny the individuality of the officeholders or of their tenures in office, any more than that of the dancers or their acts of dancing.

Another point illustrated by the example of a committee is that an individual can be part of more than one whole, even at a single instant. Any reader of this book is apt to belong to all sorts of organizations, including professional societies, clubs, committees, departments—you name it. In the case of physical objects rather than social ones, however, there are more restrictions. Except in such dubious cases as Siamese twins and colonial organisms, an organ is never simultaneously a part of two different organisms. In medical practice one might arrange for two persons to share the same heart or kidney for a while, though the example is a bit forced. On the other hand, an organ can in fact be part of more than one organ system. For example, the gonads of vertebrates are parts of both the reproductive system and the endocrine system. Certainly a part of an organ is both a part of the organ and a part of the organism, which can in turn be a part of various populations and societies.

Returning to the trichotomous classification suggested by Wiley, we may conclude that cohesion or the lack of it does not seem to imply a deep metaphysical cut. Things can gradually come unglued, fall apart, and have the components go their separate ways gradually and by insensible degrees. Such changes are commonplace and readily imagined, but it is far from obvious what it would mean to say that an individual has been transformed into a class and has lost not just its cohesiveness but its spatio-temporal restriction and the like. The difference between historical entities and cohesive individuals may be important, but it is far less profound than that between historical entities and classes. Therefore cohesiveness, though perhaps a sufficient condition for individuality, is by no means a necessary one.

Having disposed of cohesiveness, let us now proceed to consider a few other putative criteria of individuality. And let us reason so much as possible by analogy with individual organisms and other objects the individuality of which ordinarily is not contested. One very commonly invoked criterion is autonomy, or self-sufficiency. According to this view, individuals can get along independently of others. The organism may reasonably be defined as the unit of physiological autonomy, though this definition is far from adequate. But organisms in general are not fully autonomous, and this autonomy is a matter of degree. Man, in particular, is a social being. Protracted solitude means degradation and misery. Even autotrophs depend upon heterotrophs for the regeneration of nutrients. Aristotle said that a fully autonomous man would be a god. And Spinoza concluded that in this sense there is one, and only one, true Individual, namely,

God. This is definitely not the sense that is intended when we are discussing the individuality of species or organisms.

Another possibility is that an individual has distinct boundaries that set it off from everything else. And yet when we look closely even at an organism like ourselves, we are hard pressed to decide where to delimit the organism and its environment. There has been a lot of loose talk about the cavities of the body, and where the exterior ends and the interior begins. How about the gut, for example? Is the food within it inside or outside the body? Some persons would insist that it "really" is outside the body. Certainly the food in question is not within the tissues. The cavity of the gut might be outside by virtue of the continuous cellular lining that covers it. This would make the ducts of the pancreas and the liver external to the body as well. But then, what about the coelom, which (just in case you have forgotten) is the cavity with a mesodermal lining that surrounds most of the viscera. In women, the fallopian tubes open into the coelom, allowing for passage of the ova. But there is no such opening in men. So here we have the secondary body cavity outside of the body in one sex, inside it in the other. In many organisms, the body is much more open to the exterior, and likewise the boundaries between what we would call Harry and John are harder to draw. Sponges are particularly difficult; the organisms may even be made up of cells from different species or even different orders (Little, 1966). Nonetheless such boundaries are important, and in later sections we will consider them in greater detail. Suffice it to say for the moment that when we call this "individuality" we are not considering "individuality" in the sense that primarily concerns us here.

Another sense of 'individuality' means being different from other things of the same kind. We often use such expressions in an honorific sense, especially when referring to outstanding personalities. An individual person is more than just another instance of a class, and perhaps wants to be recognized as such. According to such a criterion, a pair of identical (monozygotic) twins are not two individuals. But we rarely if ever talk that way in biological discourse; we might say that the pair of twins constitute a single individual in the class of clones, but we certainly would not deny that they are individual organisms. In the case of Siamese twins, we find that this criterion tends to conflict with that of autonomy. Obviously neither of these is defining of individuality in the sense that concerns us here.

Another sense is apparent from the etymology of 'individual'—something that cannot be divided. A lot of organisms cannot be cut in half and survive, but many can, and indeed scissiparity, which is what happens, for example, when certain starfishes tear themselves in pieces and regenerate entire organisms from the fragments, is a not uncommon mode of asexual reproduction. One might wish to argue that once something becomes divided it ceases to be the individual that it was, and perhaps that it turns into two other individual organisms. We have already entertained this possibility. On the contrary, I would say that if a thing can divide and tear itself in half in a literal sense, then it is indeed an individual, for only a concrete being can engage in a real process.

The "division" of a class is something else again, and perhaps the inability to do so beyond the level of the (concrete) individuals is meant. Thus, if one starts "dividing" the class of furniture into chairs, couches, etc., then divides chairs into armchairs, thrones, etc., one cannot proceed beyond the chair upon which, perhaps, you now sit. Otherwise one has to switch to the whole-part relation, which is a different matter. If that is what is meant, then this is just another way of stating that individuals are concrete. As we go downward in a hierarchy of classes included within classes, the groups that we encounter become ever less abstract, until we reach the solid ground of individuality, and can go no further.

5

Some Definitions of 'Definition'

"When I read your remarks on 'purpose' in your *Phytogeographie*, I vowed that I would not use it again; but it is not easy to cure oneself of a vicious habit. It is also difficult for any one who tries to make out the use of a structure to avoid the word purpose."

—Darwin, letter to A. de Candolle, 24 January, 1881

The reader who has had sufficient patience to follow the above discourse has been given ample evidence that the term 'individual' is not easy to define, especially if we want a definition that is adequate for the task at hand. The impatient reader, again, might ask for what he would consider, by definition, to be a definition—namely a terse, verbal formula that encapsulates the meaning of the term. But why does a definition have to be terse? What does it mean to 'mean'? Of course I can repeat the Aristotelian adage: an individual is a "that." But it is not very informative to say that a species is an individual means that it is a "that" to a someone who does not already possesses the sort of information that has been provided in earlier chapters. For most people calling a species "a that" is at least as puzzling as calling it "an individual." We might use the distinction between "such" and "that" as a starting point, but the very fact that it works only as a starting point merely serves to underscore the inadequacy of so terse a definition.

What would our impatient reader expect if he asked for a definition of 'definition'? Could it be both terse and adequate to the task at hand? There is no easy and straightforward answer. There is a branch of philosophy called the "theory of definition" that addresses such questions, but it leaves many of them without a satisfactory answer (Robinson, 1954).

One point that might well be mentioned at the outset is the distinction between a "lexical" and a "stipulative" definition. A lexical definition, such as one that occurs in a dictionary ("lexicon"), is a kind of report on how language is used. A stipulative definition proposes ("stipulates") that language shall be used in a given way. The former is in the indicative mode. It purports to be an accurate description, and therefore can be true or false. The latter is in the imperative mode, and something that we accept or reject on the basis of a value judgment. The distinction between lexical and stipulative definition is particularly important in the present work. Describing how scientists use the word 'species' is not the same thing as advising them to use it one way or another.

Bearing these caveats in mind, let us attempt to define 'definition'. The task is made somewhat easier because there was no practical way to avoid introducing some material on definition in earlier chapters. A definition, let us say, is the criteria that justify the application of a term to that to which it refers. Wittgenstein (1921 translated 1961:61) along such lines said that "A definition is a rule dealing with signs." Already we have a problem. Wittgenstein's words as translated suggest that the definition consists of only one rule. But consider how one defines bachelor: an unmarried male. By "definition" do we mean just one of these so-called defining properties, or is it both taken jointly? And what about other criteria such as eligibility? We need to make it clear whether the definition consists of just one necessary condition, or of all that are jointly sufficient. At the very least this tends to increase the amount of words needed to define the term. On the other hand it is widely asserted that a definition ought to be such that one can substitute the definition for the word that is defined, i.e., the definiens for the definiendum, without changing the meaning of a sentence. Obviously we need more than just one of the defining properties to meet this requirement.

Note how it has been taken for granted that when we define, what we define is a "term"—in other words that definition is "nominal" in the sense that what we define are names, symbols, and the like. We define the names, not the things of which they are the names, and, furthermore, the names are the names of those things, not the names of something else, such as concepts. This position is pretty orthodox. But it stands to reason that if classes and individuals are ontologically very different, we may have to deal with their nomenclature in rather different ways. Old-fashioned nominalists said that the only thing that classes share is a name. Certainly they share at least that, though most of us would say that they also share properties. Individuals, however, can in fact be named. On the other hand, we probably wouldn't say that they "share" proper names. At least not in the sense that classes share class names. One might say that lots of people share names like 'Michael' or 'Agnes', but that is only because these are not "unique identifiers" like a person's Social Security number. Likewise the parts of an individual such as my hand and my foot "share" my name only in the equivocal sense that one can refer to them by enunciating a sentence in which that name occurs.

We might therefore say that individuals *have* their (proper) names and *share* the names of the classes of which they are members. On the other hand, language is a bit more complicated than all that, and some words would seem to be exceptions to this generalization. In particular, we have what I will call "componential sortals" for want of a better term. By this I mean a name that is shared by all the members of a kind of part of a particular individual. Good examples would be 'Englishman' or 'human being'. When we say that Charles Darwin was an Englishman we mean that he was an organism-level part of the English nation. When we say that he was a human being, rather than a dog, we mean that he was an organism-level part of *Homo sapiens*, not of *Canis familiaris*. Note that this is not the same thing as class membership. We would not say that Charles Darwin was an instance of the English nation, any more than we would say that Middlesex County is an England. Likewise, when we say that Charles Darwin was not a *Homo sapiens,* we are not denying that he was a human being, as Mahner (1993) asserts.

We must emphasize that "human being" refers to the organism-level components of *Homo sapiens*, but not to all of the components. Gametes, such as sperm cells, are components of species without themselves being organisms or necessarily being parts of organisms. One may wish to argue whether a human fertilized egg (zygote) is a human being, but hardly the sperm that failed to fertilize anything. In other words, *Homo sapiens* is not extensionally equivalent to the class of human beings.

Such componential sortals properly analyzed would seem to be neither classes nor individuals in a straightforward ontological sense, but rather a verbal ploy that saves us the work of expressing ourselves with greater ontological precision of a kind that is quite unnecessary in everyday life. However, the existence of terms that are ambiguous in this respect is a very important source of confusion. It illustrates how misleading an "ordinary language" approach to metaphysics can be.

Note that I referred to the "criteria" for the applicability of the term, and not to the "defining properties" or the like. As a matter of fact, 'definition' sometimes is taken to mean the enumeration of the defining properties, and nothing else. In other words, all definition is "intensional." Terms that have no defining properties would (by definition) have no definitions. Accordingly those who use language in this way say that ostensive definition is not really definition, but merely fixes the reference. I prefer to stipulate otherwise, and use 'definition' in a broader sense, and to alert the reader to the possibility of equivocations.

From what has been said earlier, it should be obvious that there are at least two kinds of terms that are defined otherwise than by intension: extensional class names and proper nouns. An extensionally defined class (name) is one that is defined by stating its membership and nothing more. Such a definition would consist of a kind of list, and being thus listed would be the only criterion for the name to apply. If one adds or subtracts an item, then one gets a different list and therefore a different class. This is the basic reason, as was already pointed out, for denying that biological groups such as families and species are extensionally defined classes; if somebody dies or is born, one class has replaced another.

What is truly an extensionally defined class would seem to have very little interest in the natural sciences. If the only thing that a list of items had in common was being on a list, then why would anybody bother making out such a list? A telephone book is a list, but the listees share the property of wanting their telephone numbers made public. Faunal lists are classes of what is known about the animals from a given area or other significant part of the natural world. When we encounter what purport to be extensionally defined classes or sets in biology, we are probably dealing with what is something else, both ontologically and epistemologically. It is either an intensionally defined class or an individual. One might make an exception for certain "artificial groups" in biological systematics. Say, for example, we make a group out of all animals minus toads and sponges. It would be rather like you, versus your right ear and your left foot. Such groups are scientifically uninteresting. If a group is considered artificial because it has arisen by convergent evolution, for example whales and ichthyosaurs, then it is indeed interesting to scientists, and for that very reason. But then it is obviously an intensionally defined class.

Some people may object to my calling some of these artificial groups scientifically uninteresting, so I hasten to point out that it all depends upon what we mean by "scientifically interesting." I don't mean that scientists have no proper concern for groups that are not strictly natural. When I am filing my reprints, I am pleased to use such groups as Pisces and Reptilia, because they put the reprints in an order that is convenient for locating the publications about fish and reptiles. It really does not matter that these names do not designate entire lineages. Gnathostomata and Amniota are indeed the names of such lineages, but using them when filing the reprints would make my filing system unnecessarily complicated. On the other hand, when trying to understand the evolution of vertebrates, it is groups like Gnathostomata and Amniota, not Pisces or Reptilia, that I think about.

As briefly alluded to in an earlier chapter, proper names, which are names of individuals, have no defining properties—no intensions. The notion that proper names are a kind of "meaningless mark" has been widely, but not universally, accepted since the work of Mill (1872; cf. Marcus, 1993). Hence one reason for referring above to the *criteria* that justify the application of their names. They are defined "ostensively"—by "pointing" as it were to the individual thus named. Ostensive definition has been likened to a christening (Ghiselin, 1966b). One might get some wrong impressions here, so let us consider the matter further. Suppose, for example, that instead of just bestowing a name upon a person, the procedure was rather to take a lock of hair from his or her head, attach that hair to a certificate bearing the person's name, and file it with the authorities as a means of determining who had that name. This would be a strange way to do it, and one might prefer a fingerprint or something more readily identified, but a lock of hair will do. Such a procedure would, as jargon has it, fix the reference in a rigid manner. The name could designate that person and nobody else. As here stipulated, such fixation of reference is one kind of definition.

If it seems odd that we would "point" not at the whole organism, but at a part of an organism, or even at a product of an organism, this situation is far from unfamiliar to systematic biologists. Indeed it is the rationale for the practice of type-designation as provided for in the *International Code of Zoological Nomenclature* and in the *International Code of Botanical Nomenclature*. (For an excellent discussion of type designation, see Mayr, 1989.) When a new species is described and a name bestowed upon it, a "type" is deposited in a museum. The term 'type' is unfortunate. Philosophers and even some biologists have tended to assume that the type specimen is an "example" or an "instance" of the species thereby named, that it is somehow "typical" or a standard. The word 'typology' suggests itself. Biologists often insist that types are not typical (J.M. Schopf, 1960), but this has often fallen upon deaf ears of philosophers (Heise and Starr, 1968, 1969). For nomenclatorial purposes a bird's nest used to be just as good a type as a bird's skin, and this ought to clinch matters. But the procedure hardly makes sense if species are classes. It "hardly" makes sense, because Heise and Starr tried to interpret type designation as ostensive definition of class names. To make matters worse, Colless (1970) replied that organisms, such as individual human beings, are also classes.

Now, one might argue that the ostensive definition of (the name of) an individual is really an intensional one in disguise. This is, indeed, the unquestioned assumption that has commonly been presupposed. In this connection we need to consider two additional issues. First, do proper names perhaps have an intension after all? Second, can the names of classes be defined by the sort of ostensive definition that is commonly provided for those of individuals? The first question will be answered with a qualified "no," the second with an unqualified "no."

As to the first question, we might argue that when we name individuals we imply that these individuals have certain properties, and that some of these are actually defining. Some arguments for this view can be readily dismissed. For example, if somebody is named "Betsy" we assume that she is female, but such a name might be bestowed upon a male out of accident or perversity; and in the case of certain wrasses and other sex-changing fish, a female named Betsy might become male and yet remain the same organism. So such terms may *suggest* particular properties, but they are by no means logically necessary in the way that defining properties are.

A more serious consideration is that when we name objects we generally name them as objects of a given kind. In order to provide an ostensive definition, we need to make it clear precisely what is being pointed to. For instance, in naming a species by ostensive definition, the rules make it clear that we are naming a species, not an organism. Likewise at a christening everybody knows that we are naming the entire person, not part of that person. We may believe that being alive is a necessary condition of being that person, and that after death that person has ceased to exist. But others might well disagree, and apply a proper name to a corpse as well—to somebody's "remains" as it is put. In other words, being an instance of a particular class is sometimes considered a necessary condition for that name to apply to that individual, whereas on the point of view taken here, the name would apply no matter how much that individual changed, provided of course that there was anything left of it. We run into a problem with sortals again. It is one thing to say that an individual ceases to exist as a member of a class that has some defining property, another to say that it ceases to exist as a member of any class whatsoever.

The very fact that an individual exists implies that it has properties of one sort or another. The notion of a "bare particular" in the sense of a featureless object makes no sense, especially if we take concreteness as a necessary condition for individuality. What we deny is not that individuals must have properties, but that they must have any particular property, with the possible exception of being individuals in the category of property itself.

The recent exchange between Kevin de Queiroz (1992, 1995) and me (Ghiselin, 1995b) about the putative "defining properties" of the names of lineages (clades) is really about a way to fix their reference by making them particular instances of a class. In the case of an organism name we would point at Betsy and say that "Betsy" is, by definition, "that" and an organism. Supposedly the defining properties of "Betsy" are two: thatness and organismness. In the case of a species name, we would use type-designation so that, for instance, *Homo sapiens* has two properties: thatness and speciesness. In the case of a lineage name, we would single out the common ancestor and say, for example, that "Mammalia" has two properties: thatness and lineageness. "Thatness" means of course individuality—being that particular instance of one of the three classes in question (organism, species, and lineage). If we stipulate that the name being defined can only refer to an organism, and that organisms by definition must be living, then of course we have provided a set of rules whereby we effectively fix the reference of the proper name to one and only one individual. But to treat this procedure as if it were a matter of enumerating a list of defining properties strikes me as a serious equivocation. It is highly problematic whether "being that" is a property in any ordinary sense of "property." After all, it is not something that can be shared with another individual. And the mere fact that properties somehow enter into the definition of proper names does not necessarily mean that they function as defining properties.

Furthermore, the names of taxa remain names of the taxa themselves. They are not "shared" by the component organisms or lineages. They are not even "componential sortals" such as I mentioned earlier: they are terms like 'Mammalia' or *Homo sapiens*, not 'mammal' or 'human being'. Therefore the claim by de Queiroz to somehow have broken down the distinction between classes and individuals is spurious, "just a piece of semantic trickery" as Frost and Kluge (1994:265) put it. It is another case of confusing the spatio-temporal restriction of a member with that of a class. Be this as it may, there is no disagreement between us, so far as I can see, with respect to the suggested procedures for fixing the reference ostensively (de Queiroz & Gauthier, 1990, 1992, 1994; de Queiroz, 1994; Rowe, 1987, 1988; Rowe & Gauthier, 1992). On the one hand it allows us to get away from essentialistic definitions. On the other hand it provides us with a way to make the reference of terms increasingly determinate, something which, as Field (1973) points out, seems to occur quite generally as science progresses.

A very common mistake is to confuse the defining properties of a class of which an individual is a member, on the one hand, with properties that are supposed to be defining of the proper name of that individual, on the other. Those who discuss such matters in terms of essences rather than defining properties might say that individuals have essences. For instance, if we say that Charles Darwin is the name of a person, that which has a defining property is not "Charles Darwin" but "person." And when we define the name of a species, we do stipulate that it is indeed the name of a species—the one that is pointed to. (That species, rather than some other species.) A failure to understand this distinction led some of my critics to claim that individual species must have some property that is defining (or essential) of their (proper) names. But we can see that they were on the wrong track when, for example, Bernier (1984) suggested that chromosome numbers might be defining (or essential) (Ghiselin, 1985).

This search for defining properties nonetheless raises the very interesting question of whether proper names might be looked upon as more like titles than as the mere labels that Mill claimed them to be. Mill was careful to distinguish between proper names such as "John" and descriptive names such as "King John," which he thought do in fact have what we now call intensions. It seems almost paradoxical when it is said "The King is dead, long live The King!" Much as a king might cease to be called "The King" upon abdication, it seems quite reasonable that the name of a species should not refer to the entire genus that gets produced when it speciates. But a person's title does not assert something that must, of logical necessity, be true of a particular individual. It is more like the name of a role that somebody does in fact play, but might not. This point has been alluded to already.

Many philosophers have argued that ostensive definitions are really what might be called "covert descriptions" and that the descriptions themselves provide defining properties. This position has been maintained by many modern logicians beginning with Frege (Searle, 1958, 1960; Zink, 1973; Katz, 1977; Ziff, 1977) and has enjoyed considerable popularity from time to time. As was previously mentioned, it is contrary to the view of Mill, as well as to the position taken in the present work, which is that although individuals can be described, they cannot be defined. Searle (1960:70) puts it this way: "Mill was right in thinking that proper names do not entail any particular description, that they do not have definitions, but Frege was correct in assuming that any singular term must have a mode of presentation and hence, in a way, a sense." Later in this chapter, we will examine the notion of a "description," one of several things that are quite commonly conflated with definition, perhaps in part because they begin with the letter *d*.

The usual tack in arguing that ostensive definitions of proper nouns are really descriptions is based upon the usual way of telling somebody who is referred to by a proper name. Thus, for example, if asked who Charles Darwin was, we can answer "the author of *The Origin of Species*." As a matter of contingent fact, Charles Darwin did write *The Origin of Species* and several other books. But there is nothing necessary about that. It could have been otherwise, and indeed he was hard at work on a book entitled *Natural Selection* when the arrival of a manuscript by Alfred Russel Wallace made him change his plan. Again, for a property to be defining, it has to be logically necessary, and in this case it is abundantly clear that it was not just unnecessary, but a most unlikely state of affairs.

Nonetheless, he who wants to maintain that proper names have defining properties is not apt to give up easily. One possibility is to maintain that there simply must be at least one such defining property. Another is to maintain that even if there is no one such defining property, one or a few among a so-called "cluster" of these will do. In neither case have I seen a legitimate example of such a property or group of properties. But in both cases it would seem that those who want to have defining properties for individuals are expressing a desire to have individuals behave as if they were classes. Not only that, they want both classes and individuals alike to have "essences." Further consideration of these matters will be deferred until we consider essences in detail. For the moment let us merely say that it seems reasonable to reject the notion that proper nouns have defining properties, even though ordinary language may treat them as if they did, and even though that possibility has seemed attractive to some philosophers.

This answers the first question: proper names not only do not have intensions, but we can and do define them without recourse to intensions. Let us proceed to the second question: can the names of classes be defined ostensively?

Perhaps some can, but not all of them. The thesis that some classes, namely what are called "natural kinds," can be ostensively defined was maintained by Kripke (1971, reissued 1980), and attracted a considerable amount of attention (Donnelan, 1977; Goosens, 1977; Katz, 1977; Putnam, 1977; Quine, 1977). In the present context Kripke's views are important for at least two reasons. In the first place, we have his notion about how classes are defined, and the present work is concerned with the definition of classes in general and of certain particular classes, such as the class of species. In the second place, Kripke and many others have maintained that biological species are "natural kinds" (hence classes). A lot of such claims may be dismissed as the consequence of not having been aware of the individuality thesis or not knowing much about biology. But a few authors have attempted to argue for it (e.g., Kitts and Kitts, 1979) and they have been taken seriously enough to evoke rebuttals (Hull, 1984; see Kitts, 1984). Furthermore, in much of the literature, not only in philosophy, but also in other fields such as cognitive psychology and anthropology, the notion that biological taxa, both folk and otherwise, are "natural kinds" has all too often been taken for granted. And the distinction between natural kinds on the one hand, and natural systems in the sense traditionally used by systematic biologists on the other, has been a most pernicious source of confusion.

Kripke's terminology is a bit opaque, but it fits in quite well with what we have been discussing all along. He says that (proper) names are "rigid designators." Something is a rigid designator if "in every possible world it designates the same object." If not, it is an "accidental designator." To give an example, "Charles Darwin" designates a certain organism. In the actual world, he was the author of *The Origin of Species*. But we can imagine various possible worlds in which he wrote *Natural Selection* instead, or even wrote nothing at all. His name would still designate him, and nobody else. The notion of a "possible world" is invoked, it seems to me, as a perhaps unnecessarily complicated way of saying that the connection between the name and that which it designates is logically necessary. That Charles Darwin shall designate him and nobody else follows deductively from the definition of his name. Whatever else could have been otherwise, this, at least, cannot be otherwise.

All this may seem quite simple and straightforward, at least when one applies it to proper names. But what about class names? Kripke and his followers at one time maintained that at least some class names in the sciences are rigid designators, and in so doing presented a rather daring view of how scientific nomenclature evolves. Let us compare what he says with more traditional views. Consider 'atom' for example. Etymologically again it means "indivisible" and for a long time it was believed that atoms really are indivisible. The notion of splitting an atom would seem to be a contradiction in terms. With the advance of scientific knowledge, what had previously been believed turned out to be erroneous. The class names were redefined, though some of the old linguistic labels continued to be used. This view makes a lot of sense, and not just for atoms. Oxygen means "acid producer," but then again one might say that the name just happens to be inappropriate (a misnomer).

Let us consider how Kripke would have had us view the definition of 'gold.' Somehow we need to come up with the correct cause, which suggests the appropriate scientific theory. Certainly, "all that glitters" would not do. We might define it as the most malleable and ductile of the elements, but then again people were familiar with gold even before they knew things like that. At last, however, we know that gold is the element with atomic number 79. We have discovered its definition. We knew how to identify gold all along, and at last discovered how to define it. Gold, then, was defined ostensively, as a sort of "something I know-not what," and gradually became a "something I know what full well."

Kripke's view of defining natural kind terms as rigid designators makes some empirical claims about how scientific knowledge is in fact obtained. Against those claims we might urge a variety of empirical evidence. What happens, for example, when the name in question turns out not to designate anything that exists? Or when it designates more than one class of objects. The traditional counter-example is jade, which turns out to be more than one "chemical species." In a later chapter I will show how little sense Kripke's idea makes when we try to apply it to the class of biological species. The philosophical literature with respect to these matters is seriously flawed. One reason is that the authors go on and on, using the word 'tiger' as an example of a natural kind. It never occurred to such persons that one reason why we find it hard to come up with the defining properties of this class is that *Felis tigris* is actually an individual. Another reason is that when we are evolving our language in the light of repeated scientific discoveries, we often have to deal with terms for things that we very poorly understand. The philosophers wanted to treat language as if it were something immutable—as if it were a class rather than an individual.

One outcome of the Kripkian view of the definition of natural kinds was that it supposedly provides definitions that are analytic, and hence a priori. Although these terms 'analytic' and 'a priori' are some of the most common and elementary ones in the language of philosophy, I had better explain it for the benefit of any reader who has been exposed to the muddled usage of some philosophers and many biologists, and just to make sure I am not misunderstood. Knowledge is a priori if we can obtain it without recourse to experience. Anything that has to be found out by observation, experiment, or the like is a posteriori. That means the entire empirical content of the natural sciences. "Analytic" knowledge is that which merely follows logically from premises: because it does so follow, we do not need experience to verify it, and it is, therefore, a priori. The whole of mathematics, for example, is analytic and a priori. Knowledge that does more than merely deduce consequences from posits is said to be "synthetic," and since the only way to obtain such knowledge is through experience, it is said that there is no "synthetic a priori" knowledge (contrary to what Kant believed).

At this point we might as well mention that some philosophers have maintained that our knowledge of metaphysics is a priori. Such a claim, if true, would make it hard to defend the position taken here, that metaphysics is a natural science. But it is easy to see how some metaphysicians might ensnare themselves by assuming that certain premises might be treated as if they were, or could be, known without recourse to experience. And furthermore it would be very hard to convince such persons to change their minds. If the ontology we embrace is laid down as a posit, and the beings that supposedly are its subject matter are forced to conform to that ontology, irrespective of any experiential evidence that we might bring to bear upon it, then we are in real difficulties.

Let us briefly consider how the notion of analytic, a priori knowledge might affect our view of definitions. If we say that all unmarried persons are, by definition, single, and that, furthermore, all bachelors are, by definition, unmarried males, it follows logically that all bachelors are single. We can tell that this is true without having to appeal to experience, and because it follows deductively, we can say so with apodictic certitude—it cannot possibly be false. One reason why I persisted in my belief that species are individuals was that it seemed to me that it followed from the definitions of 'individual' and 'species' as I understood those terms. For that reason it seemed that to claim otherwise was like denying that bachelors are single. However, such an argument would not have the force that one might think. Merely by deductive reasoning about definitions we cannot know that anybody is married, or male, or even that he exists. And by the same token, it is certainly not true a priori that a species exists, or, for that matter, that any material being whatsoever exists.

Now consider an ostensive definition of the name of a species. It would take the form: let "*Xus yus*" be the name of the species of which "that" is a specimen. Now, one might want to say that the act of naming in this fashion fixes the reference in such a manner that it is logically impossible for the species name to apply to anything other than the species thus named. But obviously experience is necessary to do more than tell us that a specimen is a type of a species. A great deal indeed is needed to tell us what the conspecific organisms of the type might be. Without experience we have no way of knowing whether "that" exists, or whether it is a specimen of anything, let alone a specimen of a species. In the case of natural kind terms, such as 'gold', we have the same kind of difficulties. We do not know whether there are any instances of such kinds, nor whether what we treat as one kind is really two, three, or none whatever. And if we presuppose a "something, I know not what," we have even greater difficulties.

Even so, the notion of a "causal theory of reference" that relates the meaning of terms such as "gold" to laws of nature has much to recommend it. Even though we scientists do not define natural kind terms ostensively, one of our goals is to get at something more fundamental than mere appearances. Although the notion of a cause is philosophically difficult and has often been denigrated, nonetheless the most valuable classifications in science are what may be called "etiological" ones—those based upon causes (see Carnap, 1965). In medicine, for example, the "etiology" of a disease is very important, especially the identification of the causal agent. Then one can do more than just treat the symptoms. One may be in a much better position to cure or to prevent the disease. Knowing that the causal agents are parts of an individual rather than instances of a class may be particularly useful. The underlying metaphysical reason why quarantine is useful for controlling epidemics, but not for preventing heart attacks, is that lineages of pathogens are individuals, and therefore they are spatio-temporally restricted in a way that allows us to protect ourselves.

Etiology is so fundamental to scientific classification that it is hard to exaggerate its importance. Its neglect has had so many disastrous consequences in the past that the point will be repeated time and again in the present work. But for the moment let us continue to pursue the topic of definition and ask what we have discovered when, for instance, we find out that gold is an element and that its atomic number is 79. I maintain that we have discovered some important facts about the natural world, some regularities manifested in the periodic table of the elements, and some laws of nature of great power and utility. Possessed of these we are in a position to restructure our language and system of classification, i.e., to revise the elements as Mendeleef did. Something new has emerged in the evolution of our knowledge. Nothing less has happened, but nothing more. According to the view of Kripke, however, early chemists, perhaps even alchemists with their alembics and retorts, had been seeking to discover the "essence" of gold and modern scientists have succeeded in that task.

One might therefore treat definition as the discovery of essences, and this seems to be what Kripke and at least some of his followers were trying to accomplish. Their views have been labeled "essentialist," but this is just one of many positions that have been so characterized. Essentialism, or what systematic biologists call "typology," was briefly mentioned earlier. The time has come to consider it in some detail. If all we meant by an essence were the defining properties of a name, then essences would be pretty innocuous. But there has long existed a pervasive view that essences are more than just "nominal." John Locke distinguished "real essences" from merely "nominal essences," the latter having to do with language but the former with something more ultimate. Kripke and his followers seem to have had "real essences" in mind when they identified them with the underlying laws of nature. It is as if someone were able to define the things themselves, and not just their names.

Were essences nothing more than defining properties, then, given that individuals have no defining properties, it would follow that individuals do not have essences. And yet essences have quite generally been attributed to individuals. One symptom of this is the notion, earlier discussed, that an individual ceases to exist when it undergoes some kind of change. An individual has an essence, in this sense, if it possesses the essence of its class. For example, if being alive is an essential property of a person, then upon death the person ceases to exist. Not all properties, of course, are essential. Some are what have traditionally been called "accidents," or in another jargon, "accompanying properties" rather than "defining" ones. One argument in favor of the existence of such essential properties is that they give us a criterion for deciding where an individual begins and where it ends. But is there any metaphysical necessity for having such a criterion? Definitely not. Individuals are what they are, and enjoy their existence, irrespective of anybody being able to recognize them. We decide that a person ceases to exist upon death rather than upon puberty because of the greater importance that we attach to a lifespan than to a childhood, both of which do end, but the entity that interests us by no means ceases to exist in an unqualified sense.

Although essentialism often has to do with the properties of things, there is also what is called "mereological essentialism" (see Chisholm, 1973). According to some persons, an individual ceases to exist when it loses an "essential" part, rather than when it ceases to have some property. For example, it might be said that although the Roman patriot Mucius Scaevola lost his right hand in a celebrated act of bravado, he did not lose an essential part, and remained Mucius. But maybe there was some essential part that he could have lost and ceased to be he. Such necessity would have to be more than just physical however. A heart or something like it has to be present of physiological necessity if somebody is not to die. In Roman times there were no heart transplants or mechanical equivalents thereof. But it could have been otherwise, for now it is in fact otherwise. One might want to treat the brain as an essential part. (I refrain with great difficulty from making remarks about brain transplants and certain persons who could perhaps use one!) But consider the phenomena of regeneration and the substitution of function. One can cut large chunks out of the brain, and yet it regains the functions that had formerly been localized in those places which had been removed. As a matter of contingent fact, the regenerative capacity of a human being is rather limited, but this is not even physically necessary, at least to judge from what happens to other kinds of animals that can regenerate an entire organism, say from a single arm of a sea-star.

Religiously inclined persons might want to make a part called a "soul" the essence of a person, and perhaps the desire for immortality has made this hypothesis especially attractive. Another candidate for an essential part of an organism or a species is the genome, which in its most technical definition means the haploid complement of chromosomes, but might be used in a somewhat looser sense simply to mean the genetical components of an individual. An organism that had no genes at all would be rather odd. But then again, mammalian erythrocytes (red blood cells) have no nuclei, so the lack of genes might be a physiological possibility even for a whole organism, indeed, enucleated protozoans have been found to live for quite some time. Getting rid of an organism's genetic makeup altogether might be looked upon as an extreme case of changing the genetic makeup, perhaps by substituting one piece of chromosome for another, perhaps by rearrangement. One might want to treat some properties of the genetic material as essential in defining a class of organisms, and indeed the classes of haploids and of diploids have just such defining properties. But is any organism, or, for that matter, species, the organism that is (or the species that it is) by virtue of any chromosomal properties? A convincing argument would have to provide some examples and not just insist that there must be something of the sort.

There are many varieties of essentialism and many interpretations of the word 'essence'. In addition to what might be called "Kripkian" essentialism (see Salmon, 1981) there is the Platonic variety, the Aristotelian, and any number of others. Our task of defining "essentialism" would be a lot easier if it were abundantly clear that essentialism itself has an essence, but since one of our goals is to avoid such suppositions, we had better try another approach. Essentialism at the very least may be characterized as an attitude toward definitions, and when it becomes a philosophical doctrine it tends to be bound up with theories of definition. We might also characterize it psychologically, as a kind of thought-habit or behavioral predisposition, and this is what systematists often mean when they refer to "typological" thinking. The epithet 'typological' has become almost strictly pejorative in this context: like 'anthropomorphic' or 'teleological'. At a profoundly ontological level, essentialism would seem to result from trying to make classes "prior" to individuals. The classes are treated as the ultimate reality, and individuals are merely epiphenomenal, even if one goes so far to make individuals the only entities that actually exist.

An essence is almost always taken as a kind of 'norm,' 'ideal,' or 'standard,' according to which individuals are evaluated. An individual is "typical" according to the degree to which it approaches that putative ideal. Such standards are thought to represent a desirable state of affairs, so that departures from that standard are categorized as "abnormal" or "deformed." The standards are conceived of as immutable and not dependent upon the context that applies to particular situations. They are monolithic: there is but one standard, or at least not as many as there are actual variants. Defenders of essentialism have often insisted upon the need for "norms" and it would be rash to dispose of norms altogether. Life is better with just laws and other standards of conduct. And if we are going to measure things, we are better off with a common system of units, and perhaps a Bureau of Standards. The question, however, is not whether we have norms, but the appropriateness of our attitudes toward them. In the United States, one drives on the right side of the road; in Australia on the left. So long as everybody in one place drives on the same side, this prevents accidents. But there is no ideal standard that tells us that one practice is better than another. They are just conventions.

Typology often takes the form of thinking in stereotypes, which are, indeed, a kind of norm. Often the victims of such stereotypy say that they would like to be treated as "individuals" rather than as instances of some class. The "norm" for a particular kind of thing is often the means of rationalizing acts of bigotry and worse. Were the male not accepted as the norm for the human being, then it would be harder to justify the practices against which feminists so often protest. Were the adult not the norm, it would be harder to justify practices against which children are in no position to protest. One could go on and on.

Among evolutionary biologists, the outstanding critic of essentialism has been Mayr (1959, 1963a, 1982a, 1988a, 1991a, 1991b). Yet it seems to me that Mayr has underestimated, rather than over estimated the case. Mayr from the outset has emphasized the point that essentialists ignore variation. Since the process of natural selection could not take place without variation, it stands to reason that typological thinking would make it difficult, if not impossible, to think properly about evolution. He suggested that Darwin invented a whole new way of thinking, ("population thinking") which revolutionized biology and much else besides. Mayr (1959:2) characterizes population thinking as follows:

> The assumptions of population thinking are diametrically opposed to those of the typologist. The populationist stresses the uniqueness of everything in the organic world. What is true for the human species,—that no two individuals are alike,—is equally true for all other species of animals and plants. Indeed, even the same individual changes continuously throughout his lifetime and when placed into different environments. All organisms and organic phenomena are composed of unique features and can be described only in statistical terms. Individuals, or any kind of organic entities, form populations of which we can determine the arithmetic mean and the statistics of variation. Averages are merely statistical abstractions, only the individuals of which the population is composed have reality. The ultimate conclusions of the population thinker and of the typologist are precisely the opposite. For the typologist, the type (*eidos*) is real and the variation is an illusion, while for the populationist the type (average) is an abstraction and only the variation is real. No two ways of looking at nature could be more different.

This statement, powerful as it is, tends to be taken as saying that a typologist ignores variation and little more. The ontological message is there, but it does not come through with full force. In my own writings, I have rather emphasized the point that Darwin's discovery reversed the traditional conception of the priority of classes and individuals (e.g., Ghiselin, 1974a). This sort of ontological individualism is sometimes interpreted as nothing more than a kind of nominalism or methodological individualism. It differs, however, in recognizing that there are in fact individuals at more than one hierarchical level, including the species. Hence the "radical solution to the species problem" was more than just tinkering with the traditional alternatives.

One of those alternatives was to try to define species and other taxa in a way that acknowledged the variability of species. Early in his career Hull (1965) attempted to apply the notion of a disjunctive definition to taxonomy. According to that view, a species could have a whole suite of defining characters, and no two organisms of the same species need share any one of these. The result was still essences, ones to be sure that allowed for variation, but they were still essences. Many are the variations upon this same theme. The literature abounds in assertions that amount to just about the same thing. Species are supposed to be "polythetically defined," which is another way to say that their definitions are disjunctive. Or they are said to be "fuzzy sets" or "cluster concepts" (see Zadeh, 1982). All this would work very well if in fact they were defined, or if in fact they were sets, but they are neither.

From the above quotation we can also see that Mayr was having trouble, because on the one hand he wanted to reject nominalism, and on the other hand he wanted to reject typology. So he stressed the point that organisms with their variations are the fundamental reality. Yet in his zeal to make the averages no more than abstractions, he did not do justice to the extent to which he believed that species themselves are real. After the seminal papers by Hull and me had appeared, Mayr (1976) at last endorsed the individuality thesis. Later he maintained that he had supported it all along (see Mayr, 1987a, 1987b; Ghiselin, 1987e, 1987f). It seems less hyperbolic to say that it fits in very well indeed with views that he has long advocated, and those on essentialism are a case in point. (For more on essentialism in evolutionary biology, see Sober, 1980.)

In recent years it has become popular to maintain that some kinds of essentialism are a bad thing, whereas others are quite innocuous or even desirable. Plato believed in an archetypal world, populated by Ideas, of which individuals are imperfect copies. Aristotle, on the other hand, while maintaining that essences exist, denied that they exist apart from particular things. Mediaeval jargon has it that Plato's essences were transcendent, whereas Aristotle's were immanent. Now, it is easy to see how Plato's essences could be far more pernicious than Aristotle's. And indeed some of the worst excesses of essentialism in biology have been due to professed Platonists, such as Lorenz

Oken, Louis Agassiz, and Richard Owen. For such Platonists, the archetypal world was nothing more nor less than the Divine Mind. And the study of comparative anatomy was explicitly conceived of as a kind of Mind reading. Oken's anatomical and systematic writings were loaded with numerology: he found threes, fives, and sevens everywhere. And this makes very good sense, for Oken got it, indirectly at least, from Plato, and Plato got it from Pythagoras.

In Aristotle, and in the writings of professed Aristotelians such as Cuvier, we do not find that kind of excess. And it is easy to see how Aristotle, much more of an empiricist, came up with a biology that in many ways appeals to us moderns. Yet the claim of Balme (1987) that Aristotle's biology was "not essentialist" is flatly contradicted by the texts. At least some Aristotle scholars feel that their colleagues have been going a bit too far in their revisionism (see Granger, 1987). As Pelligrin (1986) shows quite nicely, Aristotle had very different goals than a modern systematist does, and arranged his materials on a very different basis. Of course Aristotle's biology may not have been affected by essentialism in the way some persons have thought, but the essentialism is there, and it is pervasive. The idea that material objects and persons alike have a "natural place" in the world depends upon the notion of a nature, or essence, irrespective of whether it is used to explain the fall of bodies or to justify the existence of slavery. Likewise the notion that some organisms, whether seals or women, are deformed or imperfect representatives of some class, presupposes an ideal standard, even though there is no Ideal World to the inhabitants of which such individuals might be compared. And although the search for knowledge is not treated by Aristotle as a kind of Divine mind reading, for him knowledge is knowledge of essences, and a major source of knowledge is intuition. I should add that a lot of modern "Aristotelians" have been flagrant anti-evolutionists (e.g., Gilson, 1984).

Essentialism may perhaps best be looked upon as a body of attitudes or beliefs about definition itself. One might even call it a kind of linguistic mysticism. Rooted in our thought processes, and inherent in our routine use of language, it is reinforced by our upbringing. When children learn a language, and that is the first major body of knowledge that they acquire, they have to discover a body of pre-existing grammatical rules and conventions about naming. The standards and norms make a great deal of sense, much as do those that we call good manners, and the rules with which one learns to deal with the physical universe. It is very easy to conclude, or merely to assume without giving it any thought, that the various kinds of rules and norms have a great deal in common. So when one has learned a definition, one may readily confuse that acquisition with the discovery of something more fundamental than just a convention about how to communicate. The rule comes to function like a law of nature and becomes a source of authority. So when people believe that they have discovered the definition of something, they believe that they have discovered what it is, what it has to be, and what it ought to be. The formula of a definition becomes an essence. And instead of defining words, we attempt to define the things that we talk about.

In learning a language, we have the best of reasons for believing that the grammatical rules and definitions are already there for us to discover, for language has long existed and is passed on from generation to generation. But in scientific research, there is nothing analogous to a pre-existing relationship between the instruments of discourse and the natural order that we seek to discover. There are no essences in the sense of entries in some Great Dictionary beyond space and time. The language of science, like language in general, evolves and adapts to new circumstances. Terms are defined and redefined as a part of the ongoing process of research. We do not discover the definition of a previously existing term, but rather the basis for creating a new definition, perhaps by changing the sense of a pre-existing one, perhaps by coining another to replace it.

So far as I can see, essences are like anthropomorphism and teleology. They may not always be harmful, but there are better ways to deal with what interests us. We can do without them, and the simplification that results confers the same benefits upon ontology that getting rid of unnecessary hypotheses does upon the natural sciences in general. In closing this chapter, it may well be worth remarking that given the position that has been defended here, the reader should not expect a definition of 'essence' that provides its essence in a mere handful of words.

Definitions of 'Species' and Some Other Terms

"It must be admitted that many forms, considered by highly competent judges as varieties, have so perfectly the character of species that they are ranked by other highly competent judges as good and true species. But to discuss whether they are rightly called species or varieties, before any definition of these terms has been generally accepted, is vainly to beat the air."

—Darwin, *Origin of Species,* first edition, p. 49

A species can be an element in a hierarchical classification system. If so, it is a member of a taxonomic category, such that it occupies a particular level in the system. In the Linnaean hierarchy, that level or category is called "the species" or for clarity "the species category." Thus "*Homo sapiens* is a species" can be interpreted to mean "*Homo sapiens* is a member of the species category." Perhaps something else might be meant, but surely it means at least that, given the various definitions that have been presented here. But we had better take a closer look at the Linnaean hierarchy, and at hierarchies in general, before we proceed. (Recommended reading: Buck and Hull, 1966; Beckner, 1974; Eldredge & Salthe, 1984; Vrba & Eldredge, 1984; Eldredge, 1985; Salthe, 1985; Greene, 1987; Brandon, 1990; Ereshefsky, 1994; Grantham, 1995; Valentine and May, 1996.)

Often the term 'hierarchy' has been made to seem like something obscure, even downright mystical. One often wonders what is being talked about and why it is supposed to matter. But from a logical point of view, the concept of a hierarchy is reasonably straightforward, if perhaps a bit difficult to formalize. Hierarchies are a familiar part of our daily lives. Etymologically, the term means "holy rule," and the government of a church is a good example. A hierarchy has a series of levels, each with a definite position relative to all the other levels, such that any entities ranked at different levels are "higher" or "lower" than others. If A is higher than B, and B is higher than C, then A is higher than C: the relation is transitive. It is also asymmetrical, for A is higher than B does not imply that B is higher than A. (As mentioned in an earlier chapter, there are rare, and probably just apparent, exceptions.)

The notion of hierarchy is so general that little more can be said about all of them. But the very diversity of hierarchies is well worth pondering. Let us take a look at academic hierarchies, for example:

Professor	Professor Smith
Graduate Student	John Jones
Undergraduate	
Senior	Roberta Allen
Junior	Stephen Brown
Sophomore	Ann White
Freshman	James Scott

Here we have a system of levels, together with some of the entities ranked at those levels. Note that we have classes and (included) subclasses, such that, for example, undergraduates includes freshmen, and therefore if James Scott is a freshman he must be an undergraduate. That is perhaps all too obvious, and such an *inclusive hierarchy* of members of classes included within classes is not very interesting, though we might mention that such inclusion makes syllogistic reasoning possible. But consider the individuals here. The only thing that the purely *subordinative hierarchy* represented by such levels as freshman, sophomore, junior, and senior gives is their rank, or position in a sequence through which one of them has passed and all of them might. They are members of the classes, and that is it. As one moves up in the hierarchy one does not proceed from more particular to more general or from a part to a whole. On the other hand try:

University:	Big U
College:	The College of Arts and Sciences at Big U
Department:	The Biology Department at Big U
Professor:	Alphonso Urechis Smith, Professor in the Department of Biology at Big U

Here we have a whole-part, or *incorporative hierarchy*, with Professor Smith being a part of a particular Biology Department and a part of a particular College and University; the Biology Department here, of course, is a part of the College of Arts and Sciences.

There are of course all sorts of complications, and this is one reason why taxonomy is interesting from a philosophical point of view. For example, there are serious questions as to why a particular entity is given a particular categorical rank, and in what sense such assignment makes any two entities with the same rank be "peers" or "equals." Within a given whole, this may not be much of a problem, say with two associate professors in the same department having approximately the same responsibilities, privileges, and social status. But once we start comparing across departments, universities, and other hierarchically structured entities, we get serious problems of commensurability. How does Professor Smith, in the Biology Department at Big U, compare with a colleague in another department, or another university? How does he compare with a government official, especially in some other country? This can make a great deal of difference when somebody has to make decisions about protocol. And two persons might have different relative positions in two hierarchies, for example when two people belong to two different clubs.

A "pecking order" is said to be a dominance-subordination hierarchy. Each hen can peck the hens below her, but not the ones above her. In ecology there are similar rankings in terms of competitive ability. One curious feature here is that sometimes these hierarchies are said not to be transitive. Thus A might outcompete B, and B might outcompete C, but C might outcompete A. An ecological community is often said to be hierarchically structured. This because of its trophic levels. The first, or lowest, level consists of producers, which are eaten by the first-order consumers that form the next higher level, and so-on through second-order consumers and increasingly higher level ones. But the world is well understood to be a lot messier than this ideal scheme suggests, thanks to omnivores, animals that get promoted to higher rank as they get larger, and so-forth. A food web is often considered more appropriate than a food chain. In such hierarchies, the only individuals are the entire social or economic wholes (such as an individual flock of hens or an individual community) and the individual organisms that are their components. The entities, such as secondary consumer, ranked at higher levels are not wholes made up of parts, such as primary producer, that are ranked at lower levels. Now of course such wholes may exist, and often they do. The freshman class at a particular university might get organized and elect a president. The faculty are generally organized into individual departments. The point is only that hierarchies need not have such higher-level individuals. An entity referred to as a hierarchy may itself be an individual, but it does not have to be an individual. The higher levels can be classes pure and simple, as with assistant professors, associate professors, full professors, etc.

Whole-part (incorporative) hierarchies are of great interest to those who are concerned with the laws and principles that govern organized beings in general (Simon, 1962). In armies with their incorporated divisions, battalions, companies, platoons, squads, and soldiers, hierarchical organization enables those in command better to coordinate the interactions among the parts. An analogous hierarchical organization exists in the animal body and its components such as organs and segments. This is all very significant, and we shall allude to such matters later on. However, at the present moment it seems desirable to point out something that sometimes gets overlooked. The "subordination" of one entity to another in a hierarchy does not necessarily mean that an entity that occupies a lower level "takes orders from" or is the boss of, another, though perhaps one might get such an impression from some of the familiar examples. Yes, seniors may bully freshmen, but they don't have to, and I dare say they should not. The relationship can be despotic, but it can be quite the opposite. In a democracy, individual citizens make decisions and instruct those upon a "higher" level to carry them out. An elected official is supposed to be a public servant whose constituency is free to dismiss him just like any other employee.

The periodic table of the elements is hierarchical, with definite levels for classes that are treated as mass nouns. Here, passing from lower to higher does not mean going from lower status to higher status, but from lower generality to higher generality and from more nearly concrete to more highly abstract. Thus the class of halogens is more general than chlorine, and chlorine is more general than any one of its isotopes, as with students being more general than freshmen. Sometimes it is said that the periodic table is not hierarchical, but I am not quite sure what is meant by such assertions. Perhaps it is only meant that the elements are arranged so as to show increasing atomic number and the like; but such an arrangement would still be hierarchical. Anything with levels would seem to qualify as a hierarchy.

The periodic table is an inclusive, not a whole-part hierarchy, at least above the level of the individual atom. In this it differs from the modern ideal of a natural, phylogenetic system in evolutionary biology, an incorporative hierarchy in which all the taxa are individuals united by cohesive forces or by community of descent. The Linnaean hierarchy likewise has generally been treated as something that consists only of the higher levels. In the words of Simpson (1961:18), "Classification involves only groups; no entity possible in classification is an individual." If organisms are not groups, Simpson would seem to be telling us, they cannot be classified. But organisms are in fact groups, and so are their parts; so the decision as to where we must stop classifying has to be done on some other basis.

At this point we should preclude a possible source of misunderstanding. In the passage quoted above, Simpson was trying to explain the difference between classification and identification, and of course he meant "organism" when he wrote "individual." But this only shows how muddled he was, and how his philosophical confusion led him to provide the wrong solution to the problem that he was addressing. It never even occurred to him that he was dealing with an incorporative, rather than an inclusive, hierarchy. This is precisely what one would expect from someone who had embraced the set-theoretical analysis of Gregg (1954). We can classify both classes and individuals in the sense of creating a system in which all of them occupy a position in that system. And we can identify both classes and individuals by deciding where to place them in a system that already exists.

To argue that there is no good reason for assigning names to particular kinds of units is one thing; to deny that possibility is another. For most species we really do not need a formal taxonomy below the level of the species, but for some species naming such units is common practice. We do name organisms, and sometimes we even label their parts with unique identifiers, even though this is not considered a legitimate aspect of formal taxonomy. In any event our basic biological classification system has a definite place for individual organisms and their parts, and it is an integral part of a conceptual scheme that extends to lower levels. At the very least, treating the organismal, suborganismal, and supraorganismal levels together helps us to understand how classification relates to evolutionary theory.

The following is a rather sketchy list of some of the hierarchical levels that biologists consider important:
Genus
Population
    Species
    Subspecies
    Deme
Organism
Cell
Molecule
Atom
This is a list of classes, instances of which, such as *Homo sapiens* and yourself, can easily be supplied by anybody with a high school education. Among the items in the list is a class of such classes, the population, which is designated by a term that sometimes is used in different senses that need not bother us for the moment. The interesting point here is that the individuals ranked at higher levels are composed of individuals ranked at lower levels, giving a whole-part hierarchy. The individuals in such a hierarchy are never parts of more than one individual at a given higher level at any given time: a molecule is part of one and only one cell, an organism part of one and only one species. Not everything is thus constrained. Had we included a level called the organ system and another called the organ, we would have to deal with the gonads, organs which are parts of both the reproductive system and the endocrine system.

Another noteworthy point is that not everything in the universe has to be a part of any particular larger whole except for the universe itself. A subatomic particle does not have to be a part of a molecule, or a molecule a part of an organism. And within an organism, a part does not have to be a part of all the parts of higher rank. For example, a molecule that is part of an organism need not be part of a cell; blood, for example, consists largely of extracellular materials. Finally, we may note that an individual that is ranked at one level can also be ranked at some other level. The most obvious example is an organism that is also a cell. There are plenty of protozoons and protophytes that could serve as examples. Whether we call a human zygote a unicellular organism depends upon whether we think of unfertilized eggs as fitting our definition of 'organism'—a point that had better just be mentioned for the time being.

On the other hand, the claim of some persons, notably Dobell (1911) and Hyman (1940), that protozoans are not unicellular but acellular can definitely be dismissed as based upon logical fallacy. They treated the multiplicity of components at a lower level as defining of a class of wholes at a higher level. But biologists in general have not treated being part of a whole made up of many such units to be defining of the cell. Perhaps Dobell and Hyman were a bit confused as a consequence of the "holistic" metaphysics that was widespread in the first half of the present century. Be this as it may, to say that an organism cannot consist of just one cell is like saying that a house cannot consist of just one room. But the world still contains a fair number of one-room schoolhouses, and committees of one, and other such things.

Likewise, there is nothing about species that rules out the possibility of one of them consisting of but a single organism. In the case of a lot of quite familiar creatures, one would think that the last of its species would be unable to reproduce, and consider the existence of that species as having been terminated. We need not quibble here. Rather we should remember that some organisms are facultatively self-fertilizing hermaphrodites, and a species composed of them might survive even if a generation consisted of only one of such organism. This possibility raises some problems, for if we say that a species is by definition a population, and that a population must by definition consist of more than one organism, we have excluded such uni-organismal species from our definition.

As was mentioned earlier, the term 'population' has other definitions that are often used in other contexts. In addition to the broader sense given above, it also has a much narrower sense, meaning a small local population or "deme." In the ecological literature it often means a mere sample, perhaps singled out by drawing an imaginary line around a group of organisms. In textbooks it is often defined as a group of organisms living in a given place at a given time. Such an expression may point the students in the right direction, but they would have to know a lot more to apply it properly in a concrete context. In and of itself it does not tell one whether the group is an artificial kind, an individual, or perhaps just a set. It does not tell one what, if anything, makes it cohesive. It does not tell one whether it remains the same population if time passes and perhaps some of the organisms move around a little.

The above list shows only two levels below that of the cell, namely the molecule and the atom. It doesn't bother with subatomic particles and their components. Molecules and atoms are often singled out as being particularly important, and we ought to say at least a little about them. Given the supposed "maturity" of the physical sciences, one might think that defining anything so basic would be a very straightforward matter. Not at all. It turns out that dictionaries and other supposedly authoritative reference works provide definitions that contradict one another, and in ways that suggest that the authors never bothered to think things through. One noteworthy point is that the concepts of molecule and atom roughly correspond to those of compound and element, but only roughly. (For a good discussion on "element" see Paneth, 1963.) 'Compound' and 'element' are good examples of mass nouns: they do not refer to any individual, but rather to an indefinite amount, like water and gold, for example. 'Compound' is a contrary of 'mixture' and differs insofar as there is supposed to be a definite numerical proportion among the different kinds of parts of which the compound is compounded. Such definite proportions would follow from general principles if the compounds consisted of individuals that are composed of a definite number or at least proportion of subunits (and perhaps other defining properties too). And a molecule would be such an individual. 'Element', on the other hand, is a kind of substance that cannot be decomposed by ordinary chemical means. Atoms would qualify as individuals at the atomic level.

Atoms, then, are the smallest entities that can be interpreted as members of the class of elements. This seems to me a perfectly good definition, though of course it is intelligible only if one knows what an element is. The elements are defined as mass nouns, implying that an indefinite amount of gold or chlorine would quality as "some gold" or "some chlorine." If one takes a chunk of gold apart, one has to stop at a single atom, for otherwise what one has would not qualify as gold. Atoms are important to chemists and to scientists who use chemistry in their research because atoms participate in chemical reactions, form larger units, and play a major role in causing such units to have the properties that they do. There is a definite lawfulness to their structure, each with its nucleus and electrons in characteristic orbitals.

One might want to define 'atom' by describing that very same characteristic structure, thus making the definition depend upon the arrangement of the parts, rather than upon the properties of the whole. In any case we seem to have at least three different ways of explaining what is meant by the word 'atom', all of which seem to be fully consistent with modern scientific theory, and which are extensionally equivalent, i.e., the terms would refer to classes having exactly the same members. Suppose that somebody took each of these definitional formulae and used them to define three different names, say 'atom', 'mota' and 'tamo'. These terms would be synonyms, in the sense that any sentence in which one of them occurred could be rewritten without changing the meaning of the sentence. In one sense we might say that the three definitions were definitions of the same concept, since they lead us to conceive of the same class of objects, but in another sense we might say they are definitions of different concepts, since they lead us to conceive of it in different ways. But one might say that we can derive all of these from a single body of theory without contradiction, and that therefore the various formulae all have the same conceptual basis. That again would suggest that there is one concept here, not three. Be it understood that this is a purely terminological matter insofar as what 'concept' means goes, but in the present work I have chosen to make the last of these the criterion of what makes for two concepts being the same; the first criterion is perfectly legitimate but only insofar as it is derivative of the third. The theory provides an etiological basis for deciding that the identity of extensional reference results from more than just chance—that a real cause underlies the phenomena.

On the other hand, some of the definitions I have seen for 'atom' make the term seem rather equivocal. Some of these make an atom be an atom only when it has the full complement of electrons, neither more nor less. Consequently an ion cannot be an atom, for the electrical charge would necessarily mean that at least one electron too few or too many is present. Thus virtually all of the salt in the ocean would be ions rather than atoms. By no means is it obvious why it would not be better to say that elements consist of atoms which can exist in an ionized state. After all, they can exist as isotopes, which have different numbers of neutrons in the nucleus. This is not the same as suggesting that every elementary particle in the material universe ought to be made, by definition, part of an atom. We are asking here why the addition or subtraction of a part is supposed to make it change so as not to be a member of some class, not proposing to place it in some whole that does not exist.

It is common knowledge that our conception of atoms has changed through time. The notion that they are indivisible is graven in their etymology, much as the history of life is preserved in vestigial organs. The atom has had to be redefined, at least in the sense that we have changed our theory. But the notion of indivisibility does not seem to have had the sort of central place in chemical theory that would cause one to claim that the term had been "redefined" in a way that made the term do more than just evolve a little. Making atoms divisible probably had very little effect upon what objects were recognized as such by chemists.

The molecule may be defined as another minimal unit. It would be the smallest unit for which a pure substance maintains its chemical properties. Again, if one breaks it down, it becomes something else—just a part. Alternatively one might try to define it by means of structural characteristics, such as being held together by bonds. There is a problem here, because in many substances there is no definite single unit made up of a given number of atoms held together by bonds. Common salt in its crystalline form is a good example. Either one has to say that the molecule concept does not apply to such materials, or one has to say that molecules are not of a definite size. The former possibility means that there are molecule-less compounds. The latter possibility conflicts with the aforementioned minimal unit notion.

There would appear to be no good reason for insisting that a molecule must have a particular number of atoms. Nonetheless particular chemical compounds are commonly defined in terms of the number, kinds, and arrangement of atoms of which each molecule is composed. Thus the ala-phatic hydrocarbons—methane, ethane, propane, butane, etc.—have as defin-ing properties one carbon atom, two carbon atoms, three carbon atoms, four carbon atoms, etc., with each individual having a definite number of atoms as well as being held together by covalent bonds. In an individual diamond, the carbon atoms are also held together by covalent bonds, and in a certain con-figuration; but no particular number of bonds need occur in any diamond.

And indeed when we speak of "macromolecules" such as polymers, it is far from obvious that a class of such molecules need be defined as having a par-ticular number of subunits. Of course there are polymers that are defined in terms of their number of subunits. Dipeptides, tripeptides, and so forth are a good example, with two, three, or more amino acid residues respectively. But "polypeptide" specifies only an approximate number of such units. And most of us would probably not object to calling the individuals in the class of proteins—which are, after all, longer polypeptides—molecules. In thinking about an individual protein molecule, most of us would not consider its length defining. When a protein molecule is synthesized by a ribosome, one amino acid is added after another, and it makes sense to say that it is still the same individual protein molecule in the various stages of its synthesis, just as we would say that a juvenile organism is nonetheless an organism.

Sometimes the molecule is defined as an aggregate of atoms. In any case there is a problem here if the term is defined so that a molecule cannot consist of just one atom. The "noble gases" such as argon and neon do not form com-pounds under ordinary circumstances. In the context of gas laws, the individ-ual noble gas atoms function exactly like any polyatomic molecule, for what matters is the number of particles.

Not being a chemist I would rather not pass judgment on how they ought to classify. Nonetheless it should be abundantly obvious that how we define the basic units is not the sort of question for which simplistic answers will bear crit-ical scrutiny. Indeed, we may feel encouraged in seeing that the problems of defining basic units are just as much a problem for our reductionist colleagues as they are for us systematic biologists.

Now let us consider how we might define 'organism'. As a rough-and-ready definition, we might say that the organism is the unit of physiological au-tonomy. The leading idea here would be that life exists as discrete units that can get along more or less independently of one another so far as such basic func-tions as nutrition, growth, and maintenance are involved. These things are called organisms, or "individuals" in the narrow sense. Leo Buss (1987) obvi-ously had such a conception in mind when he entitled his book *The Evolution of Individuality*. 'Organismality' would have been a dreadful neologism. Yet the very idea that this property evolves suggests that there are transitional condi-tions to which the notion of an organism is difficult to apply. One need hardly be surprised that biologists have devoted considerable attention on the problem of deciding what is an "individual" in contradistinction to a "colony" (interesting examples: Erasmus Darwin, 1800; T.H. Huxley, 1852; Haeckel, 1878; Perrier, 1898; Bancroft, 1903; Mackie, 1963; Janzen, 1977). Many organisms reproduce by dividing or fragmenting into parts that become creatures much like the parent. Corals and other cnidarians provide some good examples. During such fission it may be very hard to decide at which point one is dealing with one organism or with many, for they may remain stuck together during a long and gradual process of separation. The products of fission can also lose their autonomy. In many of these creatures one finds an evolutionary series, with the products of fission tending to stick together, forming a colony of very similar zooids or "persons." In later stages the persons undergo a division of labor, with some specialized for feeding, others for reproduction, and others for defense. The result may be a very complicated creature, with the persons functioning in the same basic way as the organs of more conventional creatures like ourselves. No matter how much we refined our definition of 'organism', we would always have problems with intermediate cases in which the organism concept was hard to apply.

As was mentioned in an earlier chapter, one may have a very hard time deciding what is and what is not part of an organism, including delimiting its interior from its exterior. Many marine animals, especially tropical ones, have little plants in their tissues from which they derive at least some of their food supply. Are these parts of the animals? In some cases the animal goes so far as to destroy all of the plants except their chloroplasts—organelles which, it is thought, became symbionts and permanent parts of the plants themselves something like a billion years ago. Whether we are to regard symbiotic algae—or *a fortiori* symbiotic organisms that have become organelles, such as chloroplasts or mitochondria—as organisms is highly debatable. The definition of 'organism' in terms of autonomy makes it seem reasonable to say that they are not organisms. Autonomy, however, is a matter of degree. We animals could not do without the plants that produce essential amino acids, let alone food. To some degree, every creature on earth is a mutualist, if not an outright parasite. In the case of parasites, dependence is often accompanied by extreme degeneration. Viruses, which consist of genetical material and very little else, are often considered the extreme of parasitic degeneration, although their historical origin is by no means clear.

Tradition (I daresay not much else) has it that viruses are not "life," and yet one wonders how such a view can be justified as more than a matter of linguistic convention. In textbooks "life" is generally defined in terms of certain processes (such as growth by intussuception) that occur in the more conspicuous organisms. And yet we see that these properties disappear gradually and without clear boundaries with the degeneration in question, and any reasonable scenario for their origin must have been equally unbroken. Evidently "the organism" has become the Essence of "Life." Life is a class, and organismality is defining. But why look upon it that way? What if life is a class of processes, with tellurian life—an individual—the only instance known to us? Why not look at the (admittedly hypothetical) world in which the ecosystem consisted of naked RNA molecules replicating in a sea of energy-rich molecules as consisting of, or at least containing, life? And if the extant viruses are parts of lineages that go all the way back to the *Urschleim,* why not regard them as parts of one living world (*Imperium Vitae*)?

Be this as it may, such things as organelles and viruses are generally not treated as cells, even if, from an historical point of view, they are modified cells. But the cell, to which we now turn, has its own definitional problems. One way to define 'cell' would be to say that it is the smallest, or better the hierarchically lowest, unit having physiological autonomy. Unicellular organisms are capable of independent existence, whereas anything ranked lower has to depend upon some larger whole for its existence. This makes a certain amount of sense insofar as some cells fit the definition of organism. It also fits in with the view that multicellular organisms have evolved from colonies of unicellular organisms. The problems we have encountered with the organism would also vex us with the cell.

Alternatively, we might prefer to define the cell in structural terms. Cells have certain characteristic parts, such as the cell membrane, the nucleus, the chromosomes, and various organelles. The term 'organelle' implies a distinct analogy with the "organs" of organisms. Protozoologists seem quite happy with such Greek-root terms as 'cytostome' (cell-mouth) and 'cytoproct' (cellanus) inspired by such comparisons. In general a cell contains the wherewithal for reproduction and maintenance of the basic metabolic functions, such as DNA and the machinery for making proteins and putting them to work, plus something that separates them from each other and from the outside world. So textbooks contain a diagram of "the cell" with some characteristic features.

Naturally the world is not so neat as such a typological scheme would have it. Prokaryotic cells are much simpler than eukaryotic cells, and lack some of the "typical" organelles such as a nuclear membrane, and a lot of quibbling has gone on with respect to whether their chromosomes really correspond to the same Platonic Idea. Furthermore, many unicellular organisms have more than one nucleus per cell. Ciliate Protozoa have a macronucleus and a micronucleus. The former contains many copies of the genome and functions as the physiologically active genetic material. The latter is inactive and constitutes the germ line that is responsible for heredity in the long run. When sex occurs in ciliates, the DNA in the micronucleus undergoes recombination and regenerates the macronucleus. A given amount of cytoplasm requires a definite corresponding amount of genetic material if it is to function properly. In a larger organism the appropriate ratio can be maintained in various ways. One of these is to multiply the nuclei. Often cells are connected by pores or other connections that unite the cytoplasm. There are also so-called "syncytia" in which the cells that make up a tissue in a multicellular organism are not separated by distinct membranes at all. Other cells, for example mammalian red blood cells, have no nucleus.

If we reflect a bit on the history of the cell concept, we can see that it has changed a lot, without any need for a gross departure from the extension of the term (see Maulitz, 1971; Lenoir, 1982; J. R. Baker, 1988). From the time that Hooke observed dead cell walls in cork, and referred to the spaces among them as "cells" by analogy with those inhabited by monks in their monasteries, to the formulation of the cell theory by Schleiden and Schwann in 1838 and 1839, cells were occasionally mentioned, but their theoretical importance was not recognized. Especially thanks to Schwann (1839) they came to be conceived of as playing a fundamental role in development. But their genetic continuity was not recognized until Virchow (1859 translation 1860) propounded his famous aphorism "*Omnis cellula e cellula*"—every cell from a cell. The role of the nucleus and that of the chromosomes gradually came to be understood later in the nineteenth century.

So we might say that our conception of the cell evolved from that of a tectological (or structural) unit, to that of a developmental unit, to that of one playing a fundamental role in heredity, without ceasing to be any of these things. Given our understanding of what cells do and how they work, we can easily see that the absence of certain parts in given cases is really quite trivial if cells play the same role in theory. It is the theory that tells us what to call a cell or not to call a cell, and as our understanding has evolved, so to some extent have our definitions. And it may well be that in the future we will increasingly think of cells in historical terms, as something that has evolved and need not be defined in the traditional manner.

In delimiting individual cells, the problem is not so much where one cell begins and where it ends, but rather what is part of at least one cell and what is not a part of any cell. The cell membrane may be taken as a convenient limit. Cell walls of plants and all sorts of "extracellular matrix" and the like are then conceived of as non-cellular, things to which the cell concept does not apply. The existence of parts of organisms that are not parts of cells is not considered a problem. Thus it would seem that we have no need for a "pluralistic" concept of the cell, such that a cell is one thing in Protozoa and another in Metazoa, or one thing in the gut and another in the brain. One might say that the need to account for some dubious cases such as Protozoa with more than one nucleus calls for another kind of pluralism, but the actual procedure is not to say that a cell is one thing in small protozoans and another in large ones, but to make the "monistic" cell concept sufficiently rich to accommodate such variants.

With such preparation out of the way, we may now pass on to the definition of the species category itself. We begin with the various versions of what is called the "biological species concept," which is generally accepted by modern evolutionary biologists and plays an important role in the Synthetic Theory of evolution. Only when it has been fully explicated does it make sense to consider the various alternatives. It bears repeating that we cannot do justice to the biological species concept if we focus all of our attention upon the terse verbal formulae that pass for definitions, and thereby neglect the underlying theoretical criteria that really determine what is and what is not a species.

To be sure, it would be very nice if we could produce a terse and pithy "definition" of 'species' that accurately presented the defining properties of the term, something that really is accomplished by that shopworn example "A bachelor is an unmarried male." Unfortunately that turns out to be difficult, perhaps downright impossible. Given what has been said earlier in the present chapter, one might say something like the following: "species are the units of evolutionary autonomy." This makes a certain amount of sense in the light of the suggestion that cells are the smallest units of physiological autonomy. But it carries the virtue of brevity to excess. Indeed, if we reflect upon the definition of 'bachelor' for a moment, we see that it too is somewhat problematic, and for basically the same reasons. Usually it is only applied to a human being, and one who is old enough to marry; but how old is that, and what constitutes a marriage?

It is not too difficult to say what are the defining properties of 'species'— *sensu* "biological species"—in a few sentences without doing too much violence to the ideal of a properly lexical definition. In the first place, a species has to be a population, it being understood that the term 'population' is used in the broad sense of a reproductive community suggested above, and not in other senses such as strictly a very local one. That provides us with one necessary condition. Second, under ordinary conditions, such a population must be sufficiently cohesive to prevent its components from undergoing indefinite divergence, provided that the cohesive forces are operative. Here again, "cohesive" has been pretty well explained, although we will have more to say about that later on. We will discuss "indefinite divergence" later on too, but to explicate it briefly, it simply means that the population is held together as an evolutionarily unit. This provides us with another necessary condition. And thirdly, to differentiate the species from lower units in the hierarchy, such as subspecies and demes, we must say that it is the largest, or most incorporative, such unit. This provides us with yet another necessary condition. Together the three of them, along with the qualifications that go along with them, constitute the definition, or the totality of defining properties, of "biological species."

It seems clear that a criterion of evolutionary autonomy has been presupposed in many efforts to formulate species definitions. Dobzhansky, for example, was quite emphatic about this, and even in his earliest writings on the species concept he emphasized the role of isolating mechanisms in keeping their gene pools separate (Dobzhansky, 1935, 1937, 1940). Species, accordingly, are evolutionary units, though they are not the only ones. But we can conceive of evolutionary processes and of the units that participate in evolution in various ways. One way to think about them is in terms of population genetics. Thus, Dobzhansky said that species are "the largest Mendelian populations." This means that they are units with a common gene pool, with—to force the metaphor a bit—perhaps little "gene bays" and such forming smaller subunits. Such a conception does not entail the absolute closure of the population, any more than a pool can have absolutely no seepage of water to or from one nearby. But genetics and heredity are just one aspect of evolution; species are reproductive units as well. So there is no reason why we might not formulate the same basic thesis in terms of reproductive processes. And there are other kinds of processes that go on within species as well. There is gene flow within them. The population evolves as a consequence of natural selection and other processes.

In his earliest attempt to define the species category Dobzhansky (1935:354) writes as follows: "Considered dynamically, the species represents that stage of evolutionary divergence, at which the once actually or potentially interbreeding array of forms becomes segregated into two or more separate arrays which are physiologically incapable of interbreeding." Roughly translated this means that species are the products of a process of diversification that has gone so far as to prevent interbreeding. The notion of a population, and the difference between a population and a class of isolated organisms, were not particularly clear to him at the time. This is evident from his treatment of "ring species" in which some populations cannot interbreed directly with some other populations, even though the whole is a single, interbreeding system; for him the isolated subunits were specifically distinct (Dobzhansky, 1937:378). Nor had the more general notion of reproductive isolation yet taken the place of old fashioned ideas about sterility tests. In some of his writings, especially the first and second editions of his classic *Genetics and the Origin of Species,* Dobzhansky (1937:312; 1941:373) emended the above so as to have the species be defined as the stage, rather than as representing that stage. It was like defining "undergraduate" as a stage in education, rather than as someone in that stage. This elementary category mistake gives the reader the impression that Dobzhansky embraced a kind of process metaphysics. This interpretation reads too much into a verbal slip, albeit he did want to think of species as intimately connected to the process that gives rise to them.

Mayr clearly and explicitly treated species as populations and characterized them in terms of what keeps them apart: reproductive isolation. His earliest formal definition (Mayr 1940:256) is as follows:

A species consists of a group of populations which replace each other geographically or ecologically and of which the neighboring ones intergrade or hybridize wherever they are in contact, or which are potentially capable of doing so (with one or more of the populations) in those cases where contact is prevented by geographical or ecological barriers.

The version that is often cited, even by Mayr himself, as having been published in 1940, is actually an emended one that appeared in print somewhat later (Mayr, 1942:120):

Or shorter: Species are groups of actually or potentially interbreeding natural populations, which are reproductively isolated from other such groups.

A still later version (Mayr, 1982a:273) reads:

A species is a reproductive community of populations (reproductively isolated from others) that occupies a specific niche in nature.

Here, the term 'population' means a local population, the "groups" of populations which by definition are species, themselves being populations in a broader sense. When any two such groups are "reproductively isolated," they not only do not interbreed, but they lack the potential to interbreed. The potentiality is important, because something other than being reproductively isolated might prevent interbreeding between two component populations of a given species; for example, they might be separated into so-called allopatric populations and therefore be geographically isolated without being reproductively isolated. If they came back together—became sympatric—they and the species as a whole would resume interbreeding.

Interbreeding is a property of populations as wholes, not of organisms, and it makes a great difference. For one thing, organisms of different sizes or the same sex could coexist within a single species. Domesticated dogs are all parts of the same species, even though a Chihauahua cannot mate with a Great Dane. A similar situation can exist upon a geographical basis, and in the case of "ring species" two local populations behave as if they were two different species in a region of local overlap. By "reproductive isolation" is meant something "intrinsic" that prevents interbreeding, in contradistinction to something "extrinsic" such as a geographical barrier. Reproductive isolation does not mean just the failure to mate or produce offspring, but the incapacity to contribute to the ancestry even of remote generations within the entire species, and even that is not quite an adequate way to characterize it. The intrinsic "isolating mechanisms" are quite heterogeneous, and are commonly subdivided into pre-mating and post-mating mechanisms. Pre-mating mechanisms prevent mating from taking place, and include behavioral disinclination to mate and anatomical incapacity to do so. Post-mating mechanisms can come into play early or late. A zygote may not be formed, or it may not come to full term, or the offspring may be weak or, for any of a number of reasons, sterile. Or the products of reproduction might form their own independent population, isolated from both parental ones, hence a new species; this sometimes happens as a result of chromosomal events.

For species to be reproductively isolated is the same thing as not to have the potentiality to interbreed. Hence the two expressions are redundant when both are given as in the earlier formulations of the biological species definition, and later versions have been rendered more terse by doing away with one of these. The formal definitions do not tell one exactly what constitutes a lack of potential for interbreeding. Therefore one might miss an important qualification, which is that the populations would not interbreed under natural circumstances, even though component individuals might be crossed in the laboratory, and even though human intervention might destroy the isolating mechanism and cause two species to fuse into one.

Also not obvious, and again a source of frequent misconception, is the point that reproductive isolation need not be absolute or complete. Some crossing between organisms of different species can occur, so that the flow of genes between them need not be absolutely cut off. This may seem a real puzzle, but if we remember that the formal definition is not the real thing, it makes perfectly good sense. What matters is not that all gene flow be cut off, but that it be cut off to a sufficient degree that the species can continue to diverge instead of fusing back together into a single populational individual. In later sections we will expand somewhat upon this point.

Another matter that needs to be explicated is that species are neither classes of individual organisms nor classes of individual populations that are defined in terms of some reproductive propensity or the lack of it. In other words, reproductive modes can evolve so that species might be kept apart by one mechanism at one time and by another mechanism later on, yet still remain the same species. Indeed, it is generally understood that post-mating mechanisms evolve first, and pre-mating ones are added later, because natural selection favors the organisms that do not have sterile or inviable offspring. The notion that a species is a class of organisms defined in terms of some particular isolating mechanism is attractive to some people, especially paleontologists, who want to recognize "chronospecies" or temporal subdivisions of a given species that differ in one respect or another. A change in isolating mechanism could be used for delimiting such chronospecies.

Instead of defining 'species' in terms of what keeps them apart, one might want to define it in terms of what holds them together. Paterson's (1985:25) "recognition" concept seems to do this: "We can, therefore, regard as a species that most inclusive population of individual biparental organisms which share a common fertilization system." Paterson, however, had something else in mind, namely that species are classes of organisms that all mate in a given way. Consequently there could be no species with more than one system of mate recognition mechanisms. Within a species such mechanisms could not vary from place to place, nor could new ones evolve without "speciation" having occurred. But if one did allow for such variation and change, the result would be the delimitation of the same groups of organisms playing the same role in evolutionary theory, and therefore nothing fundamentally different. The implicit, real definition of the biological species concept would not conflict with Paterson's formula. In the next chapter, however, we will consider the possibility of just such a conflict.

We should mention some other possibilities for formulating versions of the biological species concept. An obvious choice is gene flow, which is an obvious concomitant of interbreeding. Buerton (1995:194) does something of the sort when he writes: "A species is the most inclusive unit of organisms among which there is gene circulation."

Another possibility is the kind of process that goes on within reproductive communities: selection (both natural and sexual) is one of these. For example (Ghiselin, 1974c): "Species, then, are the most extensive units in the natural economy such that reproductive competition occurs among their parts." In any species the component organisms will be maximizing their reproductive success throughout the entire population. This is the same as to make them out as reproductive competitors. Different modes of competition go on between organisms that are conspecific and ones that are not. And since such competition is a long-term affair and something that goes on at a distance, conceiving of species in terms of this process has some advantages over things like potential interbreeding. So this criterion would give us the same groups as the conventional formulation of the biological species definition would. It would not delimit an assemblage of extensionally different groups, and the groups would play precisely the same role in evolutionary theory.

But this formula too can be thought of as following from the more basic, real definition—from the fundamental theoretical principles that are presupposed and that determine the criteria of usage. Many people, including me, hesitate to accept such a definition for that very reason. And traditional logic even has a term, 'proprium', for those properties that are necessarily true of things but nonetheless are not considered defining (or essential).

Another possibility would be to say that species are, by definition, the products of the process of speciation. I have entertained this possibility in several of my earlier publications (Ghiselin, 1969d, 1974c). There is a minor problem here, because one has to do something about the first species to come into being, but this is merely a technical matter that all too often complicates our efforts at providing a terse definition. What really seems a problem is that one naturally tends to reject such a definition as circular. One traditional way to define 'circular definition' is to say that it is one in which the definiendum (i.e., the word which is defined) occurs in the definiens (i.e., the statement of the defining properties). This is not quite adequate, and it would be better to say that a circular definition is one that is not intelligible unless one already knows how to use the word that is being defined. Surely, if somebody wants to understand what is meant by the terms 'species' and 'speciation', we cannot merely say that species are the product of the process and speciation is the process that gives rise to the product. One or the other has to be explained in some other way. But with respect to species and speciation, there is no reason why we have to provide this information by defining 'species' first rather than 'speciation'. We could define 'speciation' by explaining how populations split up and become reproductively isolated, and only after having done so say that the products of speciation are called "species."

Therefore there is no reason why the definition in question necessarily has to be circular. In spite of any psychological obstacles we may encounter with this definition, it can readily be derived from the same fundamental principles that we have identified with the real definition. Again, the class of individual species that fit this definition would be extensionally identical with that referred to by definitions such as those proposed by Mayr. And such groups would play exactly the same role in evolutionary theory.

Indeed, one advantage to defining species as the products of speciation is that it emphasizes the process that gives rise to them. At one time in the history of biology we were very much in the dark about what causes species to exist and why they are important. But things are different now. Of course, species do more than just speciate. They provide their component organisms with access to genetic resources. And on a long-term basis they constitute lineages that are transformed by natural selection and other processes. What else they do is a very interesting question, and we will consider such matters when we come to discuss macroevolution. The important point here, however, is not that we can come up with yet another terse verbal formula that allows us to define 'species' without doing justice to the concept. Rather it is that as evolutionary biology evolves, its terminology becomes increasingly theoretical and increasingly linked up with notions of process.

7

SOME ALTERNATIVES TO THE BIOLOGICAL SPECIES CONCEPT

"With species in a state of nature, every naturalist has in fact brought descent into his classification; for he includes in his lowest grade, or that of a species, the two sexes; and how enormously these sometimes differ in the most important characters, is known to every naturalist: scarcely a single fact can be predicated in common of the males and hermaphrodites of certain cirripedes, when adult, and yet no one dreams of separating them."

—Darwin, *Origin of Species,* first edition, p. 424

The best I have been able to manage in protracted efforts to present a terse and properly lexical definition of the biological species is the following:

Biological species are populations within which there is, but between which there is not, sufficient cohesive capacity to preclude indefinite divergence.

This after rejecting any number of other possibilities including, perhaps too hastily, a more colloquial, jocular equivalent:

Biological species are the most incorporative reproductive populations with enough "sticktogetherness" to make them hang in there as evolutionary units.

One can see that this formula has both advantages and disadvantages. Let us briefly consider some of the advantages.

In the first place, it does suggest what are the actual or "real" defining properties rather than ones that are not defining or only dubiously so. Second, it makes interbreeding, which is of course a fundamental cohesive process, and reproductive isolation, which of course is a contributing factor, sufficiently unobtrusive as not to be misleading. Furthermore, it specifies the amount and kind of interbreeding that is meant: sufficient to hold the species together. And finally, it emphasizes the role that species play in evolutionary theory, including systematics. Species have speciated when the cohesive forces are no longer capable of holding them together. So defining the term this way makes it more or less obvious that species are the product of the speciation process.

Simply saying that species are reproductively isolated fails to capture the point that reproductive isolation is a matter of degree. It tells us neither how much isolation is necessary for two populations to qualify as species, nor how much is sufficient. Consider a situation in which two populations diverged somewhat in allopatry and subsequently came back together again and lost some properties that initially tended to keep them apart. Most systematists would treat these as having been "incipient" species. On the other hand, we have situations in which, although the "isolating mechanisms" are in place, there is still "interbreeding" in the sense that fertile hybrids are formed and a certain amount of gene flow occurs between two populations, but these are nonetheless ranked as species. Accordingly what matters is not that these populations neither actually nor potentially interbreed, for in actuality they do. Rather, reproductive isolation has evolved to the point that even in sympatry it will be maintained in spite of gene flow.

This implies that the species by no means represents the "highest" level in a whole-part hierarchy of genetical populations. In other words supra-specific, populational individuals can and apparently do exist. That they are individuals suggests that they may be able to play some role in evolutionary processes. For example, supra-specific units may provide reservoirs of genetical resources that are available to more than one species. Such considerations, however, do not affect the fundamental truth that species formation generates the hierarchy that serves as the basis for classification. But they do imply that the properties that are used in diagnosing taxa and studying phylogenies may at least occasionally be transmitted from one species to another, creating at least a measure of uncertainty.

In fact there seems to be no consensus among advocates of the biological species concept as to how much crossing can occur between species. This is one reason for reformulating the definition as suggested above. The notion that complete lack of gene flow is necessary has some following among biologists (Häuser, 1987; Willmann, 1989). Furthermore the literature contains extensive discussion on the question of when speciation is "complete" (e.g. Short, 1969, 1970; Uzell and Ashmole, 1970). The process of speciation, we may recollect, is thought to occur in stages, with initial diversification (in allopatry) leading to post-mating isolating mechanisms (a failure to produce offspring that are viable or fertile), followed by selection (in sympatry) that leads to the evolution of pre-mating mechanisms. It is sometimes maintained that speciation is "not complete" until premating mechanisms have evolved. As will be discussed later on, Mayr even suggested that speciation is not complete until the populations in question have come to occupy distinct ecological niches. However, it would seem that most advocates of the biological species concept would agree that a total incapacity of two populations to interbreed would make them, by definition, separate species, even though they lack pre-mating isolating mechanisms. Under rare circumstances, isolating mechanisms can even evolve between different populations of the same species, so the mechanisms themselves are not defining in a manner that is straightforward and simple.

Casting the definition in terms of "cohesive capacity" raises all sorts of problems. They are the same sort of problems that attend such dispositional properties as "potential interbreeding," and this one needs to be explicated. A capacity for cohesion can exist among the parts of a thing even though that capacity does not, in fact, hold those parts together. If one coats two blocks of wood with glue, one thereby endows them with the capacity for cohesion, but they do not cohere, or stick together, unless one brings them together into close physical contact. It is the same with species in allopatry: they are kept apart in a way that precludes the cohesive capacity from uniting them into a single population of the kind envisaged. Keeping them apart until the forces of biological cohesion are weakened is like letting the glue dry on those blocks of wood.

A perhaps even more serious problem with cohesion is that the cohesive forces within a species may have a very limited capacity to restrict the amount of diversification within them. On the one hand there is always the possibility of sympatric speciation, which means that a species could be transformed into two species in spite of the entire population remaining effectively in contact. Here the physical analogue would be tearing a couple of blocks of wood apart in spite of the fact that they were glued together and the glue had set. Sympatric speciation is at least an interesting theoretical possibility, and has had some able advocates, especially Bush (1975, 1981, 1982, 1994). One of the reasonable models depends upon the idea of disruptive selection. According to this model, selection in opposite directions might, if strong enough, actually suffice to produce sufficient difference to produce two separate populations incapable of interbreeding.

Whether sympatric speciation actually occurs in nature remains debatable. But the very possibility raises problems for the definition that has been suggested. The cohesive forces are not always sufficient to hold the whole together even if they are fully operative. So when disruptive selection is still in the process of turning one species into two, we might want to say that they were separate species already. I see no way of getting around this difficulty short of adding some perhaps merely implicit disclaimer such as: 'under ordinary conditions'.

Another problem with cohesive capacity is that its operation tends to be a local matter. This is the fundamental reason why it does not keep allopatric populations together. In addition the parts that are joined together to form a composite whole are generally bound most strongly to adjacent ones. And the cohesive forces that unite them may differ from place to place. So if we imagine a pair of populations A and B held into a species A-B by reproductive propensities, gene flow and whatever else may be operative, it seems likely that A and B will diverge from each other only to a limited degree. On the other hand, let B bud off another population, so that we get A-B-C, and so forth until we get a long string of populations increasingly different at the extremes: A-B-C-D . . . Y-Z. This would constitute a kind of "indefinite divergence" if one chooses to so call it. And it is not entirely unlike the sort of divergence that we get within a genus, especially if there is a small trickle of genetic material owing to occasional hybridization. It is not exactly the same, however, for indeed the parts in close proximity are linked. The problem is really with interpreting what is meant by 'indefinite divergence' and it might help to have a better term.

The cohesion due to gene flow is by no means everywhere the same, being dependent, among other things, on the vagility of the component organisms. Historically there was a long dispute over whether species are in fact held together by the cohesion of "coadapted gene complexes" and the like. Mayr (e.g., 1963a) originally emphasized just such cohesion. Then he was challenged by Ehrlich and Raven (1969), who maintained that very little gene flow actually occurs in nature. Their claims were part of an academic propaganda campaign in favor of phenetic taxonomy (Ehrlich, 1961; Ehrlich & Holm, 1962), and were argued on the basis of substantial underestimates of the amount of gene flow that occurs in nature (Jackson & Pounds, 1979; Slatkin, 1985). Among marine organisms it turns out that rapid and extensive gene flow occurs over vast distances (Rosenblatt and Wapples, 1986; Mitton et al., 1989; Shulman and Bermingham, 1995). These are creatures which are adept at getting from place to place as eggs or larvae, but slower-moving ones were under-estimated as well. For example, the possibility that much of gene flow occurs episodically rather than as a constant trickle was ignored. Be this as it may, there ensued a period in which cohesion tended to get de-emphasized in favor of a highly stable phenotype. This tended to reinforce the notion that the species is a "nondimensional" affair and its application a matter of deciding whether the local elements are distinct from one another. Emphasis shifted toward Mayr's "peripheral isolate" model of speciation at the expense of the traditional "dumbbell" model. In other words, speciation came to be conceived of more like the budding of a yeast cell than like fission into equal-sized daughter cells as in *Paramecium*.

Consequently, polytypic species with locally diversified populations become things to which the biological species concept applies only with difficulty, if it applies at all. And the possibility that a concatenation of populations might easily break up into two separate species by elimination of one or more intermediate links comes to be looked upon as a mode of speciation that is not "typical." Essentialism has raised its ugly head again. Realizing this, we had better appreciate the extent to which species concepts tend to be linked up to particular theories about how they are formed. They also get linked up to particular theories about the nature of the cohesive forces that supposedly unite them, and about whatever it is that supposedly tears them apart and keeps them apart.

There are good reasons why, in striving to formulate a proper definition, we should want to be somewhat noncommittal, or theory-neutral, with respect to what it is that forms and maintains species. For one thing, we want the question of how it is that species form to be a matter left for further empirical scientific research, not turned into something to be decided by lexicographical arbitration. The possible role of learned (rather than instinctive) behavior provides some instructive examples. Often it is said that the biological species concept is a genetic one, and that, therefore, specific status necessarily implies that any two species must differ genetically. However, the definition does not rule out the possibility that mating behavior acquired purely by learning might get passed from generation to generation and suffice to isolate populations, making them, by definition, separate species. Hence genetical change is by no means a necessary condition for speciation to occur. All sorts of cohesive forces that we don't know about may be holding species together as well.

There are perhaps even better reasons for not making some particular mechanism of species formation defining of 'species'. Nobody to my knowledge would want to define 'species' in a way that does not include all the products of speciation processes that exist, or for that matter those that might possibly exist, including both the dumbbell and the peripheral isolate mechanisms, sympatric as well as allopatric speciation processes, or for that matter even special creation. The species is defined in a way that allows it to function in evolutionary theory, but not in a way that forces such theory to be "monolithic" with respect to the kind of causality that is deemed a legitimate possibility.

And yet the critics of the biological species concept often argue that it is not broad enough to suit their needs. Their point is not that they object to drawing the line in one place or another. Rather they want a species concept so broad that every organism is an element of a species. And of course this precludes asexual clones and the like. The real justification for this claim is the supposed advantages that we would have from being able to refer to each and every organism by a specific epithet, and to do so in what seems, at least, to be a straight-forward manner. We lose, however, the advantage of having the most basic unit in systematics coincide with one of the most basic units in theories of evolutionary processes.

The reason for not treating asexual clones as parts of species is of course that sex provides the cohesion by virtue of which species evolve as units. For all we know something else may hold them together too, and such possibilities are well worth exploring. It has occasionally been claimed that asexual groups have the same basic properties as sexual ones. Such claims are erroneous, but the issues are sufficiently complicated that we shall defer rebuttal until Chapter Ten, which provides a more general discussion on sex. At this point we should perhaps merely draw attention to the fact that the definition can be cast in terms that are rather noncommittal about what the cohesive forces actually are.

Hybridization between species suggests that there is a certain amount of cohesion between at least some of them. Here we have a minor source of criticism occasionally leveled against the biological species concept. Making the amount of cohesion more explicit helps to defuse such criticism, but at the same time it brings another kind of criticism to mind. Namely, a great deal of diversity can arise within species, especially when the cohesive forces are weak or inoperative. Under such conditions the amount of diversity within a species can be, at least on the face of it, comparable to or even greater than that within a genus. The amount of difference, however, is irrelevant, for what really matters is the restraining influence that the diversifying parts may exercise upon one another.

So obviously those biologists who study situations in which the cohesive forces would seem to have had very little effect may feel that the biological species concept is not particularly useful in their own research. This is especially true among wide-ranging organisms of low vagility. Frequently the claim gets made that the biological species concept applies to animals but not to plants. Such claims would be more appealing, perhaps, if all plants, but no animals, were wide-ranging and of low vagility. But terrestrial pulmonate snails and earthworms provide good examples of populations made up of animals of limited dispersal capacity that diversify into widespread polytypic species with cohesion that is a very local matter at best. The land snails of Moorea (alas, now extinct) provide a fine example, wherein slow-moving creatures evolved so as to form a bewildering network of interconnected but diversified populations (Murray & Clarke, 1968, 1980, 1984; Johnson, Murray, & Clarke, 1986, 1993a, 1993b). The authors of the classic papers here had a very difficult time deciding how to rank the various groups that they encountered.

Time and again certain botanists have provided putative examples of situations in which the biological species concept is supposed not to apply; but time and again examination of the facts clearly and unequivocally shows that it applies perfectly well. In the case of what advocates of the biological species concept would call a polytypic species, it is a matter of presupposing that the parts of such species must be species. But to argue in this manner is the logical fallacy of begging the question. After all, the human species has all sorts of local variants, and we do not insist upon calling each and every one of them a distinct species.

So in order to make a decision as to what definition to accept one has to do more than just listen to dogs howl at the moon. One had better consider the alternatives. In many cases one can find perfectly reasonable alternatives to the biological species definition by recognizing and naming groups that are not biological species. If one wants to incorporate parts of biological species into the taxonomic hierarchy, one obvious possibility is the recognition of taxa of lower categorical rank, such as subspecies. These can even be designated by trinomials.

It is by no means unusual for local populations to be sufficiently different from other components of a species that it is worth talking about them. One very good reason is that they may form lineages that might be interesting objects for phylogenetic research. If diversification starts within biological species, one certainly wants to be able to study it. In addition to populations of organisms, there are also classes of organisms that interest the student of biological diversity. The sexes are a very good example. So there are perfectly good alternatives to calling a group of organisms a species, and before we do anything rash we might consider some of these.

This is not the best place to work out precisely what such alternatives should be. But we might at least suggest how one might proceed. Analogy is at least helpful. We have no dearth of situations in which parts break off from a whole and remain separate. People, for example, quit their jobs, get excommunicated from their churches, and graduate from college. We often refer to such persons by means of their historical relationship to the organization from which they have departed. For example, we refer to graduates by some such locution as "B.A., University of Maryland, 1989." And if a species should give off a clone, we should, in principle, be able to refer to it by naming the species from which it arose, and giving it a name (or number, or other symbol) that distinguishes it from other clones. And in fact this seems to be what we have done with HeLa cells cloned from the tissues of a woman now long-since deceased, the fantastic exercise (perhaps not intended to be taken seriously) of Van Valen and Maiorana (1991), who treated them as a species, notwithstanding.

Having discussed the possibility of applying concepts other than that of the species to particular groups of organisms, let us now consider the possibility applying a species concept other than the biological one. If somebody comes up with a better alternative, we should all be grateful, but as we shall see, those made available thus far create difficulties of their own, and not everybody is willing to pay the price.

In its most naive formulations, the species concept is left so vaguely defined that one hardly knows what it means. Species are kinds of organisms plain and simple. But so is the class of bachelors. To say that species are kinds of organisms is to suggest that species are classes, though perhaps those who say that they are kinds of organisms think that baseball teams are kinds of organisms, too. The sort of "natural history" that is generally the possession of even the most primitive tribes of humanity acknowledges the basic facts of procreation, the family, and the lineage. So when anybody tries to define 'species' so as to make them classes of similar organisms, a certain amount of accommodation seems almost inevitable, and one way or another the notion of an ability to procreate and give rise to fertile offspring gets added, perhaps as a kind of afterthought.

Definitions of species as classes of organisms that differ from one another by a certain amount are often spoken of as "morphological" species definitions. However, the characters that supposedly separate these classes are not necessarily "morphological" in any ordinary sense of that term. What is really meant is any attribute that might be predicated of an individual organism or specimen. Therefore things like physiology, behavior, or the properties of chromosomes and genes are by no means ruled out. Hence the term 'phenetic', more or less as it was used by advocates of the "phenetics" or "numerical taxonomy" that was popular in the sixties and early seventies, seems more appropriate.

The term 'phenetic' has the same root as 'phenotype' and 'phenomenon' and refers to the "appearances" of things. As a result, perhaps, of false analogizing, we got a confused notion that the issue has been between classification on the basis of phenotypes as advocated by pheneticists, on the one hand, and classification on the basis of genotypes as advocated by the evolutionary systematists, on the other. However, the problems with respect to phenotypic similarity turn out to be equally applicable to those with respect to genotypic similarity. It has never been obvious whether attributes of populations rather than individual organisms, their ranges and sex-ratios for example, might qualify as "phenentic," but nobody seems to have made a fuss about that.

Irrespective of what the similarities are supposed to be, all "phenetic" definitions of the species category (and other categories as well) suffer from the same basic problems. There is no such thing as similarity that is both "overall" and objective. The different kinds of similarities are of such a heterogeneous nature that they are incommensurable in principle. Furthermore, even if one establishes an arbitrary and subjective amount of difference as justifying specific rank, there remains the question of why one calls the taxa of such rank "species" rather than genera, subspecies, or whatever. In effect, a phenetic or morphological definition leaves 'species' incompletely defined.

In some cases the criterion for specific distinctness is patently subjective. A fine example is provided by Cronquist (1978:3): "Species are the smallest groups that are consistently and persistently distinct, and distinguishable by ordinary means." Cronquist freely admits that what qualifies as "ordinary means" can vary from person to person and taxon to taxon. It could be anything from examination with the naked eye to sequencing the entire genome. Taxa that are presently species would not have been species at a time when the techniques by which they are now distinguished were not ordinarily used. We should emphasize that the mere fact that human beings cannot ordinarily tell which organism belongs to one species and which to another does not necessarily imply that the organisms themselves cannot make that distinction and do so by means which, for them at least, are perfectly ordinary. For instance, we may not even be able to smell the odors that many animals use in finding mates. This is especially true under water. So too with sound, which is often pitched at higher frequencies than we can hear.

Subjective species concepts such as that advocated by Cronquist and others have often been called "practical" (J.S. Huxley, 1940). The term 'practical', however, is used in a rather peculiar, economic sense. Namely, the employees of museums are sometimes expected to identify a lot of specimens. Often they think of themselves more as identifiers than as classifiers (Calman, 1940). If not actually forced to do piece-work, they at least have an incentive to maximize the number of names that they can put on specimens per unit time. Anything that makes their task less burdensome will, for that reason, be welcomed. Furthermore the consumers of systematics do not always care whether or not the names that they use correspond to taxa that a serious evolutionary biologist would want to call a species, rather than a subspecies or a genus. They will be perfectly happy, for example, with the labels on their shell collections. Of course other consumers may be very interested indeed. The textbook example is the mosquitoes that are very hard to tell apart, especially as adults, but which differ in their ecology and with respect to whether they transmit malaria (J. Smart, 1940; Thorpe, 1940).

Of course such "cryptic species" have long been known, and have been described and named just like the more obtrusive ones (Dobzhansky, 1939). As electrophoresis and other new techniques have come on line, an increasing number of cryptic species are being identified. The results are particularly striking among marine animals, as has been documented in a recent review by Knowlton (1993). To give but one example of the implications, it was formerly believed that a lot of marine animals display "poecilogony," a polymorphism in the mode of development. Although there are just a few well documented instances, it turns out that many are just separate species that are hard to tell apart (Hoagland and Robertson, 1988). It should be obvious that bad taxonomy can be a disaster to just about any reproductive biologist.

It seems odd in the extreme that Cronquist (1978) would treat cryptic species as if they posed some kind of problem for the biological species concept, for all sorts of others have drawn precisely the opposite conclusion. Indeed the very terminology here has created some curious discussion. Some authors (Gochfeld, 1974) prefer 'sibling' to 'cryptic' species. But 'cryptic' seems apt, for it means hidden, whereas 'sibling' has connotations of close relationship. Although cryptic species are apt to resemble their closest relatives, there is no necessary connection here. McCafferty and Chandler (1974) object to 'cryptic' because it might be taken to mean a species of cryptic organisms: ones that conceal themselves beneath leaf-litter or something. But given the individuality of species, it should be obvious that crypticity is a property of the species, not the component organisms. Somebody who belongs to an invisible college is no more apt to be run over by a truck than anybody else.

The old notion of a "physiological" species concept has often been confused with the biological concept, but it is really quite different. The physiological species has taxa that are classes of organisms which share the ability to cross with one another. This is in fact a very limited kind of similarity and one that leads to contradictory conclusions, for the ability to cross varies considerably within populations, and a cross between one organism and a second may fail to produce offspring, whereas if it crossed with a third the result would be fully fertile progeny. With the biological species definition this is no problem, for what matters is not a relationship between organisms but a relationship between species as wholes.

The "recognition" concept of Paterson (1982, 1985) was introduced in the previous chapter, wherein it was suggested that it matters little whether we emphasize what holds species together or what keeps them apart. In the formulation provided at the beginning of this chapter, an effort was made to emphasize the cohesiveness as well as to specify the degree of separation. We need to address the question of whether the isolation and recognition concepts really are equivalent. Claridge (1995) takes the position that isolation plus recognition together give the biological concept, so for some people at least, they are not all that different.

Paterson (1985:25) defines 'species' as follows: "We can, therefore, regard as a species that most inclusive population of individual biparental organisms which share a common fertilization system." As usual, the term 'fertilization system' is not self-explanatory. Paterson (1988:69) says: "The fertilization system of a species comprises all characters that contribute to the achievement of fertilization. These characters are diverse and include such characters in the mating partners as the design features of the gametes, those determining synchrony in the achievement of reproductive condition, the coadapted signals and receivers of mating partners, and their coadapted organs of gamete delivery and reception."

Here, species are populations in the sense of reproductive communities, so there is no difference in this respect. Furthermore the definition implicitly excludes asexual assemblages. However, the definition itself leaves much about the fertilization system unstated. It is not obvious whether a species could evolve a new fertilization system, lose the original one, and nonetheless remain the same species. If not, then it would make a great deal of difference, because one might interpret such species as classes of sharers of a particular fertilization system. Or perhaps they would be temporal parts of species—during stages in their evolution comparable to the youth and adolescence of an organism.

Even more important is the question of whether a species might have two or more fertilization systems at a given time. We can imagine a situation in which a species consisted of a linear series of populations, with very different causes of cohesion at the two ends and a very gradual intergradation among the intermediates. Paterson rejected this possibility upon theoretical grounds, claiming that the manner in which animals recognize conspecifics is too strongly reinforced by selection. Furthermore he argued for his definition on the basis of claims that all the relevant divergence must occur in allopatry. He rejected the claim that once a modest amount of isolation has occurred, there can be selection that "reinforces" it. Suppose that two populations had the same way of recognizing mates but could not produce fertile offspring should they pair. He argued that if such populations were sympatric, they would not coexist, but rather one or the other would disappear. It is an interesting empirical issue, and one of great theoretical importance, how readily populations that have only post-mating isolating mechanisms can if fact coexist and if so can evolve isolating mechanisms. Although the difficulties would seem to be real, they also seem to have been exaggerated (Littlejohn, 1981; Spencer et al., 1986; Butlin, 1987a, 1987b, 1989; Carson, 1990; Howard & Gregory, 1993; Liou & Price, 1994).

Furthermore, although Paterson has had a few strong supporters (Masters and Spencer, 1989; Niven, 1989; White, Michaux, and Lambert, 1990; Masters, 1993), his species concept is generally not considered a real improvement over the biological one (Raubenheimer & Crowe, 1987; Mayr, 1988b; Verrell, 1988; Coyne, Orr, and Futuyma, 1989; Ryan, 1990). From the point of view of definitions, however, the interesting point is that Paterson seems to have made differentiation sufficient not only to keep the populations separate, but sufficient to allow them to coexist, when sympatric into a defining property of the species category. This clearly would individuate different groups than would most versions of the biological species concept. But not all of them, including Mayr's ecological version to be discussed later.

According to Paterson's conception of the speciation process, each species has a normal habitat and way of life. Selection maintains a kind of equilibrium condition (for critique see Eberhard, 1992). In agreement with Mayr and many others who advocate "peripheral isolate" models of speciation, he thinks that speciation involves a kind of genetic revolution that goes on in small populations. And it would seem that once such a revolution has occurred, we have a new population. To my knowledge he has not considered whether a single population would become a new species were it to pass through a genetic bottleneck. Whether or not Paterson ever held that opinion, it is obvious that somebody might. In that case we would have a kind of phenetic species concept in the broad sense that includes the physiological species.

Paterson's main arguments against the biological species concept turn out to be metaphysical ones. In the first place, he maintains that the notion of an isolating mechanism is teleological. On the one hand, we must acknowledge that teleological conceptions here not only are possible, but have been embraced by various authors. This was especially true during the time when the biological species concept was being formulated by Dobzhansky and Mayr, and a great deal of covert teleology affected the thinking of biologists. It was uncritically assumed that there are adaptations at the populational level, and isolating mechanisms were supposed to exist by virtue of the benefits conferred upon the species. On the other hand, there is no reason why such interpretations simply must be imposed upon the biological species concept. Paterson and his followers have insisted that the concept of a "mechanism" implies that isolating mechanisms are adaptive in the sense of being the product of the process of selection. But this is simply an objection to a metaphor that of course suggests, but does not entail, teleology. In the above quotation Paterson himself uses the expression "design feature" when he refers to the properties of gametes.

The other metaphysical point that Paterson urges against the biological species concept is that it is a relational one. According to Mayr, a species is like a brother. A species is a species only in relation to other species. My own answer has been that this is merely a result of taking the particular verbal formulae that Mayr has published too seriously. One can readily solve the problem by emending Mayr's definition to read that species are reproductively isolated from any other such groups as may happen to exist. In proposing what I call the "bioeconomic" version of the biological species definition (Ghiselin, 1974c) I pointed out that species are like organisms, not brothers, if the category is defined that way. The formulation provided here has the same advantage, since it does not imply that a species simply must have a relative that is itself a species. And unless the advocates of the biological species concept were to deny that the first species to exist was not a species, we ought to assume that the concept is implicitly non-relational.

One more point needs to be made with respect to the recognition concept. Although some of its advocates explicitly recognize the difference between defining the species category and defining species in terms of recognition, they have tended to play down the distinction and invoke epistemological criteria in favor of their views (Lambert, Michaux & White, 1987; Masters, 1993). There would seem to be some implicit metaphysics here that needs to be considered in evaluating their views.

Another heroic, if not very successful, effort to modify the biological species concept is the ecological version suggested by Mayr, (1982a:272): "A species is a reproductive community of populations (reproductively isolated from others) that occupies a specific niche in nature." Mayr did not push this redefinition very hard, and indeed from the outset expressed uncertainty as to whether the ecological criterion really ought to be treated as a defining property. In general, it has been treated as little more than an intellectual curiosity (Hengeveld, 1988; Futuyma, 1989). At first I thought that the reason for providing this criterion was that it might give systematists a criterion for deciding what to do with allopatric populations and asexual organisms. However, his real motive seems to have stemmed from a belief that the speciation process is "not complete" until the species are not only reproductively isolated, but able to co-exist when sympatric.

Although the niche concept will be discussed in more detail later on, we need to say a little about it at this point. The niche of a species as here defined is the Eltonian niche, or what Darwin called a "place in the economy of nature," and is often compared to the "profession" of a species although the term 'job' would be better. There are certain words that do have places as parts of their definitions. 'Chairman', for example, refers to a person who by definition must occupy a kind of social place. But this is not true of words in general, even though every material object occupies a place. And if we defined a cell as having a membrane and a nucleus, etc., it would seem redundant at least to add that it occupies a cellular place.

Both Mayr and Paterson have tried to make conditions that are physically necessary (perhaps not quite necessary) for species not only to exist, but to continue to exist, defining of the species category. In both cases, however, the species only have trouble coexisting when they are in sympatry. It is rather like insisting that an organism be defined, not only as physiologically autonomous, but economically as well, or even as able to survive long enough to reproduce.

Another alternative to the biological species concept is the evolutionary species concept. Although it must have had precursors of one kind or another, credit for an explicit definition is generally given to Simpson (1951). Simpson (1961:153) presents the following: "An evolutionary species is a lineage (an ancestral-descendant sequence of populations) evolving separately from others and with its own unitary evolutionary role and tendencies." A slightly different version has been provided by Wiley (1978:18): "A species is a single lineage of ancestral descendant populations of organisms which maintains its identity from other such lineages and which has its own evolutionary tendencies and historical fate."

The definition that I have provided for the biological species will probably strike the reader as very much like the evolutionary. Indeed, the same may be said for the "bioeconomic" version, as Frost and Kluge (1994) have pointed out. And for good reason. A common goal has been to make both clear and explicit the idea that species are units that evolve more or less independently. This provides some appearance of legitimacy for the claim that the evolutionary species is more general than the biological.

However, when people claim that the evolutionary species is more general, they probably mean that it incorporates asexual lineages and clones. On the one hand, this kind of generality may or may not be a good thing. For example, 'rape' would be more general if it designated all forms of sexual intercourse whatsoever. On the other hand, it is by no means obvious from the definitions provided by Simpson and Wiley that for them asexual groups are species. The reason is that the term 'population' does not mean the same thing in the biological and the evolutionary species definitions. Indeed, the word is used equivocally in the evolutionary definition, with disastrous results.

In the biological species concept the term 'population' designates an individual, a composite whole, and one with a certain amount of cohesiveness at that. In Simpson's version the populations are classes of similar organisms, with the additional proviso that there is a genealogical nexus among them. Hence it is possible to include asexual lineages. So when Simpson refers to "its" evolutionary role and tendencies, he implies a kind of cohesion that is not necessarily there. So too with Wiley, who explicitly acknowledges that species are individuals. When he says that a species retains its identity from other such lineages, he is saying that it remains that particular individual even in the presence of other such individuals. In other words they remain separate. But his species not only correspond to entire sexual populations but also to asexual lineages that are purely "historical entities" as he puts it. These retain their individuality in the sense of remaining separate.

The notions of evolutionary roles and tendencies are very vague. At most they suggest that different species do different things, and organisms of the same species do the same thing—more or less, at any rate. But this in no way differentiates the species from other categories, especially subspecies and even the most local populations.

We may conclude that the evolutionary species concept as formulated by Simpson and Wiley has little to recommend it over the biological. It tries to do too much and winds up doing too little. Its main attraction is that it suggests treating organisms that are asexual, or the reproductive biology of which makes the biological species concept hard to apply, as classes of organisms that sort of look like biological species. The problem can be solved without this maneuver.

On the other hand, there is an earlier version of, or precursor to, the evolutionary species concept which is often designated by the same name (Simpson, 1951:289): "a phyletic lineage (ancestral-descendent sequence of interbreeding populations) evolving independently of others, with its own separate and unitary evolutionary role and tendencies, is a basic unit in evolution." Here the word 'interbreeding' makes all the difference in the world. For the units in question are neither more nor less than biological species (Krishtalka, 1993). And at the time Simpson would seem to have recognized that, for he did not treat it as a different concept, but only as a matter of emphasis and clarification. One may wonder why Simpson changed his mind and abandoned his own version of the biological species definition in favor of something rather different. Apparently he fell increasingly under the spell of the set-theoretical treatment of the Linnaean hierarchy by Gregg (1950, 1954), who, although mentioning in passing the possibility that species are something else, insisted that they are classes.

For species to evolve, it is metaphysically necessary for them to be individuals, and an "evolutionary" species definition that treated them as if they were sets would be a contradiction in terms, or an oxymoron at the very best. If one wants to think of them as classes, then the set-theoretical approach becomes seductive. But since a set with different members is not the same set, one is driven to the conclusion that an individual species at a given point in time is not the same thing as a species at any other point in time. And granting that, one has to conclude that the species category viewed with respect to what obtains at a given moment is ontologically different from the species category considered with respect to what obtains through time. But the only way that an evolutionist can deal with such a situation is to treat the argument as a *reductio ad absurdum*, and spurn the set-theoretical metaphysics. However, not everybody seems to have realized this. Salthe (1985) in particular has claimed that synchronic and diachronic species are ontologically different for just such reasons, while nonetheless endorsing the individuality thesis. He may be in part responsible for recent assertions that Simpson's 1951 version of the biological species concept is somehow different from, or preferable to, the biological, or to gloss over the differences between the 1951 and 1961 versions (Eldredge, 1993; Kimbel & Rak, 1993). It looks like academic mythology in the making.

Next let us consider the "cohesion concept" of Templeton (1989:25) which defines species as: "the most inclusive group of organisms having the potential for genetic and/or demographic exchangeability." Templeton's views resemble those of Mayr insofar as he emphasizes something like Mayr's notion that species form as a consequence of genetic revolutions. But he has somewhat different views on the mechanisms of genetic revolutions and prefers to use the word 'transilience' instead (Templeton, 1980). On the other hand, he favors Paterson's views on the importance of mate recognition, hence the "cohesion" concept. Casting the definition in terms of an ability of one gene to be exchanged for another makes it roughly the same as the biological species definition. But the notion of demographic exchangeability is another matter. Demographic exchangeability includes one organism being equivalent to another even when they are not parts of the same genetical population. One asexual organism can replace another one demographically. So his groups are abstractions rather than concrete populations. Or perhaps we may say they are biological species with properties such that demographers can give "honorary" membership to certain individuals that otherwise do not fit in. Therefore it is a cohesion concept only by virtue of applying an inappropriate label and not by virtue of making cohesion a defining property.

We may now pass on to a variety of species concepts that are often called "phylogenetic" or "cladistic" because they tend to be associated with the advocates of "phylogenetic systematics" or "cladism." The founder of cladism, Willi Hennig (1950, 1966), explicitly endorsed the biological species concept, and this makes a great deal of sense given his realism and his opposition to idealistic morphology. On the other hand, as various authors have pointed out, his version of the biological species concept was not quite the same as that of Mayr (Ridley, 1989, 1990; Wilkinson, 1990; Nixon and Wheeler, 1990, 1992). According to Hennig, whenever a species speciates, the ancestral species ceases to exist. But according to Mayr, a species might bud off a new species as a peripheral isolate and continue to exist. This is, as already explained, a metaphysical problem for which no solution seems possible. Clearly these "biological" species definitions individuate quite different groups. It is well worth noting that none of the brief and terse formulae of definitions that we have encountered specify what one is supposed to do when applying the concept under such circumstances.

On the other hand, such a rather arbitrary criterion as that made by Hennig has methodological justification for the kind of task he set himself: delimiting the supraspecific lineages that are generated by sexual populations. And for such purposes the biological species may not be the appropriate unit to single out for study. When populations split up, they may have already formed separate lineages before anything like reproductive isolation has occurred, so units smaller than species are perfectly good objects of study. Indeed, lineages of organisms, and of parts of organisms, can easily and productively be studied using cladistic techniques. Species are of particular interest to phylogeneticists because they are the largest units that are generally homogenized by sexual reproduction and may reasonably be said to evolve as such, but they are not the only units of interest.

Some persons, however, have proposed that species not be the largest populational units, but rather lineages and populations of somewhat lesser incorporativeness. They have by no means been able to agree how large such units should be, and indeed often suggest that it really does not matter. In that case the term 'species' simply means a taxon of lower rank that happens to interest somebody. Rosen (1978), who gets much of the credit for the phylogenetic concept, seems to have had this sort of notion in mind. It is obvious that the defining criteria here tend to be arbitrary and subjective. When efforts are made to specify such criteria, the results are not particularly encouraging.

According to Kluge (1990:411): "The species category is defined as the *smallest historical individual* within which there is a parental pattern of ancestry and descent." Kluge compares this to the evolutionary species concept, and this makes sense insofar as there is an idea of the groups evolving separately from one another. The clades, insofar as they represent the diachronic aspect of populations, would then represent separate lineages, irrespective of whether they have the potential for interbreeding. On the other hand, it is by no means clear what to do with asexual lineages under such a species concept, for they "speciate" every generation. As Frost and Kluge (1994) point out, the effect of such a species concept is to increase the number of taxa at the species level, and at the same time reduce the amount of paraphyly. The latter has some utility for phylogenetic analysis, but the same goals can be achieved otherwise. But making every branch into a species means that a deme that 1. happens to be separated from its species and exist for a few generations, and then 2. happens to become extinct rather than reunite with its collateral branch, has to be treated as a full-fledged species. This is not what is meant by a species in most theories of speciation. Furthermore, at the present time, we find all sorts of allopatric populations that are by no means yet extinct, yet may or may not fuse depending upon what happens in the future. If all one wants to do is figure out their relationships, all well and good, but what if one wants to make decisions about conservation policy? Kornet (1993) has used what she calls "permanent splits" in lineages as a basis for delimiting "internodal" species. But such permanency has a different causal basis, depending upon whether it is a matter of intrinsic isolating mechanisms or mere historical accident. On the other hand, Kornet has pointed out to me in conversation that it would be possible to adapt her formalism for delimiting biological species and at least in that sense reconciling the two approaches. Exploring possibilities along such lines has lately become an active focus of research (Graybeal, 1995).

Eldredge and Cracraft (1980:92) try a slightly different tack:

> a diagnosable cluster of individuals within which there is a parental pattern of ancestry and descent, beyond which there is not, and which exhibits a pattern of phylogenetic ancestry and descent among units of like kind.

This was later somewhat emended by Cracraft (1987:341): "an irreducible cluster of organisms, within which there is a parental pattern of ancestry and descent, and which is diagnosably distinct from other such clusters." The "beyond which there is not" a pattern of ancestry and descent was evidently removed, since obviously there are patterns of ancestry and descent at all levels in the hierarchy. 'Cluster' is a metaphor that is not altogether clear from the definition provided, but from the text one can see that the taxa in question are definitely supposed to be populations. Thus it rules out things like parts of organisms, and asexual lineages as well.

A version that does cover asexual groups is provided by Nixon and Wheeler (1990:218):

> We define species as the smallest aggregations of populations (sexual) or lineages (asexual) diagnosable by a unique combination of character states in comparable individuals (semaphoronts).

Another effort is provided by Mishler and Brandon (1987:406):

> A species is the least inclusive taxon recognized in a classification, into which organisms are grouped because of evidence of monophyly (usually, but not restricted to, the presence of synapomorphies), that is ranked as a species because it is the smallest "important" lineage deemed worthy of formal recognition, where "important" refers to the action of those processes that are dominant in producing and maintaining lineages in a particular case.

Again, the units are lineages, not populations, so asexual groups can be accommodated. It is not obvious why the species in question have to be what they are because of "evidence" rather than on the basis of properties of the organisms and the lineages themselves. In other words, there is once more a strongly subjective aspect to this definition. One person's importance is all too often another's triviality.

Trying to have such a "minimal unit" for systematics is by no means unreasonable. In population genetics we have the deme. And in fact there has been a long tradition among systematists of naming subspecies and other parts of species. The question is, why call such things species? I see no reason why calling such units species adds anything substantial from a scientific point of view to Cracraft's (1986, 1988, 1992) outstanding work on biogeography. And there are all sorts of objections. Not the least was one raised by Stanley Salthe at the end of a lecture by Cracraft at a meeting of the International Society for the History, Philosophy and Social Studies of Biology. According to Cracraft's definition, *Homo sapiens* is not a species. There are very good reasons for not adopting that definition, and we may suspect that part of the motive for advancing it was factional politics among biologists.

Politically, the repercussions of abandoning the biological species concept in favor of smaller units might be most unfortunate, and I urge every morally responsible reader to consider it carefully. At present, 'human being' is a componential sortal, and means an organism-level component of *Homo sapiens*. The rules of nomenclature are such that, were we to have species be only parts of biological species, the name would cease to apply to all of us, though it would have to apply to at least one white person. Most people would be outraged by someone even suggesting the possibility that they are not a human being, and life might become downright miserable if one put oneself in a position of having to explain why one gave even the appearance of being a racist.

But philosophically the most serious objection we might raise is perhaps with respect to the notion of a "diagnosable" group. If diagnosability is contingent upon our ability to diagnose the group—in other words to differentiate it from other groups on the basis of properties—then there is a serious problem with respect to objectivity. A group becomes a species when we have the ability to differentiate it. If, however, we adopt an objective criterion, it would seem that very small groups indeed can be diagnosed. Sequencing entire genomes should give diagnostic characters every time a cell divides.

Perhaps driven by a sense of desperation one might consider having the species be not one kind of thing, but many kinds of things, perhaps all things to all men, women, and dogs. We might want, as some have put it, a "pluralistic" species concept. Here we do not mean that the same symbol might designate different classes, in the sense that "class" has a pedagogical as well as a logical sense. In such cases a precise intension is supplied by the context of the general term, and the universe of discourse (what is being talked about) is clear and unequivocal. With pluralistic species concepts there is no particular relationship between what the words mean and what the discourse is about, so one has no way of knowing, for example, whether or not the same thing is designated by a word in the major premise or the minor in a syllogism. And indeed steps may be taken to see to it that there is no connection at all.

The most straightforward form of pluralism in systematics would have it that there is, or even should be, no necessary connection between the language of taxonomy and that of other branches of biology, especially evolutionary theory. Accordingly, when taxonomists use the word 'species' or the name of one of its instances such as *Homo sapiens,* they do not designate anything to which evolutionary biologists or lay persons refer in their discourse. Although taxonomists name taxa, the taxa are not the names of groups of organisms, but of something else. Accordingly, the statement that "*Homo sapiens* may become extinct" does not assert anything about something of which this is possible. But what, if anything, is it that taxonomists are supposed to name? The best answer I can give is something like constellations—strictly phenomenal groups.

It seems perfectly reasonable that people might want to bestow names upon constellations and other "things" that are useful in locating objects from their own perspective. However, it seems very strange indeed that they should refrain from doing anything else. In astronomy there are class names, such as 'galaxy' and 'star', and proper names for their instances, such as 'the Milky Way' and 'Sol'. But there are not two kinds of astronomy which define these terms in different ways if they define them at all. Nobody asserts that there is one thing, the sun, which does have a name, distinct from that entity which seems to rise in the morning and set in the evening, and which acts in conjunction with the moon to produce the tides, but which does not have a name. If 'the sun' is not the name of the sun, if 'The United States of America' is not the name of The United States of America, if your name is not your name, and if *Homo sapiens* is not the name of your species, then what, if anything, is it that these words are supposed to designate? This kind of pluralism, however popular it may be, simply won't bear critical examination. It would have us talk about a metaphysical fiction rather than about the material universe.

Usually pluralism is less bizarre from a logical and metaphysical point of view. In general it is proposed that biologists can pick and choose whatever happens to suit their needs or inclinations as to how to define the category. A celebrated example is provided by Regan (1926:75): "A species is a community, or a number of related communities, whose distinctive morphological characters are, in the opinion of a competent systematist, sufficiently definite to entitle it, or them, to a specific name."

This is a mixture of populational, morphological, and phenetic criteria. By a "community" Regan says that he means "a number of similar individuals that live together and breed together." Thus he means something like a reproductive population, and in this respect his is like the biological species concept. The role of similarity in defining 'community' is not specified. But once a systematist has decided that his populations are similar, then he ranks them as species or subspecies, or perhaps as a genus, purely on the basis of "convenience." There are no rules for deciding this, and there should not be any. One naturally wonders whose view to accept when the experts disagree.

Another author who advocates a pluralistic species concept is the philosopher Philip Kitcher, who, as was pointed out in Chapter 4, has been a persistent critic of the individuality thesis (Kitcher, 1984b, 1984a, 1986, 1987, 1989). Right or wrong, he is at least consistent insofar as some of the entities that he wants to call species are hard to interpret otherwise than as sets. He suggests that a biologist ought to pick out any kind of unit that interests him and call that a species. A knowledgeable biologist might agree that even advocates of the biological species concept use 'species' in what might be called a "pluralistic" sense, but in such cases a very different kind of pluralism is involved. Like many other words, 'species' has different senses that may be equivocal. Which of the senses, that is used has to be specified, perhaps by context. In a very broad sense, it refers to a particular kind of group, but such groups can be anything, not just groups of organisms; terms are said to be "generic" or "specific" rather than "general" or "particular" for example. We occasionally hear of "mineral species," and in fact these were recognized by Linnaeus in a taxonomic hierarchy that included Lapides as well as Plantae and Animalia. But when we talk about organisms, rather than words or rocks, we infer from the context that the species in question are the species of biology. Part of the problem with the individuality thesis has been that in biology we generally do not think of "individual" as something other than an organism. That is why we so often have to use a locution such as "individual in the ontological, rather than the biological, sense." Those who want the biological species concept to be the generally accepted one do not insist that nobody should ever use the term otherwise, but only that the biological sense should be presupposed unless the alternative is specified. Hence such locutions as "nominal species" and such coinages as 'pseudospecies' or 'agamospecies' for asexual assemblages. If there is any doubt that biological species are meant, one can say "biological species" or "biospecies" instead of just "species." And we do.

Not everybody, however, agrees that the biological species ought to be the ordinary sense of species and that the others ought to be avoided. Nor does everybody agree that an asexual species ought to be considered a contradiction in terms and ought to be called "asexual 'species'" at best. Nonetheless, the reason for wanting such restrictions are simple and straight-forward enough: to avoid the logical fallacies and misunderstandings that arise from equivocation. It may seem bizarre, but Kitcher, a professional philosopher, is arguing that the language of science actually ought to be equivocal, and not that it is just a problem that we may have to put up with.

Kitcher argues in favor of an equivocal species concept by means of an analogy between the species and the gene that can readily be turned on its head. Thus Kitcher (1984b:308) gives the same definition of species that Regan (1926) did: "Species are those groups of organisms which are recognized as species by competent taxonomists." Likewise (Kitcher, 1992:131): "A gene is anything a competent biologist chooses to call a gene."

Biologists have never been altogether happy with the concept of "the" gene in the sense of what are sometimes called "hereditary determinants" in general. It originated as something very hypothetical with an assumption that the hereditary material (or whatever) consisted of corpuscular functional units. As it turns out of course, there are indeed functional units. However, these are definitely not a bunch of things like beads on a string. Chromosomes and entire genomes are inherited and transmitted, and the genotype that corresponds to a phenotype may be a level of ploidy (haploid, diploid, tetraploid, etc.). Within the chromosome there are a whole hierarchy of functional units, going all the way down to codons and base pairs. According to Kitcher (1992) genes are stretches, whether longer or smaller, of nucleic acids. But he overlooks the fact that some genes are inversions, translocations, position effects, or even deletions.

Furthermore, biologists have increasingly settled upon a particular kind of functional unit as the gene in the ordinary sense. This is the stretch of nucleic acid that specifies the amino acid sequence in a single enzyme. There are various problems with this definition. Some structural genes do not specify the structure of an enzyme. Other genes are regulatory, rather than structural, in function. Nonetheless there is a clear intent that the term 'gene' shall designate a class of functional units that participate in a given kind of process (with complications such as genes that are not transcribed).

When geneticists try to estimate the number of genes in the nucleus of a cell, they are not trying to count the number of things that somebody might want to call a gene: codons, chromosomes, inversions, you name it. Rather they want to single out the units that can be meaningfully and fruitfully compared in functional and perhaps historical terms. Likewise when it is asked how many species there are in the world, the question is not what somebody might want to call species, but how many units of a given kind do in fact exist. A pluralistic species concept renders such an effort altogether worthless. Imagine what would happen if the United States Bureau of the Census attempted to do its job with a "pluralistic" conception of "inhabitant" and instead of everybody collecting the data counting heads, one census taker were to count heads, another legs, another digits, perhaps others hairs.

The only argument that Kitcher gives in response to such objections is that we supposedly get along fine with such "pluralistic" language. It may not matter if we divide up the world in different ways and use the same word to refer to different kinds of parts. But no number of examples of problems being avoided can contradict the plain fact of problems having been encountered and crying out for solution.

But Kitcher's main thesis about genes and species is that no units are "privileged," and therefore none should be singled out as the standard intension of a term. This completely misses the point. Different kinds of units play different roles in natural processes. We want to use the same general terms for classes of individuals that do the same kind of thing and play equivalent roles in theory. The molecule, the cell, the organism, the deme, the species, and the phylum are classes that differ with respect to what their members can and cannot do, and these differences are important. If a heterogeneity of intension for 'species' is such a good thing, why stop with groups of organisms? Why not broaden the concept so that any group of cells, or even molecules, is a species?

In embracing pluralism rather than monism, Kitcher has implicitly taken the position that scientists ought to be parochial rather than catholic in their outlook. He treats knowledge itself as if it ought to be a class, or even worse, a set, rather than an individual. Some of us have a rather different attitude toward our work. It seems to me that malacology and carcinology are parts, not instances, of zoology, that zoology is a part of biology, and that biology is a part of science. An integrated body of knowledge benefits from a reasonably uniform and coherent language, especially with respect to the names of the important units. When somebody uses a word, it should be easy, not just possible, for somebody else to figure out that he means one thing and not another. If malacologists never conversed with carcinologists, and if nobody else ever read their publications, then perhaps it would make very little difference whether what malacologists call species were what carcinologists call subspecies. But of course, a scientist's work is, and should be, known to other scientists. One measure of excellence or lack of it in research is the extent to which other scientists want to know about it, especially in order to make use of it in their own research. A common language for the whole of biology is precisely what taxonomy is all about. It has been ever since the time of Linnaeus. And what about knowledge in its entirety?

Not everybody, to be sure, thinks that the unification of knowledge is a good thing. Indeed, John Dupré (1993) argues in favor of a pluralistic species concept on that very basis. He maintains that the world will be a better place if we find room for all sorts of diversity, and the equivocal use of terms is a good way to accomplish that end. However sympathetic we may feel toward the sort of liberal ideals that he advocates, it does not seem to me that the end justifies such means. Indeed, if diverse groups of people are going to live together in a spirit of mutual respect and understanding, and to work together in a spirit of cooperation and good will, they need a common language, not a Babel-like situation in which they cannot communicate their thoughts and feelings.

Since what we are looking for is not just language, but system, especially a hierarchical system of classification, it stands to reason that biologists need unequivocally defined categories with everything at each level sharing certain important properties. Nothing stands in the way of having a wealth of categorical levels and taxonomy being pluralistic in that (very, very different) sense. Nor does it mean that we cannot have ways of classifying things apart from the Linnaean hierarchy. Such classes as "parasite" are not the names of taxonomic categories or their members, but of course they play a very important role in biological discourse. A system of classes of classes of individuals is one thing, a system of classes of individuals made up of parts another. On the other hand, an incoherent mixture of the two plus whatever else somebody might happen to throw in is no system at all. It isn't even a good list.

8

Objections to the Individuality Thesis

"How absurd that logical quibble—'if species do not exist, how can they vary?'
As if any one doubted their temporary existence."

—Darwin, letter to Asa Gray, 11 August, 1860

The previous chapter disposed of many objections to the biological species concept, putting us in a better position to rebut objections to the individuality thesis itself. We have already provided answers to some of the criticisms that have been offered, but a more thorough discussion still seems justified. Here, we will consider the various kinds of arguments that have been proposed and explain how to deal with them, rather than merely rebut some of the polemics that have been published. Human ingenuity being what it is, people can be expected to come up with additional arguments in the future.

Let us begin with some allegations that more or less obviously result from just failing to understand what the individuality thesis is all about. Sometimes it is said to mean that "The species is an individual." If this were taken literally, it would mean not that the taxa (such as *Homo sapiens* and *Ardea herodias*) are individuals but that the category of which the individual species are members, or in other words that the species in general or in the abstract, is an individual. In some cases such expressions are obviously not intended to have been understood that way by their authors. When, for example, Mayr (1976) first seriously addressed the question of whether species are classes or individuals, he asked in the title of his paper whether the species is a class or an individual. Some authors, however, really seem not to have understood what is meant, at least to judge from efforts at making sense of what they say. I find it very difficult to figure out what it would mean to say that the species category is an individual. One very clever philosopher, Edward Reed (1979), did attempt to argue along such lines. He treated the species category as a kind of "sector" of the natural economy. I can understand, vaguely, how the species that are components of the biosphere might come together one way or another to form some kind of composite whole. It is rather as if, in the political economy, "labor" ceased to be just individual laborers and got organized into a kind of labor union. But such a thing would be an individual in another class, and not the species category itself.

One wonders how many people actually have had the impression that the individuality thesis means that species are organisms, and have rejected the individuality thesis for that very reason. There are few if any examples of this misconception making it into print. In a much weaker sense of making them at least seem to take on all sorts of gratuitous organismal properties, however, it seems not all that uncommon (Kellogg, 1988; Bock, 1991). The individuality thesis does of course imply that species have at least one property in common with organisms: individuality itself. Of course it does not rule out species having all sorts of other properties in common with organisms. By the kind of logic that alas all too often does make its way into print, we have the claim that it implies that species must have this, that, or the other property that organisms have, and since they don't have such properties, they are not individuals. Sometimes such claims are made by those who do not approve of certain theories that may be true, but only if species are individuals. More about that sort of opposition later.

If "individual" commonly means "organism" and is less often used in the ontological sense favored here, it also has some other meanings, and these occasionally create some misunderstandings with respect to the individuality thesis. The various notions of autonomy, discreteness, distinctiveness, and the like were discussed at some length in an earlier chapter, so we need but mention them at this time. Some of the various dictionary definitions of 'individual' were discussed in an exchange of opinions between Ledyard Stebbins (1987) and me (Ghiselin, 1987e, 1987f).

A more serious sort of confusion has been a failure to identify the individuality thesis for what it really is and to confuse it with other philosophical notions. As Mayr (1987b) points out, various authors would seem to have claimed that species are individuals long before I did. And in point of fact I had already acknowledged such predecessors as Buffon (Ghiselin, 1969d, 1974c; see also Sloan, 1986). But in retrospect it turns out that a lot of these resemblances are more apparent than real, and that the authors had something rather different from the individuality thesis in mind. Haeckel (1866), for example, thought that a population could move from one species into another, which makes little sense unless species are classes. Willi Hennig (1966), the founder of cladism, has been treated by some of his followers as having believed that species are individuals. Certainly, his views make much more sense if we interpret his writings in the light of that assumption, but he never came right out and said it. Furthermore, his followers completely missed this point until they read my paper of 1974. A perfect example is Dupuis (1984), who said that Hennig shifted from treating species as intensionally defined classes to extensionally defined classes after encountering the writings of modern logicians. I am by no means alone in thinking that Mayr somewhat underestimated the novelty of the individuality thesis (Hull, 1987; Rosenberg, 1987). After all, I grossly underestimated its importance at first myself.

Another kind of opposition to the individuality thesis comes from indignation at finding a word used in an unfamiliar sense. Having patiently listened to a very good-natured graduate student at the University of California at Irvine lament the situation, I feel moved to express a certain amount of sympathy, and if somebody can come up with a better term I would gladly change my own ways. Calling species individuals at least does not confuse anybody who knows that the term 'individual' has an ontological sense and what that sense is. And the alternatives that have been proposed thus far are, without exception, much, much worse. Mayr (1987b) suggests that instead of saying that species are individuals, we should say that they are populations. But although all species are populations, and all populations are individuals, not all individuals are populations. A monophyletic genus, consisting of several species all of which are descended from a common ancestor, is an individual but not a population. Likewise with the term 'system' as advocated by Stebbins (1987) and Löther (1972, 1991). The term 'system' has been variously and vaguely defined, and in the case of these authors it evidently means a kind of material whole. But as will be developed at some length in a later chapter, individual languages are individuals, and these are not exactly what is meant by a material object, albeit some persons might want to call them systems composed of something. Furthermore, a lot of individuals, such as wars and speciation events, are by no means physical objects. They fall under the category of process. Again, the existence of such entities tells us that we need a more general term. And 'individual' is the traditional one.

Furthermore, as I said in my response to Mayr and Stebbins (Ghiselin, 1987e), there are additional sanctions for using the term 'individual' when speaking of things other than organisms. Nobody seems to have any problem when we speak of an individual chair, whether it be a piece of furniture or an academic one. And we wouldn't say that I am sitting on a populational chair, or a systemic chair, as I write this book. In some cases, the term 'particular' works perfectly well in place of 'individual', but even here there are problems. 'Particular' is a contrast term for 'general'. It does not entail the concreteness that 'individual' does. And finally, the word 'individual' has a long history of standing in opposition to 'class'. Once made aware of this point, however, others have occasionally protested that the individuality thesis makes all sorts of things besides species be individuals. Again, this seems merely to be a matter preferring old ways of thinking to new ones. A lot of the things in question are indeed individuals, and it really matters. One of the reasons why the individuality thesis has attracted so much interest is that it has so many important implications.

Objections have been raised to the individuality thesis on the grounds that it conflicts with intuition, common sense, or tradition. One might better argue that because it conflicts with all those things, so much the better. It would not be the first time in the history of the world that intuition and common sense have proved misleading. And given the opportunities that so often result when we reject old ideas in favor of new ones, this speaks in favor of the individuality thesis, not against it. The individuality thesis does conflict with the usual "analysis" of the problem that has been provided by many philosophers. So much the worse for those philosophers! If they had really analyzed the problem, they would not have missed the boat. Among philosophers, Marjorie Grene (1989) has protested that Hull and I are not using the orthodox philosophical concepts of individuality and class. But where is the consensus among philosophers? And what is so heterodox about Aristotle, Locke, and John Stuart Mill? I freely admit that the early discussions of the individuality thesis were not as clear as they might have been (Eldredge and Grene, 1992). But this only underscores the point that unfamiliar ideas may be hard to articulate. More importantly, once philosophers have come to understand what the individuality thesis is all about, the only one who has raised any objections to such usage has been Grene. Everybody else seems to assume, and quite rightly, that the argument has been carried on in the sort of language that has long been conventional among metaphysicians.

Others are disturbed because the individuality thesis lends a certain amount of plausibility to positions to which they are opposed, or tends to weaken their own arguments. This is hardly unexpected, given the many implications that have been recognized thus far. In some cases the fears are justified, in some not. Many people find nominalism distasteful. Mayr (1970:11-12), for example, originally rejected the notion that species are individuals because it seemed to him nominalistic. Because the individuality thesis was developed partly as a response to claims based upon misinterpretations of nominalism, it makes a certain amount of sense that the real issues have sometimes been confused. Recollect that nominalists say that only individuals, not classes, are real. If species are classes, it follows that they are not real. But what if species are individuals? Then, obviously, it does not follow. The individuality of species is perfectly consistent with realism. It merely asserts that species are not classes, and is completely noncommittal about the reality of what it is that they are not. To be sure, the individuality thesis does strengthen a nominalist's hand by increasing the number of things that are acknowledged to be individuals.

I do not see any reason why an advocate of the individuality thesis simply must be a nominalist. Rieppel (1986) has made this claim, but he provides no arguments to back it up. Given his generally idealistic stance, it is easy to see why he would not be very favorable to nominalism. The suggestion of David

Johnson (1990) that my position is that of an inconsistent nominalist strikes me as quite gratuitous (see also Hogan, 1992). I do find it hard to imagine how a strictly nominal class can "do" anything, or, as he puts it, have causal force. He tries to find an example, but the things he treats as abstractions, namely ecological niches and the like, might better be interpreted as concrete. And just in case anybody wonders, it does not seem reasonable to me to say that classes exist only in the mind. If the laws of nature refer to classes, that creates some very sticky problems for anybody committed to such a metaphysical position.

Many people, conversely, would be very happy if species were in fact mere abstractions. For if they are individuals, they might have "causal force" or the ability to do things. In an earlier work, I discussed at length the notion of "organicism," the history of which goes at least as far back as Plato (Ghiselin, 1974a). Various kinds of entities, including societies, species, ecosystems, and even the entire world, have been interpreted as "super-organisms." Such organicist thinking has led to all sorts of errors, such as the notion that social evolution proceeds like the embryological development of an organism, with stages of childhood, maturity, and old age. In order to avoid such abuses, some persons have advocated "methodological nominalism" or "methodological individualism" to take the place of "methodological holism" when thinking about supra-organismal phenomena. There is some danger of turning such a program into a kind of metaphysical nominalism, and thereby coming to erroneous results. The terminology here can be misleading. If one accepts that there are individuals at supra-organismal levels, then individualism consistent with nominalism makes good sense from a terminological point of view. The methodological stricture that seems appropriate is not to extrapolate across levels without good reason, and mere possibility is not a good reason. From the mere fact that the planet Mars is an individual it by no means follows that it copulates.

In evolutionary biology the problem was fairly straightforward. Darwin understood it very well and explained it in various places in his works. Because natural selection works by differential reproductive success of individuals within populations, it can only evolve that which "benefits" individuals. But he specifically included families under the rubric of individuals (Darwin, 1859: 235-242; Ghiselin, 1969d:58). This means that although natural selection can produce adapted organisms, or even adapted families, it cannot produce adapted species. Some authors believed that groups of intermediate hierarchical level, such as local populations, might compete with each other and thereby provide a basis for "group selection" and hence adapted species. It was the often rather extravagant claims of Wynne-Edwards (1962) that led various persons, including George Williams (1966), to think very hard about such matters and re-examine the evidence. Indeed, the realization that group selectionist explanations for all sorts of things simply would not work stimulated a great deal of effort to find better ones, including my own research on the evolution of hermaphroditism (Ghiselin, 1969a). Theoreticians of various stripes have concluded that group selection will work only under rather restricted circumstances, especially if it is opposed to natural selection of the more conventional sort. Nonetheless such conditions can be specified, and they are not biologically unrealistic, at least in the view of its able advocate David Sloan Wilson (1983, 1992). On the other hand, it does seem to me that "reviving the superorganism" along lines suggested by Wilson and Sober (1989) goes a bit too far.

The reason why organisms and families can be selected is that they are much more "cohesive" than species are. They stand and fall as units, and anything that happens to demes and larger units is mainly a consequence of what goes on at lower levels. In other words, there seems to be little differential reproduction of demes as such, over and above that of organisms, within species. (As will be discussed in a later chapter, this stricture applies even to the so-called "shifting balance theory" of Wright.) In the case of genera and families and other "historical entities," it is hard to say what it would mean for them to be selected, for they have lost their cohesion. On the other hand, the very fact that species are cohesive poses the intriguing possibility that they might be subject to some kind of selection, and participate in the evolutionary process over and above the activities of individual organisms. Not long after Eldredge had proposed the "punctuated equilibria" hypothesis for macroevolution (Eldredge, 1971; Eldredge and Gould, 1972), he realized that the individuality thesis would fit in quite well with his views. It suggested that species might play an important role in macroevolution, over and above the differential reproduction of organisms within them. Thus arose the theory of species-selection, based on differential speciation and related phenomena, fueling much discussion among paleontologists especially (see Eldredge, 1985).

Occasionally the individuality thesis has been considered objectionable because it has been invoked to render such theories as punctuated equilibria more plausible. It should be obvious, however, that there is no necessary connection here. All the individuality thesis does is remove one reason for rejecting the theories in question. Furthermore the mere possibility of such phenomena implied by the individuality thesis does not mean that such possibilities actually have been realized. To establish that, one needs empirical evidence. And in general the search for empirical evidence has been disappointing.

Were the opponents of group selection, punctuated equilibria, and species selection able to discredit the individuality thesis, then they could simply write off any evidence, theoretical or otherwise. And this they have attempted to do, but without bothering to address the issues. George Williams (1992), for example, refers to the individuality thesis as a "fallacy" without substantiating that claim. Were he able to show that some premise is false, then he could indeed make out a case for a material fallacy; were he able to show that the conclusion does not follow from the premises, then he would have a case for a formal one. But he gives nothing of the sort. Likewise Dennett (1995) rejects the individuality thesis but provides no real arguments. Instead he tells us that he has read Ereshfsky's anthology, and that he agrees with Williams. And for the most transparent of reasons. Williams has been largely responsible for the notion that genes are the real "units of evolution," and Dennett endorses that view, and likewise the "selfish gene" notion of Richard Dawkins that is by no means unrelated to it. If the higher integrational levels were mere classes, then it would be very easy to make it appear that they are merely epiphenomenal, and there would be no reason for considering the alternatives.

Darwin's realization that selection is strictly individualistic led him to develop a body of knowledge that has been revived and modernized under the rubric of "Sociobiology." Sociobiologists, and others who work on that kind of problem but perhaps would rather not be labeled as such, have been rightly skeptical of adaptations that are supposedly "for the good" of anything other than an individual organism or perhaps even gene. Ruse (1987, 1988) is quite explicit about why he rejects the individuality thesis. He wants to believe in sociobiology, and the individuality thesis implies the possibility of group selection. And yet he has railed against me for treating the organism, rather than the gene, as that which is selected. Why not turn organisms into classes too, and make genes the only individuals? Why stop there?

Ruse (1988) also argues that if species are individuals, there are no laws for individual taxa, including *Homo sapiens*. As he sees it, this would mean that anthropology and various other branches of knowledge are not sciences. I would be happy to agree that when Ruse studies the human species, his work is not science. However, we can dispose of his claim merely by pointing out that it derives from a bizarre notion of what is meant by "science," which, alas, he shares with a fair number of physics-minded metaphysicians. For Ruse, laws of nature are the alpha and omega of science. Since laws of nature do not refer to any particular individual, there are no laws of nature that refer to *Homo sapiens* if it is, in fact, an individual. So for Ruse, a science about *Homo sapiens* would be a contradiction in terms. This is a purely definitional matter and treats a stipulative definition of a word as if it told us something about objective reality. If we accept his definition of 'science', then geology isn't a science either, because it refers to the earth. Likewise we have to write off astronomy because it deals with the sun. Why we cannot include the historical and descriptive aspects of what scientists do in our concept of science is a real puzzle for anybody who seeks rational grounds for such claims.

Ruse (1987) provides what may be called a "consilience" argument for the proposition that species are classes, or natural kinds. He maintains that if species are classes, then the various kinds of species, morphological and other, will coincide. The idea behind consilience is that when different lines of evidence all point in the same direction, especially unexpectedly, this may be taken as evidence that an hypothesis is true. Granting that consilience is a valid criterion for scientific truth, we may raise two very serious objections to this claim. Firstly, Ruse's basic premise is false. The various kinds of species concepts definitely *do not* pick out the same basic units. Secondly, we can turn the argument on its head and propose a compelling consilience argument in favor of the individuality of species. Quite different arguments for the individuality thesis were independently developed by Mary Williams, David Hull, and me. My own insight largely derived from the realization that species names are proper names, and have to be defined ostensively, as is in fact the practice of taxonomists. Hull was convinced by the realization that there are no laws of nature for individual species. And Mary Williams, after producing a "partial axiomatization" of selection theory, realized that species functioned as individuals in that axiomatization (M. Williams, 1970, 1985, 1989). The fruitfulness of the individuality thesis in clarifying a wide range of issues such as are discussed in the present work may also reasonably be used as a consilience argument in its favor.

Arthur Caplan (1981a, 1981b) takes a position that is not altogether dissimilar from that of Ruse. Like Ruse, Caplan wants to preserve laws of nature, yet unlike Ruse he is hostile toward sociobiology, and rather reminds us how organicist views of the state have been used by the enemies of democracy. I have warned against such dangers myself, at length and in no uncertain terms (Ghiselin, 1974a). On the other hand, treating supraorganismal wholes as more than just classes suggests that we might take a more positive attitude toward them and thereby justify conservation policy (L. E. Johnson, 1995). Yet it seems to me that as scientists our first priority should be toward establishing what is true, and cope with the ethical problems as best we can.

Both Caplan and Bock (1988) are concerned with the need for having laws of nature in biology, which is by no means unreasonable. But they would seem to be looking for them in the wrong place. In order to make species into classes, Bock and his collaborator Szalay are driven to take two rather extreme steps (see Bock, 1979, 1986, 1994; Szalay & Bock, 1991; Szalay, 1993). In the first place, they must deny that a species exists through time, so that, for example, *Homo sapiens* exists only at a single instant and is by no means equivalent to a lineage of ancestors and descendants. Second, they make them into what they call "family resemblance classes"—that is to say, what are also called disjunctively-defined classes or cluster concepts. By such maneuvers are species turned into sets and supposedly deprived of their Aristotelian essences. Therefore when Bock uses the word 'species', he is manifestly not talking about what other biologists are when they use that term.

Finally, we may add to our list of objections to the individuality thesis various claims that certain species are, in fact, classes rather than individuals. Except where such claims are mistaken, and the "species" really are individuals after all, many and perhaps all of them result from a misunderstanding of what "species" are being referred to. The individuality thesis by no means asserts that everything that is called a species, even by justly-eminent taxonomists, is an individual. At a minimum it only says that biological species are individuals. If, of course, a local population happens to be called a species, it too is an individual, but not an individual species. Or if what an advocate of the biological species concept would prefer to call a genus is called a species by his colleague, it is still an individual, even if his colleague is using a different definition of the species category rather than the biological one.

Somebody might, however, classify organisms in such a manner that the specific epithets really do designate classes of organisms rather than individuals. In that case this would be an interesting fact about taxonomic practice, but it would not be germane to the ontological issue. Arguments cast in such terms are often given against the biological species definition itself. In this case it is perhaps more obvious that the real issue is whether or not we shall adopt that particular definition. Again, an argument about the appropriateness of language is treated as if it could be decided upon a purely empirical basis.

Let us now consider the possibility that species might not be material things, but are rather attributes, relations, or even places. Such proposals may or may not contradict the individuality thesis, depending upon whether one considers the attributes, relations, or places in question to be individuals. To begin, one philosopher, Michael Jubien, suggested to his student, Scott Merlino, that species are properties. Merlino and I have enjoyed some good discussion on this possibility, but at the time of writing I am profoundly skeptical.

If biological species were, in point of fact, intensionally defined classes, then there would be a sense in which one might say that they are properties (as James Fetzer has kindly drawn to my attention). Sometimes logicians refer to what I, following Hull, have called "intensionally defined classes" and "extensionally defined classes" as "properties" and "classes" respectively. This, however, is purely a terminological matter and has no bearing upon the ontological issues, or even the logical ones.

I suppose one could construct a system of metaphysics in which species and all sorts of other things are "really" attributes, much as some process metaphysicians have attempted to populate the world with nothing but pure activity and treat things as metaphysical delusions. The question is whether such a system would be appropriate for dealing with the real universe. Because the species category and the organism are classes defined in terms of attributes, it makes a certain amount of sense to say that there are attributes called "speciesness" and "organismness" possessed by any instance of such a class. This makes a great deal of sense when we consider what is manifestly an attribute, such as "yellow," and an instance thereof—a "this yellow, here and now"—a yellowness. But with respect to *Homo sapiens* and Charles Darwin, I am utterly at a loss to say what might possibly be meant by "Homosapiensness" or "Charles Darwin-ness."

In another metaphysical system perhaps nothing would exist save the relationships among relations. Or at least it would have it that the particular objects that we call species and organisms "are" relations. David Stamos (personal communication) informs me that he has been exploring the thesis that a species is a particular kind of relational property. Again, such an exercise is well worth considering, but the difficulties would seem formidable. Relations must of logical necessity have terms; for example, something must be taller than something else, it cannot just be "taller than." So one might argue as did Mayr that the word 'species' is relational, but that is not the same as claiming that species is itself a relation. When the second child is born into a family, the first acquires the relational property of "siblinghood," and one might well say that there is an individual siblinghood that begins at this time. But the siblings themselves are not the siblinghood, any more than their mother, or their father, or both, is a marriage. When we come to discuss the relation of homology, we will consider an example of an effort to make a relation, rather than the things that stand in that relation, the basic "unit of classification." Meanwhile let us simply insist that, in the biological discourse that is actually used by scientists, species and organisms are that which stands in relations (terms), and not the relations themselves.

And then there is the possibility of equating a species or other individual with the place that it occupies. David Johnson (1990) seems to have had something along such lines in mind when he tried to equate species with their niches. He did so in the context of an effort to refute the notion that species are individuals. His argument can easily be rebutted by pointing out that niches are individuals, not classes. They are individual places in the natural economy, and by no means the abstractions he made them out to be. We must be careful not to confuse, in this context, individual niches with classes of niches, or individual economies with classes of economies, which would be the usual taxon-category mistake. A metaphysics in which species "were" their niches, or for that matter their geographical ranges, might be an interesting intellectual curiosity. Indeed, exploring such a possibility might turn out to give very important results. But unless one can come up with such an ontological system, we shall have to say that an organism or a species is an occupant of a place, not a place itself. Nonetheless such notions as ecological place are by no means unproblematic, as we have already pointed out in Chapter Two.

The possibility that species are processes, rather than wholes composed of organisms, has perhaps never occurred to anybody besides me, though Buerton (1995) may have had something of the sort in mind. I must admit that I find it rather attractive, much as I do the suggestion that life is a bogus substance for a process that might better be called living, and mind a bogus substance for a process that might better be called thinking, and the like. The category of species would then be a class of processes analogous to living, and an individual species would be an analogue of living that instantiates it, composed of lives or livings or whatever you might wish to call them.

It seems reasonable to say that the processes here are ontologically prior to the material substrata, but this does not justify identifying them with those substrata. One might try to develop an aspect theory, such that organism and life, and species and some corresponding process, are fundamentally the same thing, albeit not identical—along lines suggested by dual aspect theories of body and mind. Some further thoughts along this line will be presented in a later chapter. So I conclude that species are processual things, but only in the more restricted sense that they cease to exist as extant species when the underlying process is terminated, analogously to what goes on when an organism dies and persists only as a corpse. But this maintains the ontological distinction between the process and the substantial being, in spite of reversing the usual priority. Species are the participants in these processes, not the processes themselves. In evolutionary biology the unequivocally processual individuals are speciation events and other "evolvings."

9

WORKING OUT THE ANALOGIES

"The formation of different languages and of distinct species, and the proofs that both have been developed through a gradual process are curiously the same."

—Darwin, 1871:59

In an earlier chapter I mentioned a rather weak analogy between the species of biology and the firms of economics. Here this analogy will be explored a little further. Two others will also be considered: language and the common law. The point here is to show by means of examples that from a metaphysical point of view there is nothing *sui generis* about species. The world is full of composite wholes that have a great deal in common with biological species. Like species, they evolve, and they evolve by mechanisms, according to laws of nature that have much in common with the biological ones. Indeed, organic evolution is merely a more particular case of evolution in general. Because there does exist a body of knowledge called evolutionary economics, because there does exist an evolutionary theory of the common law, and because linguistics actually became an evolutionary and historical science before biology did, much can be learned by such comparison. This, over and above counter-examples that might be used in answering objections to the theses here maintained.

Before proceeding, however, two important points deserve strong emphasis. In the first place, these are examples of something more general, so that we are classifying rather than just comparing. The individuality thesis opened up a lot of possibilities for treating all sorts of problematic entities as individuals, and thereby considering them from a truly evolutionary point of view. Not surprisingly, Hull (1988) devoted an entire book to showing how this could be done in a detailed study on the history and sociology of science. Theories, research schools, and all sorts of other things may be treated as individuals, and function as such in theories and historical narratives. That this can be accomplished has clearly been shown, irrespective of any criticisms that might be leveled against the details of that work. In the second place, we must flatly reject gratuitous allegations on the part of our critics that the reasoning here is "purely analogical" or "metaphorical." A lot of persons refer to such interdisciplinary work as based upon "metaphor" as if all the entities being compared had in common were a figure of speech. What we are doing of course is subsuming more particular cases under a more general rubric; it is like treating chlorine as a halogen, again.

When originally introduced (Ghiselin, 1974c) the analogy between species and firms was largely didactic: it helped to explain some rather difficult points. It will play the same role in the present work, though I should mention that I have discussed it elsewhere in publications on economics and its relations to biology (Ghiselin, 1974a, 1986b, 1992). The comparison is enlightening because it gives us an example of a supra-organismal whole that may seem a bit more obviously concrete than does a species. The analogy also helps because there is an obvious parallel between the different modes of competition that may occur between the wholes and their parts. Thus there is 1. competition between species and 2. competition between firms. There is 3. competition between organisms of different species and 4. competition between productive units of different firms. And there is 5. competition between organisms of the same species and 6. competition between the parts of the same firm. Biologists have no name for competition between species, so I will call 1. "specific" competition. When biologists speak of competition between species they usually mean 3., and call it "interspecific" competition. "Intraspecific competition" is the term for 5. That there has been no name for 1. is a symptom of ontological confusion that has long existed. We will have more to say about that in a later chapter. For the present we may note that, say, a salesman competing with another salesman would always be engaged in intraspecific competition. He might compete by getting to the customers first. It would be in the salesman's interest to get to the customers first irrespective of whether another salesman worked for his firm or a competing one. But competition between its own salesmen might not be in the interest of the firm at all. At any rate, what firm one works for, and what species one belongs to, profoundly affects one's behavior. Nothing of the sort would happen were species or firms merely classes of similar objects.

Firms can be very big, like Mitsubishi, or very small, like a "mom and pop" store. They can be very homogeneous, like a chain of grocery stores, or very heterogeneous, like a company that owns stores, hotels, and all sorts of other businesses. They can change indefinitely, yet still remain the same firm. A family business might pass from generation to generation and ultimately be engaged in altogether different activities from those it began in. When a firm gets "reorganized," it might change its name but not its identity; rather it would be like a person changing her name when joining a religious order or getting married. We generally do not individuate a new firm unless the old one has really passed out of existence. But an old one might be completely liquidated and some of its assets put into a new one.

An individual firm is not "defined" in terms of its components. It does not become another thing when somebody gets hired or retires. Furthermore, not everybody "belongs" to a firm—especially the unemployed. But what happens when an individual firm gets broken up, say, as the result of anti-trust action by government? Does it cease to exist, so that the various parts of a trust are now entirely new firms? In this case it makes sense to say that the trust itself has ceased to exist as a cohesive individual, and has become a mere "historical entity." But what happens when a corporation is forced to divest itself of a subsidiary is another matter, for instance a publishing empire that continues to produce newspapers and books but no longer makes greeting cards.

To push things a little further, when a couple of editors quit and set up their own publishing house, we would be loath to admit that the original one ceases to exist. It would be even more forced to say that the firm has ceased to exist if they set up their own part-time business moonlighting on week ends but kept right on working at their original jobs. Although organisms can work for two or more companies, they cannot be components of more than one species. Therefore systematic biologists are not driven to such dilemmas. But as we shall see, they come close.

A firm is something more or less familiar to all of us, and we can easily conceptualize it as a material system, that is, an individual, the components of which are, for the most part, tangible. Thus one can think of it as incorporating owners, employees, machinery, buildings, office equipment, inventory, and the like. We run into a little more difficulty when we deal with their less tangible resources. We easily think of money in the form of currency, which we hand to somebody when we buy things. But a check or a credit card slip, or sometimes even an utterance, is evidence of an agreement to pay. A firm's resources include such things as the good will of satisfied customers, knowledge of the market and much else that might be considered "behavioral" or "informational." Nonetheless it is not too difficult to picture such things as capital and skills, along with goods, services, and the means of production, as constituting an economic system, with firms and the like as parts.

If we do think of human economic life in that way, then all we need is to see the close parallel between it and the economy of the natural world—between economics and ecology. And granted that, one might want to think of economics from an evolutionary point of view, complete with various forms of selection. In fact there have been quite a number of books and journal articles devoted to just such considerations. There really is an evolutionary theory of the firm (Alchian, 1950; Winter, 1964; Nelson and Winter, 1982). It seems curious that such theorizing has developed only rather recently, and that it has not yet attracted a larger following. I mention it here only to flesh out the discussion and to recommend it to the interested reader.

Language, on the other hand, has long been considered from an evolutionary point of view. That languages evolve and have common ancestors was generally appreciated among educated persons by the early years of the nineteenth century (Pederson, 1931). Phylogenetics was invented by linguists before it was invented by biologists, and the impressive results of linguistic research were instrumental in furthering the Darwinian revolution. On the other hand, the study of language has been and remains a citadel of anti-evolutionary sentiment. It is easy to see why antievolutionists would seize upon any excuse to accentuate the alleged differences between the class of men and angels, on the one hand, and the class of vegetables and brutes, on the other. Anything to find a difference that is essential and qualitative rather than a matter of degree. In Aristotle we find three "souls"—vegetative, sensitive, and rational, the second of which is possessed by all animals, but among animals the third by man alone. This notion was taken over by Christian theology and continued to be explicitly discussed by Linnaeus and subsequent biologists. Descartes, on the other hand, stressed the notion that animals other than man are "automata." The effort to attribute animal behavior to "instinct" and human behavior to "intelligence" was part of that same tradition. Two kinds of behavior that are particularly well developed in man have been pressed into service in order to sustain his angelicity: the use of tools and speech.

Empirical research has gradually eroded the levees that have been thrown up to keep out the evolutionary waters. The differences between learning and instinct are recognized but not taken as constituting essential differences between man and brute. Cognitive abilities in other species have become increasingly documented. All sorts of animals use tools, and some even fabricate them. Rudiments of culture are well documented. And although other primates have very limited capacity to do what we do with words, they come impressively close.

In the early 1860s, F. Max Müller, Professor of Comparative Philology at Oxford, delivered a very influential series of lectures in which he took the position that linguistics is not an historical science, but rather a physical one (6th ed. 1871). By this he meant that like physics it is concerned with laws, not particular events. He also emphasized the role of reason and the apparent gulf that separates man from the brutes. He ridiculed the search for the origin of language by means of amusing epithets. The idea that words are based upon onomatopoeia, or the imitation of natural sounds, he called the "bow-wow theory." Another alternative, that they are modified emotional outcries, or interjections, he called the "pooh-pooh theory." He suggested an alternative, which was that words arise by instinct, like a bell being struck. His views elicited spirited responses, especially by Professor William Dwight Whitney of Yale (1868; 1883;

see B. H. Davis, 1987), who responded by calling Müller's a "ding-dong theory." Although Müller apologized for any offense he might have given by pooh-poohing his contemporaries, the practice of ridiculing the search for origins by no means ceased.

Nor have linguists altogether given up the tradition in some quarters of treating evolutionary questions as unimportant and emphasizing what are called synchronic rather than diachronic approaches, as advocated especially by De Saussure (1915 translation 1959). Chomsky (1978), for example, has maintained that of course language evolved and indeed has an innate basis, but the fundamental mechanisms are universally present in all human beings, and that it is these that ought to interest linguists. The position should be familiar to anyone who has tried to convince molecular biologists that the job will not be finished once they have solved the puzzles about such "universal" features as "the" genetic code.

In this connection it is appropriate that we draw the reader's attention to the similarity between efforts to define "language" and efforts to define "life." In both cases, we are dealing with a single thing, whether it be that remarkable capacity of human beings to communicate, produce poems, and the like, on the one hand, or whether we deal with the equally remarkable capacity of all tellurian organisms to acquire energy to feed, grow, and the like, on the other. Because we have but one exemplar of each class of individuals, it is very difficult to define the class to which it belongs. We cannot simply enumerate the properties common to all known languages or all known taxa, because we have no way of differentiating between that which is true of them of necessity and that which is merely contingent fact. So when linguists search for universal features of language, the list may consist largely or even entirely of historical curiosities—comparable to the fact that all mammals have hair. And in attempting to understand life, we realize that the genetic code is "universal" among tellurian creatures. But how much of that sort of thing again is an historical accident? Were we familiar with "life" on other planets, we would be in a much better position to address such questions.

In spite of all that, the parallels between phylogenetics and historical linguistics have received considerable attention from linguists. At the university of Jena, the linguist August Schleicher (English Translation, 1869) discussed the parallels in an "open letter" to his zoologist colleague Ernst Haeckel. Schleicher compared language families to species, dialects to races, and idiolects (the languages of individual persons) to organisms. He also stressed the difficulty of separating languages from dialects and noted that Darwin stressed the same problem with respect to species and varieties. (See Taub, 1993; O'Hara, 1996.)

Above the species level, the tradition in linguistics has been even more strongly genealogical than it has in systematic biology. At what seems at first to be paradoxical, this is reflected in their classifications by less concern for consistency in the use of names for the categories (see Ruhlen, 1987). The apparent paradox vanishes when we realize that the taxa are purely genealogical units, or strictly monophyletic taxa. Categorical rank is not much used to specify "degree of similarity." The system, in other words, has no "paraphyletic" taxa and is thus "cladistic" rather than "evolutionary." Furthermore the phylogenetic techniques applied have largely been in advance of those used by biologists. The stress upon "shared innovations" as the basis for inference seems to have been common practice among linguists long before it was championed by Hennig (1950) and his followers ("cladists") among systematists. A revival of interest in such matters is manifest in a recent collection edited by Hoenigswald and Wiener (1987). Linguists also have produced some fine examples of the sort of processual techniques that are anathema to anti-evolutionary, or pattern, cladists. Knowing how speech is produced can help one to decide which changes are more likely to occur (Wang, 1987).

Language being so familiar to all of us, one might think that it would have provided useful hints for solving "the species problem." In fact, however, the inspiration seems largely to have gone in the opposite direction. Language is differentiated from dialect in terms of mutual intelligibility, which is the analogue of the fertility test, not of the biological species concept. This leads to much the same problem as biologists have had with polytypic species. Some very divergent linguistic entities are connected by intermediate dialects with no clear break between them. Some linguists (e.g., Murdock, 1953), have explicitly invoked an analogy with the biological species definition to argue that this presents no problem. But others have expressed dissatisfaction with having Chinese be one vast language, in which speakers of Mandarin cannot communicate with speakers of Cantonese except in writing.

Driven by such considerations, Chomsky (1980:218) embraces the kind of misplaced nominalism that is not unfamiliar in the literature on systematic theory, carrying his metaphysics to its logical conclusion but failing to recognize a *reductio ad absurdum* of his own making:

> Even when we speak of 'an organism,' we are engaged in idealization and abstraction. One might, after all, study an organism in the world from a very different point of view. Suppose we were to study the flow of nutrients or the oxygen-carbon dioxide cycle. Then the organism would disappear in a flux of chemical processes, losing its integrity as an individual placed in an environment. The 'furniture of the world' does not come prepackaged in the form of individuals with properties apart from human intervention. . . ."

So if you ever encounter a cobra in the jungle, just think about ecology and it will evaporate, and rest assured that fangedness was bestowed upon it by your thought processes and mine. Given a language definition comparable to the biological species definition, we don't have to go to such preposterous extremes. Chinese really is a single language. And although Latin speciated thanks to the isolation of Rumanian, it would be perfectly reasonable to say that French, Spanish, and Italian are really the same language, thanks to such intermediates as Provençal. Or perhaps they have only recently speciated, or are still in the process of doing so. Evolution is such that we only expect the intermediate stages to make it difficult to decide whether we have before us a species or a subspecies.

On the other hand, the matter of loan words provides a good example of what the analogue of "interbreeding" is not—just any "lateral transfer" whatsoever across lineages. The German language tends to accumulate new terms by means of coinages that generally involve compounding its own roots, and borrowing from other languages is less common than, say, in English. But what is borrowed is words, not the grammatical apparatus that has to exist if they are to be turned into speech. For example, different languages have different systems of "genders," and English loan-words in French that were originally neuter have to be made either masculine or feminine: *le dancing*. The pronunciation tends to change as well. What Americans call a crocodile is *Il cocodrilo* in Italian.

The English language is not a fusion product of French and German, analogous to an allopolyploid such as the famous *Raphanobrassica,* with its cabbage-like root and its radish-like leaves. Rather, it is a Germanic dialect that became isolated and diverged enough from the others to warrant status as an independent language, and has continued to diverge from them ever since. A major cause of that diversification has been the incorporation of much French vocabulary since 1066. But the language remains a Germanic one and is easily recognizable as such. Furthermore, although it has undergone much dialectical diversification, English has never undergone the analogue of speciation. The language of *Beowulf,* the *Canterbury Tales, Hamlet,* and *Huckleberry Finn* has changed a great deal without ceasing to be one and the same individual language. Historical linguists retain the kind of phylogenetic system that they developed before the biologists developed theirs.

The mechanisms of linguistic evolution have both similarities and differences from organic evolution that can be quite intriguing. For example, one can readily think of analogies with sampling error, such as genetic drift and the founder principle. Just suppose that a small group of people were to leave their original homeland and establish a new community. By chance a lot of the words, especially uncommon ones, would not be part of anybody's vocabulary. So people would coin new ones, and this would tend to produce rapid change. The rate of loss of words, and the consequent coinage of new ones, should be a function of the size of the community. And for obvious reasons the more common words should be lost less readily than the rare ones. Therefore linguistic drift should be very similar to genetic drift, and the theories that deal with the one should have much in common with the other.

Of course a different kind of inheritance system creates somewhat different mechanisms and patterns of evolution. The invention of writing also complicates matters. It favors the existence of a more complicated vocabulary and a larger pool of terminology, with the capacity for greater long-term survival of words. The Chinese language is largely integrated by its remarkable writing system, which allows persons whose speech is not mutually intelligible to communicate, and something of the sort exists in other languages. Literature provides us with a richer received vocabulary, and dictionaries provide a disincentive to coinage when one feels the need for a new term.

If we work out the analogies between life and language in any detail, it becomes quite striking that the categories (levels) do not coincide. The closest we can say is that the organismal level corresponds to the "idiolect" of a certain person. Higher taxa, such as dialects and demes or races scarcely correspond at all. Furthermore, there are many bilingual persons, and of course still more who are fluent in more than one language, so that the same organism can participate in the evolution of more than just a single tongue. The ability of persons to opt for the use of one language or another, and to affect which language will be used by their children and others, means that a kind of artificial group selection takes place. This in addition to any selection that may occur at the level of the word or the phoneme. Sympatric languages and dialects can and do coexist, but the usual tendency is for all but one gradually to disappear within a given area. The cost of learning more than one language, and the cost of deciding which of them to use, are two good reasons for one of these becoming the lingua franca and finally the sole means of communication. If one dialect of a given language were to become the dominant one, the speakers of a dialect that is intermediate between others might switch to the "standard" form, breaking the chain of dialects and thereby producing a kind of "sympatric speciation" that has no good precedent in organic evolution.

If we try to move "downward" from the level of the idiolect and the organism, we may find ourselves somewhat puzzled. The parts of an organism are organs and cells and the like, and the so-called parts of speech, such as the noun and the verb, are classes of things that we would not look for at that level. (I suppose that at a lower level one might invoke the structural gene and the regulatory gene.) Furthermore, an organism's language is brought into being by quite different processes than its body is, and there is no way that we can equate the genes and chromosomes as transmitters and developers with anything in language. What we can do at that level, to be sure, is find some analogies between some of the functional units that have to do with the encoding of "information" and the transformation thereof into a genetic or linguistic product. So we might look for parallels between phonemes, words, and sentences, on the one hand, and nucleotides, codons, genes, or whatever, on the other.

Here things really begin to become complicated, because the units being dealt with are not just parts of organisms and idiolects, they are also parts of lineages of which the organisms and their idiolects are likewise parts. And at any given time these lineages are composed of more than one organism, and more than one idiolect. Following an analysis I published some years ago (Ghiselin, 1981a), let us consider the sub-organismal units often called genes and chromosomes. When we speak of "a" chromosome, we may note that it may be an individual chromosome in an individual organism. However, in diploid organisms there are two "sets" of these, each of which is sometimes called a "genome," in each cell in the body (with certain exceptions of course). All of these would be individuals. However, when fertilization occurs and a zygote is formed, there follows a series of mitotic cell divisions, as a result of which lineages of cells and chromosomes are formed. Hence an organism's Y chromosome can be a lineage. Furthermore, when meiosis, fertilization, and genetical recombination occur, the "homologous" chromosomes pair up and engage in various other activities with the consequence that every chromosome throughout the species is a part of a larger, populational unit, such as, for example, the population that consists of all human X and Y chromosomes. And such units are lineages too, for they are parts of species, and they evolve.

To extend this analysis downward in the integrational hierarchy and do the same for genes would be a simple matter were it not for the fact that the term 'gene' has been used in so many ways and, as was noted in an earlier chapter, so equivocally. But we can make do for the moment with the sort of "gene" that codes for a single protein. Just about everybody accepts that sense of "gene" these days. Such a gene would occupy an individual place, called a "locus," in an individual chromosome. But, exactly as with an entire chromosome, each gene in a diploid cell exists as one of a pair. It also is part of a lineage within an organism, part of a lineage that extends backward to its ancestors, and part of a population of genes that pair at meiosis. It is meiosis and recombination that generate diversity, and, together with crossing, turn the genetic complement and the organisms as well into a unitary population.

We need not belabor what goes on with respect to nucleotide triplets and the like. But we do need to emphasize how readily things get oversimplified, and then muddled, when people who don't know their molecular biology very well start trying to do metaphysics with such objects. When one looks for "units" here, one finds all sorts of things that might qualify as such, and it is far from clear what sort of role they might play. In fact, many of the "units" here are not "substantial" individuals but configurational and contextual ones. Genes are said to exist as "allelomorphs" or "alleles" which are often defined as "versions of the same gene." In this case, what makes them the same is that they are corresponding elements of a population and a lineage, but not identical in all respects.

Some alleles, however, are not versions of anything. They are "deletions"—places in the chromosome where some material has been eliminated. This makes a great deal of difference. The notion of selfish genes, so dear to doctrinaire sociobiologists, sounds more than just a bit odd when we talk about selfish chromosomal deletions. Other "genes" are the result of changes in the position of genetical material within the entire chromosomes. These include inversions and translocations. In the former case, the order of components is reversed, and in the latter, a piece of genetical material moves to a different place, either in the same chromosome or another one.

It is a pretty straightforward matter to see how the analogy with language fits in at this point. The meaning of a sentence or larger piece of literature depends not just upon which elements are there, but upon which words are left out, their order, and upon all sorts of contextual matters. However, there is another source of confusion here. When we are talking about genetics, we are concerned with parts of an organism that, in effect, enable an organism to construct itself, and to construct other organisms like itself. The constructed organism is a message. It is like a sentence, or better, a whole conversation. The organism's idiolect, however, is not a message, but rather the "machinery" that allows the organism to generate messages. So when we discuss the genetic code, we are concerned with something like the rules of grammar, not the encoded message. But that genetic code is also used to encode a kind of message that somehow contains the instructions for producing a complete organism. One might say, following Chomsky, that there is also a universal grammar which, like the genetic code, long ago ceased to evolve, and allows the formation of idiolects within the context of languages.

Thus envisioned, one might think of languages as populations of linguistic wholes made up of grammars and vocabularies and perhaps something more. The parts of languages therefore are not populations or organisms, although organisms are parts of speech-communities. They are not parts of discourse either, because discourse is a language product rather than language in this particular sense. A language is a whole made up of rules and conventions that organize discourse. Literature is one example of such organized discourse. A language evolves by virtue of its rules evolving, and perhaps a lot else besides. Indeed, cognitive psychologists have strongly criticized the notion that language consists of nothing more than a bunch of rules (Casson, 1983; D'Andrade, 1995). They point out that it, like other cultural entities, includes "cognitive maps" as well as rules. One does not have to know much about cognitive maps to realize that maps in general represent particular (individual) pieces of territory.

Language, of course, is one of many "cultural" entities that are somehow supposed to be "decoupled" from their material (largely biological) substratum.

Lately many anthropologists have wanted to define "culture" in terms of its underlying rules rather than in terms of its products and its activities (Durham, 1991). They seek to disassociate culture from any material objects or even from the activities that are often referred to as "cultural." Consider what this view implies, for example, with respect to art. Painters, their painting of pictures, the paintings themselves, and the contemplation thereof are, by definition, not to be considered parts of culture. If one tries to identify such parts, it would have to be things like individual "how to paints"! And yet in the art world it often happens that a painting is inspired by a statue or rendered into an engraving. Iconography and technique get transmitted from one artist to another, and art evolves as this goes on (Gombrich, 1977, 1989). There is a considerable literature on the evolution of objects of art, directly inspired by work on organic evolution, that traces changes in various kinds of artistic creations that form populations and lineages (Steadman, 1979). I only allude to studies on the evolution of literature and literary genres at this point (Ghiselin, 1980a).

Traditionally anthropologists have more or less taken it for granted that culture is not to be divorced from its products, but rather that such products, together with those rules and conventions and perhaps certain other things, are the components of culture. In its *locus classicus* the term is defined as follows (Tylor, 1871, vol. I:1):

> Culture or Civilization, taken it its wide ethnographic sense, is that complex whole which includes knowledge, belief, art, morals, law, custom, and any other capabilities and habits acquired by man as a member of society.

Anthropologists, we should emphasize, have an even greater difficulty in defining "culture" than biologists do "species," and the interested reader may wish to consult a very scholarly compendium by Kroeber and Kluckhohn (1952). So even here we can see an emphasis upon ways of behaving rather than the behavior itself. And yet, it is behaving as a member of society, and this is something that tends to get overlooked. Culture functions. It functions at the organismal level by integrating the personality. It functions at the social level by integrating its participants into an organized community. It is a part of our mental and social lives, a body of behaviors governed by rules, not just the rules according to which that behavior is governed. And the particular objects that embody that culture, be these organisms or objects d'art, or whatever, are participants in that activity. No, the dance is not the dancers, but whatever else it may be, at the very least it is the dancing.

If anything might be thought of as consisting of rules, it is law in the juridical sense of that term. We must emphasize that law is not just law in the abstract, conceptualized as a mass noun. Rather it exists as individuals, wholes made up of parts, including acts of legislation, judgments, and entire systems of law. It is by no means difficult to put law into an historical context and treat it as something that evolves. To do so was to break somewhat with tradition, which tended to conceive of it as a product of reason, or as something that reflects the Cosmic Order among things. So the "history" in question might not be the sort of thing that would appeal to a real evolutionist. It was partly to combat the then very popular idealistic conceptions of law that the American pragmatist Oliver Wendell Holmes wrote his classic book on *The Common Law* (Holmes, 1881; see Wiener, 1949). As Holmes put it, "The life of the law has been not logic, but experience."

In recent years, an evolutionary theory of the common law has emerged which, like the evolutionary theory of the firm, invokes a kind of natural selection (Priest, 1977; J. C. Goodman, 1978; Hirshleifer, 1982). The basic idea here is that if people are dissatisfied with the law, they litigate, and they continue doing so until they are satisfied (Cooter and Kornhauser, 1980). Accordingly, the decisions of judges are not so much a matter of providing intelligent guidance, as one might tend to suppose; the fundamental mechanism is fortuitous variation and selective retention—a kind of selection.

But if we treated the law as nothing more than the rules, we would fail to do justice to its workings and its participants. The law is an institution, and an institution is not an institution unless it is instituted. In other words, the persons whom it governs must be disposed to act in accordance with its rules, to enforce them, to sit in judgment, perhaps to legislate. The institution cannot exist unless those who are governed by the law participate in it, perhaps engage in legal activities, become incorporated in juries, litigate, plead, and settle. The law is a whole, of which courts and juries, decisions and precedents, are integral parts.

The question then raises its ugly head, of whether biological species are not just "bodies" of rules, and indeed whether the same might be said of organisms themselves. This is precisely what sociobiologists like Richard Dawkins would have us believe. It hearkens back to the old anti-evolutionary dictum of Samuel Butler that a hen is just an egg's way of making another egg, except that we are driven to say that a rule is just a rule's way for making another rule. Forget about the hens and the eggs, let alone the chicks, the cocks, and *Gallus gallus*. If you want a good example of ontological nihilism, here it is. The kind of notion that we entertain here is of course the attempt to "reduce" organisms to their genomes, or more precisely to a kind of message, or set of instructions, or whatever, that is inscribed in the genome. In such a scheme one might also want to include rules that are acquired through experience, especially those having to do with behavior but not leaving out other aspects of physiology. An advocate of such views might acknowledge the existence of such gene products as ourselves, while nonetheless maintaining that these are not the "real" components of species. Rather, organisms and species would be a sort of container, vehicle, or life-support system for the more fundamental reality that is scriven in the genes.

Calling organisms "vehicles" was one way of using a misleading metaphor in support of a dubious metaphysical thesis. More pernicious, because less obvious, was the choice of the term 'replicator' for genes and other things of which copies are made. In English, the suffix -*or* is used to indicate that the root to which it is attached falls under the ontological category of an agent. As with the -*er* in 'copier', it means that which does the copying, not the sheet of paper that we copy when we feed it into a piece of office machinery. For the (passive) patient, in contradistinction to the (active) agent, we use other suffixes. In ordinary metaphysics we recognize that text gets created and modified by its author and editor, and transmitted by its copyist, perhaps using a copier. Had Dawkins (1976, 1982) called his "replicators" "replicanda," his metaphysics would have been somewhat more transparent, but people might not have taken him so seriously (Ghiselin, 1987a). As to the selfish chromosomal deletions, Dawkins (1982:164) admits that there is an economic advantage to getting rid of some stretches of DNA, but insists on treating the deleted materials as replicators. The very notion of something that no longer exists being able either to replicate or be replicated makes no sense at all. The only thing Dawkins can do to avoid intellectual bankruptcy when faced with a *reductio ad absurdum* is to embrace the absurdity.

Thus the arguments for taking entities such as genes as more fundamental than anything else are far from compelling. The choice at best seems to have been quite arbitrarily dictated by metaphysical taste rather than any substantive intellectual claims. Rather than attempting to answer it directly, let me apply what was just said to this particular context. Organisms and groups of organisms (including both sexual populations and societies) are integrated and organized as functional units. Their components function together in a coordinated fashion. If they are to do so, it is a necessary condition that there shall be a certain regularity and consistency with respect to how their activities are carried out. At the organismal level, genes do all sorts of things, among the most important being the control and direction of development, especially as mediated by cellular activity. Groups of cells, including those that form organs and organ systems, are likewise coordinated in their activities, but so too are groups of organisms. This is especially, but not exclusively, the case with respect to organisms that join together in reproduction by sexual means. For here the organisms not only have to get together, but their separate contributions to the processes that organize the next generation must be compatible as well.

So although from a logical and metaphysical point of view it is erroneous, or at least misleading, to say that an organism or a species *is* a "coadapted gene complex," nonetheless organisms and species do in fact possess such entities. This out of physical, or biological, necessity. The genetic material provides resources that organize the organisms and the species of which they are components. Organisms and species are organized beings, and therefore they may be spoken of as organizations, in a broad but not altogether metaphorical sense. They have ways of conducting their activities that are closely analogous to those of other kinds of organizations.

One might argue quite legitimately that an organization is not the same thing as its practices, procedures, and routines. It would also seem not to be the same thing as the material embodiment of the "instructions" for carrying out such conduct, nor the instructions themselves, even though the instructions may be contained in material bodies that are components of the larger organizational whole. But then again, it is not the matter either. What then, is it? In trying to answer that particular question, metaphysicians both amateur and professional have been driven to all sorts of extremes. One approach is to say that the organization really is the matter after all. Another is to pick some other category as the substrate. Yet a third is to invoke some mysterious principle or even occult quality. All of these solutions fail. The categories do not exist in separate universes. They are not parts, but aspects, of reality.

10

WHY DO SPECIES EXIST?

"Without sexual crossing, there would be endless changes, & hence no feature would be deeply impressed on it, & hence there could not be *improvement*."

—Darwin, Notebook E, p. 50, November 1, 1838

We are now in a better position to discuss such topics as why species exist and what holds them together. The basic answer is biparental sexual reproduction. That is not quite as empty as saying that somebody is rich because he has a lot of money, but it comes close. Given the biological species definition, sex, or at least something functionally equivalent to it, is definitionally necessary for the existence of species. But what sex does is far less obvious, say, than what the heart or lungs do in our own bodies. Indeed, the reason (or reasons) why sex exists is widely considered the greatest unsolved problem in evolutionary biology. Several books have dealt with this problem at considerable length (Ghiselin, 1974a; G. C. Williams, 1975; Maynard Smith, 1978; Bell, 1982). But subsequent discussion, although it has narrowed down the possibilities, has by no means led to a consensus (Michod and Levin, 1987). Fortunately, we can address some of the more interesting issues without having a definitive solution in hand.

In order to lay the groundwork for the discussion to follow, we need to clarify a few concepts and correct some misconceptions. Biparental sexual reproduction (amphimixis) is not the only kind of sex, nor is sexual reproduction the only kind of reproduction. Sex, in the only relevant sense used here, means genetical recombination ("mixis"), something that occasionally occurs quite independently of reproduction (in ciliates for example). Sexual reproduction may or may not involve the formation of sperm and eggs. Fertilization, with the union of sperm and egg, is usually a sexual process, but it does not have to be; where the sperm do not contribute to the offspring of the eggs that they fertilize, "pseudogamy" is said to have occurred. To make matters even more confusing, "parthenogenesis," or the production of embryos from unfertilized eggs, may or may not be accompanied by genetical recombination of the maternal chromosomes. "Amictic" parthenogenesis, without recombination, is a form of asexual reproduction. When recombination occurs without fertilization (mictic parthenogenesis), it is functionally equivalent to self-fertilization. Such "selfing" sometimes occurs in hermaphrodites, but, contrary to what one might expect, as a general rule self-fertilization occurs only if a mate is not available (it is facultative) and many, perhaps most, hermaphrodites cannot self-fertilize.

"Sex" thus defined includes various phenomena that are sometimes called "parasexual" and are somewhat different from what we find in metazoans and metaphytes. Well known in bacteria, these involve a more limited exchange of genetic material across lineages. It is important that these processes are functionally equivalent to the more familiar kinds of sex, because it was long assumed that very little sex exists among "lower" organisms, and that it is something which has only evolved in the last billion years or so. Sex among microorganisms has been difficult to detect for various reasons, not the least of which has been that the culturing techniques are biased in favor of rapidly reproducing organisms, and sex not only slows down reproduction, but it occurs where conditions for rapid reproduction are unfavorable (see Lan and Reeves, 1996). It has taken even longer to establish the presence of widespread sexuality among viruses (Hu and Temin, 1990; Strauss, Strauss and Levine, 1990; Chao, 1992).

Many species in a wide range of taxa have both sexual and asexual reproduction. In this case it is fairly easy for sex to become increasingly infrequent and ultimately to disappear. Once it has been lost, it is highly unlikely for it ever to re-evolve, and almost certainly impossible after a substantial amount of time. There is no evidence that sex has ever originated within any known group of organisms, and in all probability it arose before the common ancestry of all extant life.

Sex that alternates with non-sex is quite common under certain ecological conditions, such as those that characterize ephemeral pools of fresh water. Here again it is easy to overlook the sexual organisms, and these have often turned up after diligent search. Probably sexless groups of organisms do not last very long evolutionarily, though they may continue to exist for a few thousand or even a few million years under favorable ecological conditions. Although it has been claimed that certain groups, such as the order Bdelloidea in the Rotifera, have existed for a long time and undergone extensive evolutionary change, the so-called evidence turns out to be logically fallacious. The organisms may have mictic representatives that have yet to be found, and their existence in the past has by no means been ruled out.

All this is highly relevant to our philosophical discussion. For one thing, it simply is grossly misleading to downplay the importance of sex in the manner that Hull (1988a) has. Organisms may get along quite nicely without it for quite a number of generations, and indeed there is no species of organism that I know of or can readily imagine as a viable entity that does it every cell generation. The mere fact that it occurs only occasionally and is not always conspicuous hardly means that it is unimportant. Just because not all of the customers that we observe in a bar are engaged in sexual intercourse we can hardly claim that none of them at least hope to do so later on.

A second point that needs to be emphasized is that, contrary to what often gets claimed, phenetic assemblages of asexual clones and the like (agamospecies) are not functionally equivalent to biological species. They may indeed coexist with species and occupy what might be considered the equivalent of the niche of a species, but even here the comparison is superficial. Both agamospecies and species with periods of sexual and asexual reproduction occupy a rather limited range of habitats and niches that can be fairly readily characterized. They flourish where the natural economy is undersaturated and where there is a premium upon rapid reproduction. They are much more common in fresh water, especially temporary bodies of fresh water, than they are in the sea; and in the sea they are characteristic of estuaries. Another example is the "weeds" that flourish in places that have been disturbed by human activity and the like.

According to one hypothesis, sex exists because once the environment gets saturated it becomes less important simply to maximize reproductive output and more important to produce diverse offspring that can occupy a broader niche. This "tangled bank" theory (Ghiselin, 1974a; Bell, 1982) may not provide an adequate explanation for the prevalence of meiosis, but at least it fits a real correlation. Although this particular hypothesis may or may not be the correct explanation, there should be no doubt in the reader's mind as to how well established the facts are. Both Bell and I provided extensive reviews of the older literature on parthenogenesis, and subsequent reviews have, if anything, strengthened the evidence for such correlations (E. D. Parker, 1979; Cuellar, 1986; Ghiselin, 1987c; Vrijenhoek, 1989). There is plenty of evidence that species and clones are quite different ecologically (a few recent references: Bierzychudek, 1985; Vrijenhoek, 1989a, 1989b; Avise et al., 1992; Ricci, 1992; Petren, Bolgr & Case, 1993).

Likewise the suggestion that clonal assemblages have the genetical variability that characterizes species will not hold water. It turns out, for example, that the rather diverse "populations" of parthenogenetic animals owe their diversity to multiple origins from separate sexual ancestors. (Hebert & Crease, 1983; Hebert et al., 1988; Crease et al., 1989; Hebert & Beaton, 1989; Havel et al., 1990; Quattro et al., 1991; Browne, 1992; Stewart, 1993; Dybdahl & Lively, 1995; O Foighil & Smith, 1995.)

There are other possible ecological explanations for the existence of sex, most of which invoke a kind of unpredictability of the environment. In general, these do not fit the empirical data very well. However there is some reason to believe the hypothesis that both parasites and their hosts benefit from rapid genetic change that allows the parasite to overcome the host's defenses and the host to evolve new ones (review in Seger and Hamilton, 1987). There are also non-ecological explanations. Sex plays an important role in DNA repair, as was suggested by Dougherty (1955) and later by Bernstein (1977) and further elaborated by him and his collaborators (see Bernstein, Hopf and Michod, 1988). Recently a number of workers have tended to favor the idea of Muller (1932) that sex helps get rid of deleterious recessive genes. These notions are akin to an older idea that sex provides a kind of rejuvenation (see Bell, 1988). One very reasonable possibility is that sex originally had DNA repair as its main, and perhaps exclusive, function, and has become increasingly prevalent as it has taken on additional functions.

None of the hypothesized mechanisms by themselves seem sufficient to solve the "paradox of meiosis" which tells us that an organism in effect throws away half of its genes as a necessary condition for sexual reproduction. A parthenogenetic female contributes twice as many genes to her offspring as a mictic one does. But this is not the only thing that seems a bit paradoxical about sex and reproduction. I mention only two other paradoxes, which may be related. On the one hand we have the "paradox of variation"; genetic research turned up more variation in natural populations than had been expected. This led to a lot of ad hoc efforts to account for such variation, including the notion that it was essentially neutral. Another, less well known, problem is what may be called the "paradox of dispersal"; the larvae of marine invertebrates spend a lot more time swimming around in the plankton, and thereby often traveling great distances, than seems really justified in terms of contemporary ecological theory (see Palmer and Strathmann, 1981; Strathmann, 1974). Perhaps there is more of a premium on not being quite identical to one's neighbor than has generally been acknowledged.

One hypothesis for the existence of sex is indeed that it homogenizes species. Darwin seems to have been the first to entertain that idea, although he did not make it part of his published system. A more recent idea is that it tends to break up adaptive or perhaps non-adaptive combinations of genes. However, if one then asks about gene combinations in general, one may wonder why those that happen to be adaptive in a given niche are not increasingly reinforced by selection. If selection favors a given phenotype, one might expect a combination of stabilizing selection and canalization to produce an optimal genome that is resistant to change. Under such circumstances the genome might become stabilized to a point that it "congeals" (Turner, 1967; Maynard Smith, 1977). Indeed it has been suggested that something very much like this is the general rule, necessitating "genetic revolutions" to get the genome back in a labile condition. More about that later. For the moment we need only stress that given the lack of foresight to the evolutionary process, if sex does break up complexes of genes, theory would have it that this happens because they have ceased to be advantageous, or perhaps because this breakup is an incidental byproduct of something else.

Whenever the raison d'être of sex is discussed, we need to emphasize the extent to which teleological thinking has distorted our thinking about such matters. At the present time sex seems much more problematic than it did half a century ago. This is because in the light of the rediscovery of the Darwinian principle that selection is largely organismal, it would seem that selection can readily produce species composed of adapted organisms, but it probably cannot produce adapted species or other adapted populations. At the time when the Synthetic Theory was being formulated, people simply took it for granted that sex exists because of its long-term advantage in promoting evolutionary flexibility. Isolating mechanisms were supposed to exist in order that there will be species; gene flow was supposed to exist in order to hold the species together; and species were supposed to exist in order that there will be more diversity in the world. But unless we can figure out some way in which such flexibility and the like would lead to one organism having greater reproductive success than another, and especially since calculation seems to give the opposite result, we are hard pressed to treat reproductive isolation, or gene flow, as something selected as such.

The issues may be a bit more tractable if we ask by way of comparison why it is that language and languages exist. Of course the capacity to engage in conversation may reasonably be inferred to have been favored by natural selection because it allowed our ancestors to communicate. It is manifestly but somewhat less obviously used as an aid to the memory, and to thinking, and even as a kind of plaything. By no means can we be certain which if any of these functions was the original one. But the communicative function seems to have predominated for a very long time, and that is all that matters for working out the analogy that interests us.

It would be curious indeed if language were used only in discourse between mothers and their children, analogously with what happens in asexual reproduction. If children conversed with both parents, but nobody else, we would get a kind of population rather than just lineages, but in any case it seems likely that the language could change very rapidly without the interlocutors having much difficulty in understanding each other. Of course we do converse across lineages, especially with those who are close by and with whom we spend a lot of time. If everybody talked only to his neighbors, as has been more or less the case throughout much of human history, it really would not matter if words and grammar became increasingly different and mutual comprehension ever more difficult as distance increases.

So all one needs to produce languages is diversification plus some way of rupturing any cohesive forces that tend to link them together. But what makes for cohesion? It would seem that cohesion is the incidental result of the need for compatibility among speakers if they are to be able to communicate. The emphasis here is on compatibility rather than things like gene flow, and this is very important. If something such as a word happens to "flow" from one group of speakers to another, it is not doing so "in order" to hold the language together. But such flow may result in homogeneity, for the words will tend to compete with one another and be subject to a kind of selection, one that might be called "unconscious artificial selection" following the Darwinian terminology. People learn that certain ways of expressing themselves get the desired results. If a word happens often to be misunderstood, perhaps because it is a homonym, people coin another or opt for some synonym. The same kind of selection obviously occurs between languages, with people opting to speak one language or another, or to learn or teach it.

Having a common tongue is obviously advantageous in any society. We can easily see why it is particularly important within a business enterprise (firm), or in a nonprofit organization such as a university. It is rather like the advantage of having money, or a common currency. Barter is very costly in terms of time and effort, and travelers can expect to have to pay a few percent to change dollars into yen or yen into dollars. Obviously an international firm can do business in local currencies and transact it in local tongues, but such a firm needs to have some employees who speak at least two languages, and that is costly. Closely similar languages, especially those that use the same alphabet, are of course more compatible.

Within a firm or other organization, the specialists may use different jargons when speaking among themselves and thus allow a certain amount of linguistic diversification without creating serious problems of compatibility. The people who sell a product may not need to know how it is made, so they need not talk about that, and they need not expend the time and effort that it would take to learn how to do so. The diversification is an inevitable consequence of the division of labor, which helps make the firm more productive.

The employees within a firm might have a variety of "tools" of a given kind, say computers. They might want a mix of large, stationary computers and small, portable ones for different tasks. However they want "compatibility" among these things—programs that run on both the large computers and the small ones, for example. Such "compatibility" is not quite the same thing as "integration," which means linking things up into a unit that functions as a system of interacting and interdependent subunits. Compatibility makes it easy for parts to function together. Integration makes such functioning together become a realized state of affairs, as when everybody participates in creating and using a common database.

"Coadaptation" is something else again; it means the possession of features that allow different things to interact effectively. We plug electrical appliances in using cords with "male" and "female" parts that are coadapted to each other. Coadaptation may be a necessary condition for achieving compatibility, but it is not compatibility itself. When I take my portable computer abroad, I may have to use an adaptor to make my power cord compatible with the sockets. My computer is rather effectively integrated into my professional life, largely thanks to its portability. I have adapted by learning to use it, and it was adapted by its manufacturer to meet the needs of users like me. We are therefore coadapted—adapted to each other. The programs that I use are largely compatible with one another, with some annoying exceptions. Although the commands for my word processing program are similar to those for my bibliography program, they are by no means identical, and that causes an occasional mistake.

Within the computer world, there are easily recognized (ontological) individuals that consist of more than just a computer or a user. A computer and a user form a productive unit, much like a spider and her web. A group of computers may be linked up physically to form a network, and users themselves are incorporated into various associations. Some of these groups fall along lines of professional interest. Others, however, fall along lines of compatibility, for example those who use IBM compatibles or Macintosh. This sort of compatibility gives rise to a kind of cohesion, rather like that which tends to hold species together. If we think of it in these terms, rather than in terms of something flowing through a network like a computer virus, we may perhaps find a better way to conceive of species and their components. So too if we are looking for integration, we may expect to find more of it in the computers and their users than in the composite wholes that they form.

Such analogies have some currency because there has been a great deal of talk about programmed genetic information, coadapted gene complexes, and the like. There are all sorts of analogies, good, bad, and indifferent, and disanalogies too. As one might expect, some of the talk is dreadfully loose. When one thinks of an organism, perhaps a unicellular one, as something like a computer, some obvious comparisons come to mind. The genetic code with its nucleotide bases and codons has much in common with the binary logic and ASCII code, for example. All too often, loose talk conflates the code itself with that which is encoded. One might reasonably compare the chromosomes with a computer's hard disk. People speak glibly of the properties of the phenotype being the product of "programmed genetic information." Yet even if one grants that this makes a great deal of sense, it is by no means clear what the detailed analogies are. My own hard disk, for example, contains, among other things, an operating system, a word processing program, and text files. Does the chromosome complement contain nothing but programs, or might we be rewarded by searching for something like a database? Species look somewhat like networks that contain a lot of databases with minor variations on a common theme and make them readily available to their participants.

When doing "light" metaphysics, it is easy just to take it for granted that the program is the final cause of the whole system, whereas the text that gets edited is merely epiphenomenal. The program remains more or less unaltered, whereas the texts get continuously revised. By illogical reasoning it is concluded that the program is somehow more important than the text, or that the genomic soft copy is somehow more important than the hard copy that gets phenotypically expressed by the epigenetic "printer." The program evolves through successive upgrades, but not quite as rapidly as the documents do, so the programs are the "real units" of evolution. From there we may proceed to infer that the purpose of text is to provide computers with something to do, and that the people who use computers exist in order that the computers will continue to be manufactured in ever-increasing numbers.

For the moment, however, let us forget about the metaphysics of sociobiology and consider the interconnectedness of things. Whenever organisms or parts of organisms interact, they have to do so in a manner that is conducive to the functioning of the wholes that are involved. Here mutual compatibility is fundamental, and the parts may require considerable coadaptation. This strongly limits the possibilities for modification either of the wholes or of the parts. If one alters the arrangement of parts in an organism or in a machine, function is apt to be adversely affected. This is obviously true with big changes, such as reconnecting the aorta so that blood flows into the oesophagus, or sending 220 Volt AC to a chip in a computer.

Early in the nineteenth century, Georges Cuvier used this kind of reasoning to develop one of the most persuasive arguments against evolution. The parts in an animal are so intimately interdependent that changing anything would be downright fatal. Although Cuvier obviously went too far, there is a good point here, and a legitimate principle as well. Such reasoning nonetheless still provides the basic argument against evolution by what are called "saltations" or "macromutations." The "hopeful monsters" of Richard Goldschmit are about as hopeless a cause as one can imagine. Evolution has to occur by little steps, with a great deal of adjustment being necessary to maintain the functioning of the whole. This is a law of nature, and like other laws of nature, one that has its *ceteris paribus* clauses, provisos for ordinary conditions not inevitably being met, and the like. The same basic principle tells us that the rate of change will be slowed down by "frictional" forces that result from successive adjustments taking time.

The interconnectedness of things suggests that a change in one part will bring about causally dependent changes in another that may be maladaptive. They may even be sufficiently maladaptive to prevent a given kind of change. Darwin elevated such considerations to a principle of correlated variation, something suggestive of modern notions about the selection of pleiotropic genes (i.e., those that affect more than one character). One of his examples was selection for white coat color in cats giving rise to domestic varieties that are deaf as well. Evidently there are modifiers and variable penetrance here, since his informants told him that some white and blue-eyed cats have normal hearing. Such constraints definitely tell us that the genetical changes that permit evolution are not "random" in the sense of a phenotype with one genetic basis being just as probable as any other. Darwin generally said "fortuitous" rather than "random," and he had good reason to do so.

At any rate we can see from the same sort of consideration that the entire system is apt to be thrown into disarray in proportion as the parents of an organism have diverged in a way that creates incompatibilities. The usual examples of such incompatibilities are drawn from chromosome abnormalities such as result from the crossing of organisms with different chromosome numbers, but it is easy to forget that the disadvantage need have nothing to do with chromosomal function as such. Something can go wrong with the developmental mechanisms, or with the functioning of even the mature offspring.

It is noteworthy that both animals and plants have the capacity to select a mate that is neither a very close relative nor a very distant one, but rather something in between. The Japanese quail is a well-known example (Bateson, 1980, 1982). Selection would seem to have favored optimal levels of inbreeding and outbreeding (Shields, 1982; Bateson, 1983). Under conditions with a modest amount of dispersal, one would expect the optimal mate to occur among neighbors, but not the closest ones. One would also expect the organisms to have a fair amount of adaptation to local circumstances. Marine invertebrates that travel great distances as larvae are among the organisms that may have less adaptation to local circumstances and greater tendency to outbreed. In any case the reproductive community will tend to consist of mutually compatible organisms, although what makes for compatibility may vary from place to place. Obviously anything that made for incompatibilities among organisms in close proximity to one another would tend to be eliminated by selection.

Mayr took Cuvier's argument against evolution, translated it into genetical jargon, and proposed that the parts of the organism are so rigidly interdependent as to preclude local differentiation and to hold the population in a condition of evolutionary stasis. The expression "coadapted gene complex" was the result of habitual use of genetical jargon, even though Mayr specified that the coadaptation includes the functioning body of the adult organisms. Indeed, he even used such considerations to maintain that the target of selection is the phenotype, not the genotype. There was an obvious inconsistency here. Furthermore, there was a serious unclarity with respect to what it is that is coadapted, and it turns out that Mayr meant not the population, but its component organisms. So in effect his argument was virtually identical with Cuvier's. The body of the animal was frozen (congealed) into an immutable and unvarying pattern, with a stasis imposed by something intrinsic to the organism rather than resulting from the conditions of existence. This is a typological conception of organic diversity, but not of the usual sort. It tries to be materialistic, with an essence that is immanent rather than transcendent and somehow built into the chromosome. And it does not rule out variation and evolution altogether. Rather, variation in important characters is rare, and evolution occurs only under extraordinary circumstances.

Mayr had decided that many species are homogeneous over a wide geographical region. Unable to explain the homogeneity by means of either gene flow or natural selection, he had to try something else. Whether this maneuver was justified is hard to say. However, there are all sorts of alternatives. For one thing, the putative homogeneity may have been overestimated, and even if not, it may have been based upon an inadequate sample. It is quite possible that a certain range of morphologies is optimal within a given niche, and that there is no opportunity for occupying a different one.

Another kind of explanation for stasis was proposed by Sewall Wright. He suggested that organisms come to occupy "adaptive peaks" which might perhaps better have been called "adaptive ruts." The organisms might be stuck with a certain morphology or physiology simply because any variant would ordinarily be slightly less fit than the "standard" one. To get something both new and functional, it would be necessary to pass through maladaptive intermediate stages. One way to conceptualize this kind of phenomenon is in terms of the "cost of retooling." In the United States of America we still make use of the old English system of measurement rather than the metric system. What deters us from changing is not the undesirability of the result, nor is it the physical impossibility of doing so, but that it would cost so much to convert. And a population evolving by natural selection being blind, it cannot sacrifice immediate advantages in favor of greater returns in the long run.

Wright's "shifting balance theory" has been highly influential but remains largely unconfirmed (Wright, 1931, 1932; see Provine, 1986; Moore and Tonsor, 1994). It is of interest in the present context because it suggests some of the ways in which population structure may affect evolution. According to the theory, species are broken up into semi-isolated subunits (demes) that are adapted to local situations. Important evolutionary change occurs within such local populations when, due to a combination of sampling error and selection, the population evolves first in a direction of somewhat lowered fitness, and then, due to selection, it changes adaptively but in a different direction, and to a higher level of fitness than previously. Thus it passes from a lower adaptive peak, through an intermediate state that is even lower, to a higher adaptive peak. Once the new adaptive peak is reached, the population, for whatever reason, expands, and the new adaptive peak comes to be occupied by the rest of the species.

Now, whatever the merits of this hypothesis, it obviously has been much misunderstood, for it is often taken as an example of "group selection" playing an important role in evolution, especially macroevolution. This impression is reinforced both by Wright's assertions and by his use of the term "interdemic selection" for one of the stages in his model (see Wright, 1982). However, it is clear (Wright, 1977) that whether or not what he refers to is interdemic, it definitely is not selection, of demes or of anything else. Rather it is migration. It is migration of individual organisms from one place, and one deme, to another. The organisms in question do not form a unit that competes with other demes, but serve as the agents of gene flow within the species.

In Wright's model, therefore, the demes are neither the active agents nor the passive patients that the terminology would seem to imply. However, the model suggests the kind of role that species and demes may play in evolutionary theory, and the sense in which one might say that they "do" something. They have properties that make it possible for certain kinds of events to take place, and for the participants in those events to do something. Like road signs, they are "affordances" of what goes on. They structure populations, with important consequences. Such roles are by no means unimportant, the point here being only that we must not misconstrue what they or other organizations do. When a school provides an opportunity for students to learn and teachers to teach, that is about all that we can reasonably expect of it.

According to Darwin (1859), selection favors diversification within species. According to Mayr (1992b; 1994a) it does not, or rather such diversification as it favors gets swamped out by intercrossing. One point upon which most evolutionary biologists these days would agree that Darwin was obviously wrong and Mayr obviously right is that the advantage to such diversity is not sufficient to produce separate species. However, it seems to me that Mayr has gone too far. He says that such polymorphisms as occur between the various morphs within a species of butterflies that mimic more than one model are rare. One might want to say that the existence of two distinct sexes in so many animals and plants gives a kind of diversity that is not representative. But even so these are big differences. What about the little ones? It is quite possible that a little diversity is a good thing, but that there can be too much of a good thing. Furthermore, from his empirical research on the Cirripedia, Darwin had good reason to believe that the species that he had studied manifested a great deal of phenotypic variation, and that, furthermore, such phenotypic variation has an adaptive character. He did not know enough about genetics to deal effectively with the problem, but we certainly can say that many organisms, especially sessile ones like barnacles and plants, have such adaptive phenotypic variability. So it is clear that ecological diversification within species does occur, and it may play a more important, perhaps preadaptive, role than Mayr has been willing to acknowledge.

For any number of reasons, stabilizing selection may account for stasis over and above any putative role that lack of genetic lability might play. If we want to test such hypotheses, we might reasonably took to situations under which the conditions of existence definitely have changed. The trouble is that when we find evidence that, say, an insular biota that has filled an ecological vacuum undergoes considerable diversification, perhaps with shrubs becoming trees and finches coming to live like wood peckers, we cannot rule out the possibility that these are exceptional creatures that have undergone a kind of genetical revolution. Such revolutions have been part of Mayr's peripheral isolate model. The animals get isolated and undergo inbreeding, sampling error, intense new selection pressure and the like, which "melts" the "frozen" genome, decanalizing it and allowing it to change.

This model has been variously elaborated, and although controversial it retains an enthusiastic following (Carson, 1982, 1987, 1989; Carson and Templeton, 1984). Gavrilets and Hastings (1996) have removed some of the theoretical objections to its plausibility but rightly emphasize that in so doing, they by no means resolve the empirical issue of how often speciation has occurred by that or other mechanisms. Furthermore, one wonders to what extent the putative stasis cannot be otherwise explained, especially since the notion of a genome that resists change has been extrapolated to justify the proposition that the diagnostic features of taxa are something special. We shall explore this possibility when we come to discuss macroevolution. For now the important point is that metaphysical considerations at least seem to be playing a leading role in this discourse. Biologists prefer a certain kind of causality and perhaps see more of it than is really there, or at least they tend to overlook the other kinds.

The metaphor of a "gene pool" suggests something in which organisms are immersed, rather than something of which some of their parts are parts. Nonetheless there is a definite notion that a species has a gene pool, and that the organisms that make up the species have access to it. Carson's (1957) metaphor would have it that a species is a "field"—a "field for gene recombination." That tends to suggest that it is an adaptation, though probably Carson thought solely in terms of an effect.

The individuality of species has led various biologists, most notably Eldredge, to ask the question of what, if anything, species might do (Eldredge, 1985; 1986; 1989). One possible role, or function, after another has been found not to withstand critical examination. They are not adapted, and they are not components of ecological wholes. Those are things that organisms, or that some local groups of organisms (especially families) do.

Eldredge (1986) denies that species play an economic role, but only because he has defined "economics" in a narrow sense. As I see it, economics is the science of resources, and whatever else genes are, they certainly function as resources. If one does view genes as resources that supply the organisms with the wherewithal to make offspring that are like themselves but also different from themselves, then one might refer to a "gene fund" as certain Russian authors have. Or perhaps one might analogize what a species creates to a "market" in which genes are exchanged for the mutual benefit of the participants. Perhaps "trading network" would be even more apt (Landa, 1994).

Species are of great significance to evolutionary biology because they speciate, become transformed, and go extinct, but so far as organisms are concerned, their only important role would seem to be the provision of those genetical resources. If this is to happen, the component organisms have to possess various kinds of compatibility with one another. Since the linkages may have a very local character, the result on a geographical scale may be a quite heterogeneous network, and one that is "cohesive" in only a very restricted sense. But the intercompatibility that exists among the parts creates a network that has many important consequences, including the capacity to occupy an extensive territory and a range of habitats. If, for example, part of the species becomes extinct on a local basis, immigration and genetical recombination allow recolonization, and a kind of supra-organismal regeneration of the lost part with genotypes appropriate to the local circumstances. Therefore it is clear that, as Eldredge has observed, species "do" at least one other thing: they resist extinction.

11

Objective and Subjective Systems

"This classification is evidently not arbitrary like the grouping of the stars in constellations."

—Darwin, *Origin of Species*, first edition, p. 411

Acquiring knowledge, whether by learning or any other means, is a process. So too is maintaining and using it. Consisting as it does of various behavioral dispositions, beliefs, and the like, and serving as a resource, knowledge bestows some kind of privileged condition, or state, upon a knowing organism. We might even want to attribute the possession of it to machines or other nonbiological objects, though to most of us this would seem to go to far.

The possessor of knowledge has traditionally been called the knowing "subject," and that which is known by the subject has been called the "object." The ontological status of the knower is sufficiently straightforward for the purposes at hand as not to pose difficulties for the present discussion. We do need to address the problem of how knowledge (which is the subject matter of epistemology) relates to the ultimate reality that may or may not be known or knowable (which is the subject matter of ontology). However, in discussions of scientific knowledge in general, and of systematics in particular, the ontological and the epistemological aspects have so often been confused that we can hardly avoid a lengthy discussion on such matters. Furthermore, the ontology and the epistemology alike tend to get bound up with psychological considerations that are irrelevant if not downright pernicious. It was not without reason that the paper in which I first mentioned the individuality of species was entitled "On Psychologism in the Logic of Taxonomic Controversies" (Ghiselin, 1966b; cf. Kusch, 1995).

Scientists set themselves the task of evaluating evidence with respect to material objects, processes, and other entities. Their theories are thought at the very least to make claims about such things, and these claims can be true or false. Such claims are the result of judgments—decisions about what is true or false. Not all judgments can legitimately be called true or false. In value judgments, for example, we decide what is good or beautiful, rather than what is true. A common pitfall occurs right here, however, when someone confuses a value judgment with another kind of judgment in which values are involved. If John tastes wine and finds it unpleasant, his decision that the wine tastes bad is a value judgment. If Mary, watching him, observes an expression of disgust, infers that John does not like the wine, she has made a judgment, but hers is not a value judgment. Where John made a judgment based upon values, Mary made a judgment about John's values, based upon the evidence of John's facial expression. John's judgment was subjective, Mary's was objective.

This kind of distinction, however elementary, is exceedingly important when we discuss the philosophy of classification, for classification systems are said to be subjective and objective, or artificial and natural. Or they may perhaps be a mixture. When they do their research, scientists want a classification system that tells them something about the things classified (objects of classification) and says nothing about the classifier (or subject). This is true even when they are studying themselves. However, that is not the only thing that they want, and the (subjective) desire for classification systems that are aesthetically pleasing or practically expedient often enters into their decisions. We shall return to this qualification later on, but for the moment it is perhaps worth observing that there need be no contradiction.

Some authors (e.g., Cowan, 1955) have maintained that there is no difference between a subjective classification scheme, or one that is artificial (in the sense that we create it), on the one hand, and an objective classification, or one that is natural (in the sense that we discover it), on the other. To rebut such claims, I long ago proposed the example of a ranking of ink-blots in the order of their obscenity as something that was purely subjective (Ghiselin, 1966b:212). Although such an arrangement might be useful for some purposes, it clearly tells us something about what is in the mind of the beholder, not about the inkblots themselves. Less obvious is the point that "intersubjectivity"—a favorite criterion for scientific legitimacy—is by no means a synonym for objectivity. The mere fact that you and I agree that one ink blot is more obscene than the other would not justify a claim that we are dealing with the properties of the things classified rather than with the properties of classifiers. Our minds are just dirty in the same way.

Knowledge would seem far easier to place in an ontological category than truth, which might be considered a purely logical concept that has no place in any legitimate system of ontological categories. It does not seem necessary to go any further into this particular question here, except that we need to clarify a very basic point about truth insofar as it relates to objectivity in scientific knowledge. This is because, again, epistemological criteria keep getting dragged into such discussions in a manner that leads to serious errors.

With respect to what the truth is, we may adopt the view of Tarski, and simply say that, for example, the proposition "This is a fossil" is true, if and only if, this is a fossil. Such a definition allows one to adopt a position of "absolute realism" with respect to truth. By this is meant that what is true has nothing—absolutely nothing whatsoever—to do with what anybody knows, believes, or opines. In taking this position one is apt to shock certain philosophers, especially those who have swallowed various versions of conventionalism or relativism. At a seminar that I presented in Berlin on October 20, 1987, the Polish philosopher Stefan Amsterdamski asked me how I could take such a position, and I replied that it was the only consistent position that a scientist like me could take in the conduct of his actual research. When we make claims about the real world, these are indeed claims about the real world, and there is nothing about the truth of such claims that would be any different even if nobody existed. At the time I gave that seminar, for example, I was involved in a research project on molecular phylogenetics, attempting to discover the branching sequences in genealogies, which correspond to events that took place literally hundreds of millions of years ago. How could any scientist's beliefs or opinions possibly affect whether or not chordates have a more recent common ancestor with echinoderms than they do with flatworms? The cladogenetic events responsible for the relationships happened well over half a billion years ago, long before there were any scientists, and before there was anything that could "think" well enough to have what most of us would call an opinion about anything.

There may be a problem here, because it is easy to mistake a theory of truth for a theory of knowledge. One can have a theory of truth without any commitment to the notion that the truth is knowable, or even to the notion that any proposition is true. But of course the notion of truth would be useless were there no relationship whatsoever between what is known and what is the case. Usually the kind of realism that I espouse with respect to truth is considered a "correspondence" theory, to contrast it with "coherence" and "pragmatic" theories. Coherence means that the elements of knowledge hang together in a perhaps only vaguely specified way. Pragmatic approaches to knowledge emphasize its utility in problem solving, so that theories are mere tools, hence the roughly equivalent term 'instrumentalism'. An advocate of a correspondence theory generally takes the position that true bodies of knowledge are in fact coherent, and that they are useful in solving problems, but only by virtue of their correspondence to what is true. Coherence and pragmatic usefulness are therefore not defining of 'truth', although they may be treated as legitimate evidence for identifying truth. Confusion of definition and evidence is very common in the philosophy of systematics, and here is a good example of where such confusion may occur. It is difficult to maintain a coherence or pragmatic theory of truth without presupposing a correspondence theory, especially when the distinction between definition and evidence is clear.

To head off some perhaps well-intentioned criticism, I should add that the sort of realism advocated here should not be confused with realism in various other senses, some of which are discussed in various places in the present work. In particular we need to make sure it is not confounded with the sort of "naive realism" that treats what goes on in our heads as identical to what goes on in the "external" world. It should be reasonably obvious that our thoughts about ice are not cold. The notion that the brain contains a little model of the world is very attractive to the naive mind, and in fact there has been a great deal of neurophysiological research that favors the existence of "cognitive maps" and "representations." But any such "isomorphism" can hardly be perfect. So when one favors a realist view with respect to our cognition, it does not have to be that of a "naive" realist. One can be a "critical" or a "common sense" realist and quite sophisticated about such matters (Reid, 1785).

It does not seem reasonable to say that our beliefs or opinions can have absolutely nothing whatsoever to do with what is true about the material universe. Nor does it seem reasonable to say that the truth about that universe, if such truth does in fact exist, is altogether and without qualification unknowable. What that relationship may be is of course highly problematic. For purposes of this discussion we need not go into such epistemological issues in any detail. However, it behooves us to consider that mishandled epistemology has been a serious source of error with respect to the issues that are important, and especially how ontological considerations help to clarify matters. Psychology, it turns out, can be of some use here.

Jung (1921 translation 1971) divided human personalities according to various "psychological types." He believed that he could classify people as to whether they were thinking or feeling, sensate or intuitive, and introverted or extraverted. These are normal variations of personality, and there is nothing pathological about them. Except for remarking that Jung's book is well worth reading, we might as well skip all but the introvert-extravert dichotomy. These terms are part of popular culture. Therefore virtually everybody thinks he knows what they mean, and very few actually do. (When first told that I am an extravert, I remarked, "Great, let's have a party," but that is not what the term really means.) An introvert is one who is constitutionally disposed to regard everything from the perspective of the subject, whereas an extravert is one who is constitutionally disposed to regard everything from the point of view of the object. (Some psychologists have changed the definitions and even the spellings of the Jungian terms, so watch out for equivocations here.)

What this means may be exemplified by a simple little test. Ask someone whether we should study psychology: A. better to understand ourselves, or B. better to understand those around us. Introverts will tend to answer A, extraverts B. As it turns out, the overwhelming majority, something like ninety percent, of professional psychologists are introverts. We zoologists and, I suspect even more so, geologists, though I have no hard data, are mainly extraverts. Extraverted persons interested in behavior are more apt to become ethologists working in zoology departments than psychologists working in psychology departments. Whether one is an introvert or an extravert surely must influence the kind of research that one does. For example two psychologists, Shapiro and Alexander (1975) have written a wonderful book entitled *The Experience of Introversion*. For a couple of introverts to write about the subjective aspect of introversion is a fine example of the sort of thing that probably would not even occur to an extravert! The very term 'experience' means very different things to introverts and extraverts. For an extravert, an experience might be something like walking down the street and encountering a friend; for an introvert, it would be how one felt and what ran through one's mind when that happened. In confirmation of my own diagnosis as an extravert, I should perhaps reveal that my basic reaction was not to ask what this told me about my own inner life, but to seek to apply the dichotomy to the history of science and the biographies of its practitioners. (According to Jung, both Darwin and Cuvier were extraverts.)

Psychology is mainly an outgrowth, or a "speciation product," of philosophy, and one might expect a lot of philosophers to have been introverts. This is indeed the case, but in fact there have been quite a few exceptions. Plato is a classic example of an introvert, but Aristotle is just as good an example of an extravert. I do not hesitate to suggest that many features of their philosophical systems can tell us a great deal about their personalities, while warning against any effort to make those systems "nothing more" than "projections." More importantly in this connection, those scientists whose philosophy is like that of Plato are more likely to be introverted, whereas those whose philosophy is more like that of Aristotle are more likely to be extraverted.

Another test for introversion and extraversion is to ask how many items there are on the following list:

1. one's self
2. the universe.

Perhaps just about any intelligent person who thinks a bit before responding will answer "One." But an introvert might at least feel like answering "Two." It is very easy to set one's self up as something other than that of which one is, in fact, an integral part. An organism is, indeed, set off from its surroundings by epithelia and other barriers. But its body is "permeated" by its medium and various other causal influences. A statocyst detects the direction of gravity without recourse to anything outside of the organism's body.

It is very easy, however, to misconceive of the self as if it consisted of a thinking and feeling being, encapsulated within a kind of cavity and communicating with the outside world via devious channels. Carried to its logical conclusion, the self would be sort of a little man, or homunculus, sitting within the big man's head, watching what goes on outside by television. Such has been, indeed, one of the most common forms of "modeling" on the part of philosophers. Descartes, in his *Traité de l'Homme,* suggests how a blind man might locate an object that he cannot see by using a pair of sticks and a sort of natural geometry. A great deal of the neurophysiology of sensation has been built up through this sort of reasoning, much of it quite legitimate when it does not get out of hand. Of course stimuli have to get received, transmitted, and processed if an organism is to make use of them, and of course they do.

The drawbacks of this perspective have become painfully apparent when philosophers attempt to treat it as more than just that—a perspective, or a point of view. Merely granting the assumption that there really is an inside and an outside calls for a certain kind of answer to the question of what gets in, and how. If once we grant that our thoughts about ice are not cold, the way is open to say that all we know is thought and its components. Bishop Berkeley, as is well known, denied that anything but thought itself exists. He maintained consistency by populating the world with God's thoughts as well as our own, effectively dealing with the problem of what happens to things when we cease to think about them. Given that all that is known by "us" is our own mental life, the inevitable, most parsimonious, extension of this line of reasoning would be solipsism: forget about "us" because only "I" exist. Everybody else is just a figment.

In Kantian terms, what we know is only "phenomena," or appearances. "Noumena," or things in themselves, are unknowable. Most of us are willing to grant such a distinction, but the question remains of what conclusions might be drawn from it. In order to think, we of course have to think with something, be it thoughts or whatever. I cannot imagine what it would be like to put an ice cube in my head and integrate it directly into my thought processes. Obviously, inner experience has to be very different from the things experienced. Modern neurobiology has provided a great deal of useful information about how what goes on inside our brains relates to what goes on in the rest of our bodies and outside of them. We therefore need not grant that we have no idea whatsoever about how things as experienced relate to things as they are. But even were we to grant that we really know nothing about such entities, we have perfectly good reason not to draw extreme conclusions.

For one thing (and this tends to get overlooked by persons who treat the self as something other than an integral part of the rest of the universe), we function as material bodies that are components of the rest of the universe and its parts. It is not that we are like spectators at a dance; rather we are like dancers dancing. It goes on within us.

The justification of some kind of realism on a biological basis has a long history, and it has long been argued that if human beings and other animals are to exist, they need reliable information about their environments. Before the Darwinian revolution, the usual assumption was that God had endowed us with various mental powers and perhaps a certain assemblage of innate ideas. The limitations of our understanding could be explained on the basis that God had not seen fit to endow us with more than was necessary for the conduct of our daily lives. Therefore it stood to reason that the human intellect is an imperfect instrument. Substituting natural selection for God provided the same kind of answer, and the result was evolutionary Kantianism (see Campbell, 1974). The temptation to become more or less a naive realist under such circumstances gets disappointed if one is critical enough to realize that an accurate conception of the world is not necessarily favored by natural, let alone sexual, selection. All that selection favors is whatever happens to maximize reproductive success. Yet in spite of such limitations, it still follows that there has to be some kind of connection between our knowledge, on the one hand, and that which we use it for, on the other. To claim that there is absolutely no connection whatsoever hardly makes sense.

The "empiricist" tradition in philosophy and psychology has systematically rejected any "innate" component to behavior, to the point even of rejecting inherited dispositions, such as personality traits. At most one is supposed to start out with certain capacities, such as the ability to reason and to respond to stimuli. We leave the details aside in order to get at the travesty in question. Namely, knowledge has been thought to consist of entities called "ideas" that come to us as simpler "units" via sensory channels. Locke (1690), in his *Essay* (Book I: Chapter 1: paragraph 15) puts it quite nicely: "The senses at first let in *particular* ideas, and furnish the yet empty cabinet, and the mind by degrees growing familiar with some of them, they are lodged in the memory, and names got to them." The metaphor of a cabinet suggests an enclosure in a way that another very common one does not: the blank tablet or *tabula rasa*. In either case, however, the knowing organism is conceived of as something passive, and remote from where the action goes on.

We need but mention here how false and misleading such a conception is to somebody who understands how animals relate to their sensory worlds. On the one hand, their sensory apparatus is spread out all over their bodies, and the information that comes to them is structured from the very beginning. On the other hand, they do not stand idly by while things just happen; if nothing else, they explore their environments. Psychologists, such as James Gibson (1966) with his "ecological optics," have gotten a considerable amount of intellectual mileage by taking a more realistic view of things.

In such an empiricist scheme, the source of knowledge is often said to be "sense impressions," but this hardly constitutes an exhaustive list. It leaves out, again, anything that derives from the activities of the organism. It also leaves out various other things that are experiential but not mediated by the senses. For example, in addition to observing the transparency of alcohol, smelling it, and tasting it, I can apprehend its effects upon my physiological and emotional condition as it perfuses all the tissues in my body. Likewise I can classify it and other substances, such as caffeine and theobromine, by comparing such effects. One might wish to argue about whether we should just stretch the definition of 'sense impression' to salvage that kind of empiricism. Nonetheless, the point stands that we must avoid a naive picture of a little man inside our heads watching a television screen. The homunculus is a superstition.

Gilmour (1937, 1940) advanced the position that what we classify is not organisms, nor is it chemicals, nor anything of the sort, but rather the "sense impressions" that supposedly provide us with knowledge in general. In this he managed to obtain a few supporters, particularly among pheneticists, and aroused sufficient interest as to draw some fire from realists among systematists. In this sense, a "realist" is opposed to a "phenomenalist," who believes that the object of knowledge is "phenomena," or appearances, rather than the things themselves (noumena). As Gilmour (1940:464) puts it, "the object which we call a chair consists primarily of a number of experienced sense-data such as colours, shapes, and other qualities, and partly of the concept *Chair* which reason has constructed to 'clip' these data together." His position on individual organisms is quite intriguing (op. cit. 466): "the individual is a concept, a rational construction from sense-data, and that the latter are the real objective material of classification." These constructs, he says (466), are "grouped together into a large number of different classes, based on the possession in common of certain sense-data or attributes." It is not surprising that Gilmour considered all classification to be artificial, or at least designed for a special purpose. His notion of a "natural classification" is even more revealing (472): "A natural classification is that grouping which endeavours to utilize *all* the attributes of the individuals under consideration and is hence useful for a very wide range of purposes." This notion of what is natural is sometimes called "Gilmour natural," though approximately the same view, without the phenomenalism, was presented by John Stuart Mill in his *Logic*. Perhaps the most revealing thing about Gilmour's definition is the word 'endeavours', which makes the criterion of truth a subjective matter. For most of the scientists I know, just trying to achieve one's goal is not the same as succeeding.

A phenomenalistic approach was very popular among logical positivists. Gilmour cites an obscure paper by H. Dingle, but a phenomenalistic slant is noteworthy in the writings of more mainstream philosophers, such as Rudolf

Carnap, as may be seen in his *Der logische Aufbau der Welt* (written 1922–25, translation 1967). The goal here was to get around the problem of the unknowable "external world" by treating reality as a sort of logical construct out of experience, albeit, in the case of logical positivists, something that was intersubjective rather than personal. A material world was not posited, and even individual organisms would have to be treated as constructs as well. As Mahner (1994) has pointed out, the remnants of such phenomenalism are still with us, and we need to clean house. (For further animadversions against phenomenalism, see Woodger, 1929.)

The empiricist approach treated knowledge as something that was put together out of certain "elements" or "objects." In the case of Locke, the basic elements were simple ideas, and these were then assembled into compound ideas by the mind, which was also capable of the more complex operations of abstracting and relating. Alternatively it might be said that knowledge is made up out of "objects" or "facts." Bertrand Russell went so far as to develop a theory of "logical atomism" in which the universe was populated by elemental truths. Consult his intellectual autobiography (Russell, 1959) for a readable account of this and other curiosities.

Whatever the details of such empiricisms, the consequences for those who have taken that kind of position have been very serious indeed. To begin, we get a travesty of how knowledge is acquired, with a vague notion that corpuscles of Truth bash into the retina, make their way into the inner sanctum, and rattle around until they get stuck together into larger particles. Popper (1972) calls this the "bucket theory" of the mind, his metaphor stressing the passive nature and simplicity of the process. The result is a kind of naive inductionism, one that leaves out all the active powers of the mind and their manifestations—things like formulating questions, constructing hypotheses, and doing experiments.

Often such naive empiricism takes the form of claims that all legitimate scientific knowledge has to be obtained through "direct" experience with the objects known. One of the attractions of phenomenalism, of course, is that we supposedly have "direct" knowledge of phenomena. Most of the advocates of naive empiricism are probably too unsophisticated to have given that much thought, however. In all probability they mindlessly maintain that "seeing is believing" and that is that. They believe that we have direct knowledge of material objects in our visual fields, even though they also believe that perception is mediated by light rays, photoreceptors, nerves, and other things. And they give privileged epistemological status to what they believe is "directly" perceived, even though optical illusions are common knowledge. The popularity of such naive empiricist views among scientists seems to derive in large measure from the putative certitude that attaches to "direct" evidence (for critique see Hull, 1967, 1968; Shapeare, 1982).

One consequence of the empiricist tradition is that the process through which the truth is reached in such a knowledge process has been conceptualized as basically a matter of correlation, something that in the psychological realm has often been called the "association of ideas." One has to be very much a novice in science not to realize that a correlation does not necessarily mean a relationship of cause and effect. But the "Humean" tradition in the philosophy of science has been to dismiss the very notion of a cause, at least in the sense of denying that there is a necessary connection between one event and another that follows it. Nonetheless causality retains its following among scientists, and a few brave souls among philosophers have been trying to refurbish it somewhat (Cartwright, 1983, 1989). In such a climate it is easy to see why "etiological classification" would be viewed with suspicion if not outright hostility. And since evolutionary events and processes are both remote from the senses and causal in nature, these have had to bear the brunt of attack upon the part of epistemological nihilists.

Yet another consequence of naive empiricism has been the assumption that knowledge is acquired in the form of corpuscular units that have objective existence in the material universe. To a naive biologist this accords to some degree with experience in familiar branches of knowledge. After all, the world is full of atoms, molecules, cells, organisms, and other things. So why not the phenomenal raw material of classification? Bertrand Russell's "logical atoms" have already been mentioned. In the history of taxonomic theory, the same basic notion manifested itself as what was called the "unit character." At one time it was actually claimed that the characters of taxonomy have a one-to-one correspondence with the genes of genetics. Even quite early on in the history of genetics, such conceptions clearly became untenable (see Stern, 1949). Pleiotropic effects, meaning that one gene affects many characters, obviously are incompatible with such a notion of correspondence between soma and germ, which in turn is hard to justify unless one believes that the chromosomes contain a "blueprint" of the mature organism if not an outright homunculus. And as more and more got learned about genetics, the notion of a monolithic gene became ever less tenable.

The very idea of a "character" also turned out to mean all sorts of different things. This notion will be discussed at considerable length in a later chapter. For the moment let us stick to the present theme, however, and point out how such epistemological notions tended to impoverish the range of evidence that was deemed legitimate in systematics. It was explicitly asserted, in the first place, that the only evidence in systematics consists of characters and, in the second place, that characters are, by definition, the "observed attributes of specimens." Note that this is supposed to be an exhaustive list; anything that is not an attribute of a specimen must be stricken from it. Upon the premise that nothing else is possible, Colless (1967) propounded the notion of a "phylogenetic fallacy" that supposedly underlay the reasoning of those who wanted to base classification on genealogical relationships. His claim being a universal negative proposition, it was easily refuted (Ghiselin, 1969b), although the mere fact that the claim was refuted has had little effect upon its popularity. Various properties that might be used are by no means those of individual specimens, but rather those of populations: the sex-ratio and the geographical or stratigraphic range, for example. And not all evidence need consist of properties of specimens. One might also use as evidence propositions of a more general character rather than properties of individual specimens: laws of nature, for example. In a later chapter we shall consider how such inference is in fact carried out.

The heyday of naive inductionism in systematics was way back in the 1960s during the vogue of a controversial movement called "numerical taxonomy" or "phenetics" (see for example Sokal and Sneath, 1963, Sneath and Sokal, 1973). Phenetics has been treated, together with "evolutionary systematics" and "cladism," as one of the "three schools" of taxonomy, following a terminology introduced by Mayr (1965). However, phenetics has been little more than an historical curiosity since the early 1970s, and virtually nobody calls himself a pheneticist these days (Sneath, 1995). Advocates of phenetics denied that biological species and phylogenetics are knowable, and as an alternative proposed that classification should be based upon what they called "overall similarity." At the same time they were developing some computer-based techniques for the processing of systematic data. The techniques in question obviously had their limitations, but many of them are still useful for certain purposes. Although the use of computers has never been an issue, every effort was made to give the impression that it was. The real controversy had to do with the metaphysical claims by means of which the pheneticists attempted to sell their product.

Their negative epistemological claim was that biological species and the genealogical relationships among them are unknowable. Anything that involved "inference" rather than "direct proof," as they put it, had to be excluded from the domain of legitimate science. Because this claim, even in very weak form, would have abolished virtually the entire content of the elementary textbooks in physics and chemistry, let alone biology, it had to be watered down and made a claim about the certitude, or the reliability, of various kinds of knowledge. But once one has admitted that "unobservables" are knowable at all, the choice becomes a matter of whether the risk of error is a sufficient deterrent when deciding whether to put one's career on the line.

But what really took the wind out of the pheneticists' sails was the emptiness of their positive ontological claim. Namely, something called "overall similarity" was supposed to exist. Not only did they claim that overall similarity exists, they claimed to be able to measure it. My own initial foray into the philosophy of systematics (Ghiselin, 1966a) involved a rather crude but nonetheless successful attempt to discredit this notion. Although my arguments were challenged by James S. Farris (1967; cf. Ghiselin, 1967), Farris ultimately acknowledged that I was quite correct. Furthermore he broke with the pheneticists and became one of the leading cladists (Hull, 1988b). I note that the philosopher Nelson Goodman (1970, reprinted in Goodman, 1972) drew some conclusions about similarity very much like my own. His essay is well worth reading, even though, as Hull (1992) emphasizes, the individuality thesis wreaks havoc with some of Goodman's other ideas.

The notion of overall similarity, like that of absolute motion, appeals to common sense. The groups that we deal with in everyday life have a lot of features in common that are not shared by other groups. If we go to the grocery store, we find that the bananas are placed together with oranges, apples, and other things that have a great deal in common—far more than just the name 'fruit'. They share lot of important and interesting properties, and as also may be said, for example, of bakery goods, these are quite a diversity of properties. On the other hand, fresh fruits and vegetables are apt to be put near each other in one place, canned fruits and vegetables in another. And the canned things share more than just the property of being canned. Once one tries to decide whether a canned peach is more similar to a fresh mango or a canned tomato, one begins to have difficulties, and not just because the tomato is botanically a fruit and gastronomically a vegetable.

If we ask what it means when we say that two things are similar, it could mean simply that they have properties in common. However, an incredibly broad range of properties would satisfy this criterion. For example, two things are similar if an omniscient and omnipotent being might conceive of them. So maybe we had better go to the opposite extreme and consider identity. Two entities are identical if and only if both have exactly the same properties. But if "two" individuals really shared all properties, then they would have to be the same individual. If there are two individuals, they must differ insofar as one is this and the other is that. So when we speak of identical items, such as identical twins or identical engine parts, we mean that they share all of a limited set of properties, and not all properties whatsoever.

When we say that one thing is "similar" to another, we mean that it shares at least some of such a limited set of properties. In geometry, identical triangles have the angles and sides the same, whereas similar triangles have the corresponding sides proportional but only the angles the same. And it makes perfectly good sense to say that in terms of a given list of properties, item A has more in common with item B than with item C, and in that sense is "more similar" to B than to C. But this conclusion is strictly dependent upon what properties are admitted in the set, and how much "weight" is attached to them. Consider the following items:

ABCDEF
ABDCEF
AECDBF

One might say that the middle and the bottom items were "equally similar" to the top one, because they both differ in two "characters". The middle differs in having D instead of C at the second locus and C instead of D at the third, whereas the bottom differs in having E instead of B at the second and B instead of E at the fourth. However, one might want to say that the middle differs from the first in just a single "character"—CD has reversed its position, or become inverted, in its entirety. The fact that the list has to be somehow circumscribed necessitates that similarity must be "over some" not "overall." And it has to be circumscribed in a definite way if the similarity in question is to be an objective one.

This argument by no means affects the legitimacy of "over some" similarity. Of course we may rank a series of objects in terms of some criterion that they do in fact possess, and often this is very valuable, especially since one important attribute tends to correlate with another. Nobody objects when we list the various aliphatic hydrocarbons in the order of their molecular weight: methane, ethane, propane. . . . The molecules become increasingly larger, and have lower boiling and melting points as one goes from lower to higher molecular weights. Methane is more similar to ethane than it is to propane. The isomers, of course, make matters increasingly complicated, and that suggests the limitations of such schemes. And such "over some" similarity is "objective" in the sense that it expresses a purely objective statement about the natural world, unadulterated by properties of the classifying subject.

However, when we go beyond that and try to include more than just one kind of property, we encounter great difficulties. Although we may be able to rank things with respect to a given kind of property along a linear scale, each kind of property has to be measured in its own terms. But there is no common yardstick, according to which all things can be measured; different kinds of things are incommensurable in many respects. When one animal is just a little different in size than another, but of very different color, how are we to quantify the difference? We should note that not all scales in terms of which things are measured are "extensive" or made up of equivalent units (such as kilograms of mass). Rather they are "intensive" units that merely represent positions of things relative to one another. The Mohs hardness scale, for example, merely states that one mineral scratches the ones that are softer than it and is scratched by the harder ones; it is not made up of "harder thans."

This is not to say that a good classification cannot tell us a great deal about the similarities and differences among things. Anybody who looks at the periodic table of the elements can hardly fail to be impressed with the extent to which elements close together resemble one another in many important respects. But this is not because there is any "overall" similarity among elements. Rather, the arrangement corresponds to a few causally important structural properties of atoms.

One might object that scientists do seem to "measure" the amount of similarity among objects. Of course they seem to do so, but such appearances are deceptive. In general what is measured is a group of similarities and differences of a certain kind, in a particular sample, for example, various kinds of genetic distance indices. This is by no means "overall" similarity. One might also argue that organisms placed far apart in the taxonomic hierarchy become increasingly dissimilar in all sorts of things, and that this is a kind of overall similarity. Let us briefly consider what this really means.

The farther apart organisms are in the taxonomic hierarchy, the greater the number of generations since common ancestry. And as time passes, various changes occur that make the organisms different from their common progenitors—of course. So naturally we would only expect that whenever we want to find two organisms with two traits in common, the probability of finding them increases with ever closer propinquity of descent. Obviously, the common ancestor had to be identical to itself. And yet the various diverging lineages change in many and different ways, to different degrees, and at different rates. Particularly when there have been shifts in habitat, such as from the sea or fresh water to the terrestrial environment, a large number of changes have occurred quite rapidly. So too with changes in locomotion, as when flight evolved in birds, bats, and insects. Changes under such circumstances have affected different parts of the body and different aspects of physiology. Efforts to find a common meterstick for these changes are foredoomed to failure.

Genetic distances too are meaningful only as indices of particular kinds of changes. Consider the sort of "distance" one gets by DNA x DNA hybridization, which has enjoyed some popularity in the study of genealogical relationships. DNA normally exists as paired strands. If it is heated up, it melts, or dissociates into separate strands. If cooled off again, the separated strands reassociate into pairs. Such pairing depends upon the ability of each residue to pair up with its complement in another strand. Within a single chromosome the pairs are essentially identical, and the association is very strong. But DNA from different organisms will also pair, albeit more weakly because of minor differences in the structure of the molecule. Such differences accumulate gradually in the course of evolutionary time, and as time passes the accumulating differences make the reassociation ever weaker. The strength of the reassociation can be used as a rough index of the closeness of the relationship, or the time since common ancestry. The technique works, because as time passes, the proportion of homologous sites that are identical decreases. And it works in much the same way that any other timing device does that depends upon a change in something from one physical state to another at a constant rate. The dating of rocks and carbon residues by means of the proportion of isotopes is a case in point. Like some other techniques, DNA x DNA hybridization has been strongly criticized as depending for its validity upon a "molecular clock"; the change from identical to nonidentical even when averaged over many sites is not so nearly constant as one might hope. Nonetheless it works. If what we are measuring is the effect of a shift in the proportion of identical to nonidentical residues, however, this is a pretty weak and limited sense of "similarity" and not a measurement of how similar overall, or even genetically, the organisms are.

The effort of Dupré (1993) to rebut such arguments is basically a diversion. Recasting the argument, I would certainly agree that I am at this moment more similar to myself yesterday than to myself as I was when I was twelve years old, and that I differed even more when I was born. And given what everybody knows, it is easy to predict many of the similarities and differences. But am I as I am now more similar to myself as I was when a zygote, or to some other adult organism? Pick your yardstick.

We should mention that such problems of commensurability are by no means unique to the taxonomic sciences. Early in the nineteenth century, Jeremy Bentham tried to develop a "hedonistic calculus" which would allow us to decide policies on the basis of "the greatest good for the greatest number," as he called it. He soon became aware that the project was not very promising because of the difficulty of comparing incommensurables. In some cases it has been possible to get around such problems, since what often matters is not anything that corresponds to an objective relationship between incommensurables, but rather preferences upon the part of people and other animals. A lot of work on motivation has been done by finding out how much electrical shock a white rat will endure to obtain food, water, sex, and other things. In economics the study of "value" has been replaced by the notion of indifference. One finds the price at which the consumer has no preference for one good or its alternative. Although the choices made by the rats and people in the conduct of their economic affairs are subjective, they can in fact be studied from an objective point of view.

It is abundantly clear from their writings that the pheneticists had a most naive conception of what is meant by "objective." Thus Sokal and Sneath (1963) quote a dictionary: "'free from personal feelings or prejudice; based on facts; unbiased'." This they say is closely related to repeatability in the sense that different scientists processing the data in the same way would tend to get the same results, but they do not say what the relationship is. Like many folk they seem to have had the notion that any earnest seeker after truth would be properly rewarded if he just kept the faith. Bias was treated as an attitude, a feature of one's psychological condition. How easy it was to gloss over the amount and kind of bias that has evolved in the human sensory apparatus, not to mention the amount that derives from education and the exercise of the intellect. We are, after all, large diurnal organisms. We have color vision, but not in the ultraviolet range. Much of the sound in our environment passes unperceived, and our sense of smell pales into insignificance beside that of a dog.

Sneath and Sokal (1973:28) later admitted that I had a legitimate point about "over all" versus "over some" and opted to restrict their notion of relationship to certain kinds of characters. They redefined (1973:29) "phenetic relationship as similarity (resemblance) based on a set of phenotypic characteristics of the objects or organisms under study." But this left the basic criticism unanswered, for there is plenty of incommensurability among phenotypic characters. Indeed the very notion of a phenotype by that time had become so muddled as to render it worthless in this context. Originally, the term 'phenotype' had a clear and precise meaning in transmission genetics. It distinguished between inherited traits that are expressed and those that are not, which is something that really makes a difference in selection both natural and artificial. Later it came to be used in distinguishing between the germ and the soma. As language degenerated, the visibility of the phenotype became one of its defining properties. So by the time the pheneticists got hold of it, it became virtually synonymous with that which can be seen. What one does about the properties of chromosomes, such as their number and morphology, under such circumstances need not detain us here.

It should be pointed out, however, that Mayr (1965), in reacting to the pheneticist onslaught, at first maintained that an evolutionary classification is based upon genotypic, rather than phenotypic, similarity. Obviously, genetic similarity has exactly the same problems that somatic similarity does. Wrongly alleging that my views were the same as Mayr's, Ruse (1973) accused me of inconsistency in this matter. The whole topic would have been much better handled had everybody realized that the issue was not phenotypes, but phenomenalism. There is no way that subjective "resemblance" can be turned into objective similarity.

Furthermore, the demise of the pheneticist metaphysics by no means brought an end to the naive inductionist "model" of how systematists do their research. Mayr, again, made a strategic mistake at this point that has not yet been rectified. Especially in his didactic publications, he continued to depict the process of classification as the grouping of specimens together into what he called "phena," which were then evaluated and used as the basis for "taxa" (Mayr, 1965, 1969, 1981; Mayr and Ashlock, 1991). Anything but admit that the specimens were gathered and, from the outset, studied with definite theoretical questions and methodological principles in mind, even when he said as much when criticizing the naive inductionism of the pheneticists.

To make matters worse, Mayr dragged in the old notion of classification by division in this most inappropriate context. Aristotle, in *De Partibus Ani-*

malium (642b, Chapters II and III), pointed out that animals have to be classified by using many differentiae. If one "divides" groups into successively smaller units—so-called "downward" classification—one gets a hierarchy that does not cut nature at its roots. Rather, says Mayr, one should do "upward" classification, putting things together into increasingly larger units on the basis of shared characters. The point that artificial systems result when one divides the flowering plants, for example, first on the basis of sexual condition, then on the basis of number of petals or something, is all too obvious. We shall have more to say about artificiality in classification in the next chapter. But the very idea that systematists must either work "upward" or "downward" is a preposterous travesty of what they actually do, which is repeatedly re-evaluate the classification in terms of new evidence.

To make error unanimous, the third faction, or cladists, went along with the same basic, naive inductionist model, even though they accepted their leader Hennig's idea that only shared derived characters are to be admitted as evidence. And this often in spite of claiming to be followers of Karl Popper, whose hypothetico-deductive approach was precisely the opposite of what they advocated (e.g., Wiley, 1975). They seem to have been provoked into claiming to be Popperians by Walter Bock (1973).

As I have pointed out (Ghiselin, 1969d, 1984c) such inductionism in systematic biology is out of line with modern concepts of how scientific truth is obtained. Especially for someone like Popper, not to mention virtually any modern natural scientist, an hypothesis is a question that is to be tested by means of experiments and other procedures. It is what we start out with. But when many cladists and a substantial number of other persons use the term, the "hypothesis" is not a question, but an answer. It is not one or all of the logically possible trees that are to be subjected to the relevant criteria of evidence. Rather, it is the tree that best agrees with their criteria of acceptability. The very idea that there is something hypothetical about the original data set seems never to have crossed their minds. Nor has it occurred to them that science in general consists of tentative hypotheses that have to be revised again and again in the light of further investigation. Somehow we must begin with "fact" and stick to it.

One wonders why in the world such a preposterous travesty of scientific method would be so popular. The answer would seem to lie with the manner in which most people are educated. The discovery of new scientific truths is a truly creative process, one that requires the exercise of the imagination as well as the ability to evaluate arguments. Virtually the whole of pedagogy, however, is a matter of becoming acquainted with knowledge that already exists. No creativity is necessary because nothing is created. So whenever a student is presented with a "problem," it is a problem for which the solution has already been worked out. Indeed, the very fact that a problem is presented implies that a solution exists.

Much of what gets presented to students at all educational levels is mere vocabulary, and the way in which we learn words is apt to prejudice us with respect to how we learn everything else. This is especially the case with taxonomy, one of the main functions of which is to provide us with a language. When very young children learn their native language, they face the daunting task of learning how to use the rules of grammar and to apply the appropriate names to things. When we reflect upon how much is involved, it is a most impressive accomplishment. On the other hand, the child can take it for granted that criteria for calling something by a given name are already there to be found out. When they get older and take biology courses, they learn a bunch of "definitions" for organs and taxa alike, and it is unlikely that they will be aware of much difference. One dissects a frog, and finds out how to recognize stomach and liver. So too one learns how to tell a frog from a salamander.

When one identifies a specimen, one decides which group it belongs to. In this process its "distinguishing marks" or "characters" play an important role. For purposes of identification one wants characters that are conspicuous ("salient") and easily made out. Identification manuals naturally emphasize such characters, as do the "keys" by means of which organisms may be gradually narrowed down to the appropriate species or other taxon. Even if one does not get the impression from such an exercise that the characters so used are "defining" of the taxa, one still is apt to get the impression that classification is "based upon" such characters. This is true even though the features that are given as diagnostic of taxa are not necessarily the same ones that are used in the keys.

If the people who teach systematics and write the text books are confused about such fundamental matters, one might wonder about everybody else. And in fact it turns out that such confusion is rampant among those who deal with the anthropology and the psychology of classification, including the developmental aspects. This point will be developed in the next chapter, but only after a bit more has been said about natural and artificial systems.

12

NATURAL AND ARTIFICIAL SYSTEMS

"Our classifications will come to be, as far as they can be so made, genealogies; and will then truly give what may be called the plan of creation."

—Darwin, *Origin of Species*, first edition, p. 486

As was suggested earlier in the present work, a natural system is something that we discover, not just create. It represents an aspect of objective reality that has real existence in the natural world. As Plato puts it in the *Phaedrus* (Jowett Translation), the "principle is that of division into species according to the natural formation, where the joint is, not breaking any part as a bad carver might." The simile, which is widely invoked, derives its power from the concreteness of an individual organism and its individual parts.

It bears repeating that this definition of 'natural system' does not necessarily mean that there is one and only one way in which to classify organisms. There are all sorts of natural systems, some of which are genealogical, such as those of historical linguistics, and others that correspond to the nomological order, such as the periodic table of the elements. We must not forget that an individual can be part of more than one whole and a member of more than one class. In modern biology, to be sure, the term 'natural system' has come to mean the individual genealogical nexus of which every living inhabitant of our planet is an integral part—either the nexus itself or its representation. But this is rather like Saint Thomas Aquinas calling Aristotle "The Philosopher."

The idea that "the" natural system is a genealogical one is largely due to Darwin, the great reformer of taxonomy, who explained himself quite lucidly in *The Origin of Species*. However, as previously noted, his has not been the only interpretation, and the main alternative deserves some consideration. This is a view that has been strongly influenced by Aristotle's discussion on classification by "dichotomy," which means repeatedly subdividing a large group into two smaller ones, using a single differentia for each division. Say one divides animals into footed and footless, then bipedal and multipedal, then feathered and featherless. It puts man together with the chicken, rather than with some of the quadrupeds with which he would seem to have much more in common. So, according to Aristotle, what one wants to do is find an arrangement in which there is agreement among many differentiae. For Linnaeus this provided the crucial distinction between the natural system and artificial ones. His classification of plants had such an artificial character, and was based upon the number and arrangement of various sexual parts. One could easily sort out plants into quite arbitrary groupings, which at least would allow one to attach a name to just about any plant. In other words it was more of an identification scheme than a classification system. Having an arrangement that often separated objects that obviously belonged together was the price one paid for ease in identification.

John Stuart Mill's analysis of natural classification led him to conclude that natural groups are those which have a large number of important characters in common. This he took to imply that they owe their resemblances to some underlying law of nature. When one tries to apply this criterion to a hierarchy, it has to apply at every level. When we deal with evolutionary trees, we are often hard pressed to find something that unites all the members of a group but is unique to that group. The Mollusca are a notorious example. Once one has subtracted the characters that are shared with their relatives such as annelids and arthropods, little remains other than the characteristic gill (ctenidium) and the tongue-like, tooth-bearing radula. The most primitive mollusks lack a shell. They may have branched off before a shell had evolved, or it may have been lost. But then again, a shell may have originated more than once during the history of the phylum. Nonetheless it seems reasonable that what we treat as typical of the mollusks was not all acquired simultaneously. The branches are "marked" by evolutionary changes, but not many changes occur between each node.

So if there are lots of characters common to a group but not found elsewhere, it probably means one of two things. One possibility is that the group persisted as a single lineage that underwent a lot of change over a long period of time. Another possibility is that the record is incomplete. If so, there were repeated branchings, but we do not know about them. When groups have split up rapidly, they tend to have few characters that unite them, and therefore the branching sequences are harder to establish. This is why a lot of molecular studies give peculiar artifacts, goofy trees with early branches tending to associate with neighboring ones. This is one reason for preferring a "total evidence" approach that combines both the molecular and the more traditional data (Eernissee and Kluge, 1993).

As was mentioned in the previous chapter, the need for a rich and diverse empirical basis has been confused with the notion that one ought to classify from the "bottom up," putting species together, then genera, then families, and so forth. One has the same problem at every level. Mayr has accused the cladists of basing their classifications on "single characters," but that allegation hardly can be taken seriously. In the first place, a single diagnostic feature is only the minimal amount of evidence for a group to be natural. It is generally accepted that the tree is more likely to be true in proportion to the number of diagnostic (uniquely derived) characters shared by any group. And in point of fact the diagrams published by cladists often have many diagnostic features for a given branch. Furthermore, a single diagnostic character (such as an inversion in a chromosome) may indeed be quite compelling evidence if there is good reason to think that the probability of its evolving more than once is negligible. And finally, every taxon shares the diagnostic features of all the taxa of higher rank in which it is incorporated: all mammals not only have mammary glands, they have vertebrae, notochords, deuterostomy, true coeloms, multicellularity, and much else in common.

The notion that one must work from the "bottom up" toward higher levels is an instance of psychologism in logic. Even were it true that, temporally speaking, systematists always work from the bottom up, this says nothing about what follows logically from the premises. In adding up a column of figures, it does not matter whether one starts at the top or the bottom of the column. And indeed there is no reason why some insight about possible relationships or the interpretation of the anatomy might not lead one to propose certain higher-level relationships and see if one has a natural group before working out the relationships within it. This is just what has happened in many cases. The tunicates created a sensation when the discovery by Kowalevsky (1866) of a notochord in their tadpole larvae led to their removal from Mollusca to Chordata. Because the early developmental stages have long been believed, and with good reason, to be quite conservative, phylogeneticists have traditionally proceeded by examining these for criteria that might allow them to divide up the animal kingdom into large groups. And this endeavor has been reasonably successful. The division of animals into those with spirally-cleaving versus radially-cleaving eggs, and into Protostomia and Deuterostomia, still helps to reveal broad relationships, some difficulties notwithstanding.

Even were these methodological or epistemological claims correct, it still would not affect the basic ontological point, which is that the criterion of naturalness is the correctness of the tree. When groups can be diagnosed by means of a large number of characters, it is because they are natural; they are not natural because they can be diagnosed by a large number of characters. It is often possible to diagnose artificial groups by means of a large number of characters. Richard Owen (1866) introduced the term Haematotherma for the "warm-blooded" vertebrates, namely birds and mammals, while explicitly stating that the group was artificial. Enough properties correlate with the homeothermy that it is hardly surprising that in recent years some authors have even suggested that this group is a natural one (Gardiner, 1982). It is a rule with few exceptions that important changes in function bring about a large number of correlated changes in structure. Birds and mammals have insulation, though the differences between feathers and hair are by no means inconspicuous. Critics of the putative naturalness of Haematotherma have shown the value of using a wide range of evidence, including fossils, molecular data, and physiological considerations (Gauthier, Kluge & Rowe, 1989; Eernisse & Kluge, 1993).

At this point it no longer seems desirable to defer introducing some technical terms that describe different kinds of taxa. Roughly speaking, a "monophyletic" taxon is a single branch of a phylogenetic tree. In the strict sense that has been accepted by cladists, a monophyletic taxon incorporates the common ancestor of the group and all of its descendants as well. A broader sense has it mean that it incorporates the common ancestor, but not necessarily all of the branches of the tree. Ashlock (1971) proposed an unequivocal terminology in which "monophyletic" was retained for the broader sense and "holophyletic" was introduced for the narrower sense. From the point of view of etymological clarity and the avoidance of equivocation, this has much to recommend it. Ashlock's proposal has not been widely followed, evidently because of academic politics. However, it seems to be catching on in some quarters (Foote, 1996).

A taxon that is monophyletic in the broader sense of incorporating the common ancestor but leaving out some of its descendants is said to be "paraphyletic." A paraphyletic taxon, such as Invertebrata (animals without backbones) is rather like "non-British history" and reflects the parochiality of the speaker rather than the natural order of things. The contrast term 'polyphyletic' means taxa that consist of separate branches of the phylogenetic tree: they do not incorporate the common ancestor. A taxon consisting of birds, bats, pterosaurs, and insects is obviously polyphyletic. Its members share a property, wingedness, and various other properties that go along with it, but not a common ancestor.

We should mention in passing that there has been much discussion about precisely how such terms as 'monophyletic' should be defined. There is also a lot of pedantry involved, such as saying "the latest common ancestor" rather than just "the common ancestor" even though that is obvious. The interested reader may want to consult some technical literature (Hull, 1964; Nelson, 1971; Sober, 1988, 1992) or a general work that discusses such matters (Wiley, 1981; Ax, 1987; Mayr and Ashlock, 1991; Minelli, 1993). For present purposes what I have said will suffice. Our concern here is with the concepts of naturalness and artificiality, especially as they relate to individuals. Strictly monophyletic taxa, being entire branches of phylogenetic trees, are readily interpreted as individuals. They are "clades" whereas polyphyletic and paraphyletic taxa are "grades"—respectively polyphyletic grades and paraphyletic grades. A certain amount of abstraction is involved in our conception of both kinds of grades, and this indicates that they are classes. On the other hand, the two kinds of grades are somewhat different in ontological status. A polyphyletic grade is obviously a class, having at least two individuals as members, which are members of the class by virtue of sharing a trait, supposing of course that the grade is intensionally defined. On the other hand, there is generally an assumption that some kind of genetic relationship or more than just an abstract connection exists, and it is a class of parts of an individual. If we try an organismal analogy, it is like both of an animal's ears.

A paraphyletic grade might be a single branch minus one part or minus several. Again, a certain amount of abstraction is involved. Speaking of an organism again, we might abstract an animal's ears from the rest of its body, giving us two taxa—"ears" and "other." On the other hand if we push this analogy a little further, we see that the abstraction we apply does tend to cut nature almost literally at its joints. We tend to consider the head and the rest of an animal as two different individuals in their own right, parts that make up a whole which is the entire body. Here, "head" and "other" are differentiated functionally and anatomically. We have abstracted, but upon what seems to be a more "natural" or less "artificial" basis. Our ability to "abstract" within individuals has received very little attention in the literature on classification. But it seems to me very important. It suggests that our classifications tend to be of mixed character. Having such a mixed character may or may not be a good thing.

Gradal classification has become increasingly unpopular with the passage of time. Or, as the saying goes, there has been increasing preference for vertical rather than horizontal classification. There have been few defenders of polyphyletic grades. These seem to be persons who conceptualize their taxa in terms of characters rather than lineages, though it is not true of all such persons. Among the architects of the Synthetic Theory, Julian Huxley (1958) was particularly enthusiastic about gradal classification, perhaps because of his mystical notions about progress. However, the main advocate of polyphyletic grades was Simpson (1945, 1961) who, as pointed out earlier, treated all taxa, species as well as higher ones, as classes. Mayr concentrated his attention mainly on the species level, and, so far as the higher categories are concerned, went along with Simpson for some years until changing his mind. Simpson did some remarkable work on mammals and seems to have been deeply impressed by the fact that their diagnostic features seemed to have evolved gradually, and with considerable evidence of parallelism. An ancestral mammal would not have had the "defining properties" of the class of mammals. So, as he saw it, the sensible thing was to leave it out of Mammalia, which meant that for him the correspondence between phylogeny and the taxonomic hierarchy was very loose (Hull, 1964).

One way he defended this maneuver was to redefine 'monophyly' in a way that made it virtually equivalent to 'polyphyly', leading to a great deal of confusion. A second thing he did was to present a criterion of minimal mono-phyly, one that evolved somewhat with his thinking on such matters. It was a sort of "close enough for government work" rule. The amount of polyphyly that was allowed depended upon the taxonomic rank of the ancestral group or groups. This can perhaps best be explicated by a hypothetical example, which is patently not true, but seems apt as an illustration. Unless the hemichordates have been misinterpreted, it seems unlikely that echinoderms are even the sis-ter group of chordates, the heroic efforts of Jefferies (1979) to make us modi-fied echinoderms notwithstanding. But just by way of illustration, let us assume that there might be a much closer relationship between Chordata and Echino-dermata than anybody has dared propose. One might say that a taxon is mini-mally monophyletic if it is derived from two taxa of the same, but not higher, rank. Thus, the phylum Chordata would be minimally monophyletic if some were descendants of the phylum Echinodermata, others from the phylum Hemi-chordata. Or one might tighten up the criterion somewhat, and say that in order for a taxon to be minimally monophyletic, it has to derive from a taxon of the same rank, but can be derived from any number of taxa of lower rank within that taxon. For example, Chordata could have a common ancestor within Echin-odermata yet be independently derived from different classes, some being closer to sea-cucumbers (Holothuroidea), others to sea-stars (Asteroidea), oth-ers to crinoids (Crinoidea), and so on. Or one could tighten it up a bit further, and allow for separate derivation from orders within a class, and so forth.

Another alternative when one finds evidence that the common ancestor does not have the "defining" characters is to say that the group is "polyphyletic" and not recognize it. This was the procedure of Manton (e.g., 1977) and her fol-lowers (Anderson, 1973) with respect to the phylum Arthropoda. The possibil-ity that the arthropods are a genealogically heterogeneous group, related, say, to different classes of annelids, is still viable. However, there is no good evi-dence for it, as has been pointed out upon more than one occasion (Ghiselin, 1974b; Patterson, 1978). So far as we can tell, it does seem that many of the characters that are "typical" of modern arthropods have evolved more than once, so that if Arthropoda is monophyletic, the common ancestor must have had a pretty "wormy" appearance. Breaking up the phylum and raising the rank of the component subphyla to phyla merely sweeps the problem under another rug, namely a higher category. In other words, turning Arthropoda into a super-phylum does not make it disappear.

Others prefer to "solve" the problem by leaving organisms that lack the putative defining properties of a monophyletic taxon in a paraphyletic assem-blage. We might call this shoveling the garbage into a closet where it can rot out of sight. Doing so seems most plausible when dealing with a monophyletic group the extant representatives of which are quite distinct from any other ex-tant group, and where the closest relatives are little known or unfamiliar. This has been done with the so-called "mammal-like reptiles" which a real evolutionist would perhaps prefer to call "reptile-like mammals." These animals form a lineage that diverged very early from the one leading to the ruling reptiles such as birds, other dinosaurs, and crocodilians. It gave off a long series of branches that became extinct as the animals became ever more like modern mammals. The extant subclasses illustrate the same phenomenon. The Monotremata persist as the highly specialized duck-bill platypus and Echidna. Although these animals have hair, their thermoregulatory capability is rather imperfect. They have mammary glands, but the milk flows out of multiple pores rather than single ducts. And of course they lay eggs. One might want to define "mammary gland" so that monotremes don't have them, and to include viviparity as an "essential" feature of mammalhood, or otherwise find excuses for treating them as reptiles rather than mammals. If we did so, however, the relationships would be obscured. It seems not unreasonable to put animals that have begun to be transformed into something like modern mammals together with those that have gone much further along that historical pathway. Acknowledging our ancestry in this manner makes it easier for the classification to epitomize the actual history of our lineage.

Some authors have suggested that taxa such as Mammalia and Aves should be reserved for the portion of a clade that has extant representatives, when dealing with groups that have a long fossil record (Rowe and Gauthier, 1992). Supposedly the lay person is not interested in fossils, and we can ignore the earlier branches. But consider what happens with Aves, which is supposed to be roughly equivalent to the birds in vernacular terminology. Creationists are desperate to find any excuse for denying that there are transitional forms between taxa. If Aves consists only of extant birds, their common ancestor, and all of its descendants, but not of more remotely related fossils, then they will tell the world that *Archaeopteryx* is not a bird. Even a fossil that was virtually indistinguishable from the common ancestor of the extant birds might not fall under Aves. Furthermore, there might be problems with supposedly fossil groups that turn out not to be extinct after all. The deep sea mollusk *Neopilina* that was dredged up from the deep sea shortly after the Second World War fitted into a class that had been thought to have been extinct since the Devonian. Fortunately it falls well within the phylum. But suppose the next living fossil mollusk to turn up belongs to a very early branch. What then?

Although there are various ways in which one can reduce the number of organisms that are placed in paraphyletic grades, it may be difficult or inconvenient to do so. Sometimes there is an ancestral assemblage from which several strictly monophyletic groups have been derived, but within which there is no clear pattern of relationship. Provisionally retaining such assemblages in the formal classification system makes a certain amount of sense. The case becomes stronger when the group is readily recognized, and roughly corresponds to some vernacular term. The basal Vertebrata, or fishes, were traditionally placed in a class Pisces. Relationships among the fishes are still controversial, but it is clear that the lineage that led to terrestrial vertebrates gave off many side branches. It is good to have names for all the monophyletic assemblages in this group, but for some purposes such grades as Pisces are very useful. Fish are almost exclusively found in the sea and in fresh water. Ichthyologists are happy to study them and write books on them, even though the group has given rise to some nonfish. And many of the branching sequences are not of great interest to nonspecialists.

As a malacologist who specializes in opisthobranch gastropods, I am deeply interested in how the paraphyletic order Cephalaspidea ("bubble shells") gave rise to several other orders that are strictly monophyletic, such as the Anaspidea or "sea-hares" that are familiar to every neurophysiologist. As long as the paraphyly of Cephalaspidea is openly acknowledged, it probably will not inconvenience many people. The experts will work with trees instead of taxa anyway. However, I would insist upon putting the ones that are the closest relatives of the other orders in those orders, so as to reveal, rather than to conceal, those relationships that are obvious only to an anatomist.

Paraphyletic taxa are particularly useful when filing one's reprints, especially since so much of the literature is organized along gradal lines. Groups like Pisces and Reptilia are perfectly serviceable for this purpose, whereas some natural assemblages, for example, Tetrapoda and Amniota, would be little more than excess baggage. Furthermore, the way in which vertebrate zoologists collect and store their specimens makes it quite reasonable to continue having a group of specialists called "ichthyologists" even though Pisces is not strictly monophyletic. They collect organisms in the same general kinds of habitats, and use the same basic techniques to preserve and study them.

Likewise with the Gastropoda, the traditional division into the subclasses Prosobranchia, Opisthobranchia and Pulmonata is convenient, even though the first group is paraphyletic and the other two form what is believed to constitute a natural (strictly monophyletic) group called Euthyneura. People who work on gastropods tend to specialize in one of these three groups, and they have some good reasons for using them as the basis for organizing their collections. Prosobranchia is largely marine, and the collections consist mainly of shells. Opisthobranchia is almost exclusively marine, but since the animals have largely evolved into shell-less slugs, the collections consist almost exclusively of specimens preserved in alcohol. Pulmonata has a very few marine representatives but is mostly a freshwater and terrestrial group with a modest proportion of slugs.

Having some paraphyletic taxa makes it easier to bring the scientific terminology into line with the vernacular. Of course it is better if this goal can be accomplished by having the terms be equivalent to strictly monophyletic taxa, so that (granted that they are indeed strictly monophyletic) "animals" would be equivalent to Metazoa, "insects" to Hexapoda, and "spiders" to Araneae. (Better, roughly equivalent, since the vernacular terms are componential sortals.) People will no doubt continue to use the term 'fish' for anything that might be called "seafood" and aquatic animals in general, and there is no way in which such usage can be rendered scientific. But the gradal Pisces is at least an improvement.

If we opt for paraphyletic grades such as Invertebrata and Pisces, then the system is, insofar as the taxa are grades, at least noncommittal as to the branching sequences. In which case it is of no use for anyone who needs that particular kind of information. On the other hand that particular kind of information can be clearly and unequivocally expressed in the form of a tree-like diagram. And since the tree does the job perfectly well, the arguments for a strictly genealogical arrangement are by no means compelling. But I freely admit that not everybody would agree with me. One reason is that so many systematists are steeped in the tradition of thinking in terms of groups and characters instead of trees and evolutionary modifications. Having learned how to do phylogenetics on the basis of common sense historical reasoning before I got around to reading Hennig (1950) late in 1964, I found his cumbrous terminology more of an impediment to the understanding than anything else. On the other hand, my early difficulties in explaining to my colleagues why the marine Pulmonata, with their "opisthobranch" characters, are perfectly good pulmonates would more easily have been overcome had I simply been able to say that the characters being invoked were what cladists call plesiomorphies. Almost paradoxically, it turns out that Hennig did a great service to systematics, but he accomplished that by showing how to get the right answers in spite of asking the wrong questions. It was like finding an easy way to do long division with Roman numerals.

Such considerations illustrate the point that biological classification as it now exists is a mixture of the natural and the artificial, but it also suggests that such a mixture is not altogether a bad thing. To reiterate what has been said before, a classification provides a sort of epitome, or overview, of what is known. Like the abstract of a scientific paper, it need not give all the details. It also helps to provide a kind of language. Some degree of abstraction and outright oversimplification may be called for when the goal is easy communication and information retrieval. But there is no reason why the names of taxa cannot be the names of lineages, or at least of parts of lineages.

Nor is there any reason why we have to treat subjectivity and artificiality as necessarily undesirable in classification. Darwin's basic example of a subjective (or as he put it "arbitrary") classification was the grouping of the stars in constellations. These are patterns that result from our position at a particular time and place but are very useful in orienting ourselves as we navigate. They serve a very different function from the classification into galaxies and other objective individuals. And a perfectly legitimate one. What we do not need is the sort of superstition that manifests itself as astrology.

Another comparison that might be made is with maps (O'Hara, 1996). Like biological classification schemes, maps in a sense "represent" the territory that interests us and have a practical value. For that very reason, a lot gets left out of them, and they often emphasize the "salient" features of the landscape— the ones that attract our attention. It is difficult to carry a lot of geographical information around in one's head. In learning geography, we emphasize the major features and salient landmarks and expect to use reference works for fine details. A great deal of detail is left out for the very reason that it distracts our attention from what really interests us. But ideally we want a variety of maps with various scales and showing us different features. So a geologist planning a field trip might want a small-scale road map, a large-scale road map, and a geological map as well.

Given such a perspective (which is in fact an economic one), it makes a lot of sense that we would prefer to have a kind of general reference system that only roughly corresponds to the detailed knowledge that is accessible if we need it. Changing the names or boundaries of taxa provides the same sort of inconvenience that happens when people change the names or limits of towns and streets. Furthermore, if we are going to have intelligent discussion, or even ask directions, it is crucial that when one of us uses a word, the other has some idea of what it is to which that word might refer. It stands to reason that we would like to have a biological hierarchy with relatively stable names and a convention for deciding which ones we should expect one another to know. So the idea of having obligate ranks and making, for example, the phyla and classes stand as a kind of marker for such purposes makes a great deal of sense. And having a few paraphyletic taxa or other elements of artificiality may be a reasonable price to pay for such convenience.

The map analogy may remind us of the old adage that the map is not the territory. It would seem that our existing systems are a mixture of artificial and natural taxa, of classes and individuals. Such paraphyletic taxa as Invertebrata are wholes of which a part has been abstracted. They are like The United States of America except for New Mexico. So the artificial taxa in question are classes defined extensionally by enumerating their members. They coexist within a system in which there are individuals (natural taxa) and intensionally defined classes (categories). Does this erode the distinction between class and individual? After all, non-New Mexico occupies a definite position in space and time. No, but it reveals a source of possible confusion. In discussing a whole, we can always single out one of its parts and call all the rest "other": "this" is an individual, "other" is a class. Furthermore, it remains the case that neither individual has defining properties. And when we talk about definitions, it is of course the case that we are defining the names. But the names are the names of the designata, not of the system or the map. The names of the individuals are ostensively defined and have no defining properties. "Otherness" is defining of the artificial classes here, though one might wish to consider it a way of enumerating the elements of a set or extensionally defined class. The mere fact that one may enumerate artificial taxa this way, or that such taxa may exist in mixed or semi-artificial systems, in no way weakens the individuality thesis, as was pointed out in an earlier chapter.

If scientists' classifications may be mixed, or semi-artificial, one wonders about the classifications that lay persons use in their everyday lives. This is a big topic. Over the past thirty years or so, there has been a remarkable growth of interest in what psychologists call "categorization theory" and in what anthropologists call "folk taxonomy," the latter of which has become part of a broader enterprise called "cognitive anthropology" (D'Andrade, 1995). In all these areas, the revolution that has occurred in the philosophy of systematics during the same period has been suppressed, so that even today the individuality thesis is the subject of a deliberate, if perhaps well-intentioned, conspiracy of silence.

The "cognitive revolution" in psychology had at least one obvious merit: proposing a reasonable alternative to the kind of behaviorism that had long dominated the academic scene. One reason for that revolution was the invention of computers. Even if computers do not think, they do something like it, and that made asking what goes on inside the brain seem much more intellectually respectable, as well as promising of interesting results. The trouble is that computer science has come to us loaded with metaphysical assumptions both overt and covert. One assumption that should be obvious by now is that computer science treats the world the same way that the computer treats the data—as a set. It stands to unreason that the universe consists only of classes and that individuals may be left out of consideration. The cognitive psychologists started out with the same sort of naive and simplistic travesty of perception and epistemology that the numerical pheneticists did. But unlike the systematists, the cognitive psychologists were not particularly critical and adopted notions about classification that were already discredited by workers in other fields. Progress in the field does go on, the naive presuppositions gradually being repudiated, but slowly and in response to criticism that is internal, due largely to experimental results that conflict with theoretical expectations (Smith and Medin, 1981; Medin and Smith, 1984; Keil, 1989).

The leading assumption has always been that when organisms classify, what they are doing is putting similar things together. The very notion of etiological classification was completely overlooked until fairly recently (Young, 1978; Mandler, Bauer and McDonough, 1991). When it was found that some of the classes that people singled out seemed not to have clean definitions, it was proposed that the taxa in question are "fuzzy" or disjunctively defined. A fuzzy set supposedly has ill-defined boundaries. However, it is not always clear that a class is really as "elastic" in its inclusion as proponents of such a notion might want us to believe (see Jones, 1982). It is perfectly possible that the same verbal symbol will be used in a broader and a narrower sense, depending upon the context, just as it might have quite different meanings altogether. For example, if one asks a farmer how many animals he has on his farm, he probably will not include himself, his wife, and his children, even if he minored in zoology when studying for his degree in animal husbandry. Nor would he count the fleas and cockroaches. The trouble is that people are very good at recognizing the context in which words are used; computers are lousy at that sort of task. Cognitive psychologists have become increasingly aware of the dangers of overlooking context in their work on categorization (Medin & Schaffer, 1978; Roth & Shoben, 1983).

A search for good examples of disjunctively defined classes with no single group of necessary and sufficient defining properties was disappointing, and subsequent research has tended to reject the notion. It turns out that people can indeed classify using disjunctively defined classes and tend to do so under experimental or natural situations in which they have no idea of what the real order is. But given a choice, they generally prefer the neat, conjunctively defined classes of conventional logic (Medin & Wattenmaker, 1987). Interestingly enough, there is an ontogenetic shift in the direction of defining properties (L.B. Smith, 1979; Keil and Batterman, 1984; Keil, 1986a, 1986b).

Cognitive psychologists have come across some material of great interest from a metaphysical point of view. Keil (1979:160), for example, has presented a tree-like diagram that purports to show how people classify the world in terms of a basic ontological pattern. I have simplified it a bit and recast it as a hierarchy of the same logical form (Ghiselin, 1987b:76):

All entities
    Atemporal structures
    Temporal structures
    Things that occur in space
    Events
      Intentional events
      Non-intentional events
    Things that can be sensed
      Things with no mass
      Things with mass
        Non-cohesive things
          Masses
          Aggregates
        Cohesive things

Non-substantial forms
Substantial forms
  Artifacts
  Living things
  Plants
  Animals
    Non-human
    Human

As I have pointed out, this is a pretty heterogeneous mess from an ontological point of view, and it suffers from the problem that it presupposes that there is just one, monolithic hierarchy. But more importantly, it confuses real ontological categories and various kinds of classes. One wonders to what extent such a psychologist's scheme is a projection of his own metaphysical confusion, and how much it is that of the persons whom he has studied. Human beings are not very good "intuitive statisticians" (see Peterson & Beach, 1967), and it may well be that they are not very adept at "intuitive metaphysics" either. On the other hand, one result of research in cognitive psychology has been that people perform better if the tasks are designed so as to take advantage of their real abilities. A particularly interesting result of Keil's work is that people recognize real category mistakes. Children are apt to laugh if one asks them, "What color was the fight?"

The ontological changes that occur with respect to cognition can be very enlightening. And here is where individuality enters in, for it relates to the well-known stage theory of Piaget. Namely, there is a supposedly "concrete operational" stage in human cognitive ontogeny in which children think in terms of particular things rather than classes. Piaget (1971) went overboard and created a preposterous ontological scheme in which the world is populated by classes rather than individuals (see Ghiselin, 1986a). It would appear that the ability to think in terms of wholes and parts appears earlier in human ontogeny than does class-membership thinking. On the other hand, the capacity to do whole-part logic seems to atrophy as some persons get older, much as does the capacity to learn foreign languages. Piaget seems to have made a very common mistake. Just because younger persons are socially subordinate to older ones, we can conclude nothing about the legitimacy of anybody's thought processes. In fact there seems to be considerable variation in human mentality. Some people are better at one, some at the other. That is one reason why I am an anatomist rather than a mathematician. For those of us who deal better with concrete thinking, translating into whole-part terms often makes it easier to solve a problem. That is precisely how Venn diagrams work and why they are so popular.

Markman (1973, 1979; Markman & Seibert, 1976) found that children can solve certain kinds of problems if they conceptualize them in concrete terms. So if one asks them if all of the ducks remained in the house, they may be stumped, whereas if one asks them if the whole family remained in the house, they can cope with the problem. A family of course is an individual, and this is precisely the point. It would seem that children conceptualize better in terms of wholes and parts than in terms of classes and members.

Markman's colleagues have largely ignored her most admirable work, and one naturally wonders why. It is probably not because she has focused so much upon the pedagogical implications of her research. Nor is it for lack of points that it clarifies. Rather it would seem that the ontological message has not been fully appreciated, as may be seen from an examination of her recent book (Markman, 1989). She continues to treat both inclusive and incorporative hierarchies as if they were "natural kinds" and has created a neologism, 'collection', for what roughly corresponds to an "individual" in the sense that has been used here. Her treatment of mass nouns as a kind of compromise between inclusion and incorporation is a case in point. She has found that there are more mass nouns at higher levels, but evidently has not noticed that this is true only of inclusive hierarchies, not of incorporative ones. As terms become more general, it makes a great deal of sense that class names will refer to an increasing number of things that do not come to us in definite amounts. Hence terms like the mass noun 'furniture' and its included 'chair'. But such is not the case with respect to whole-part hierarchies, which are more incorporative and in that sense "general" at higher levels. We do not use more mass nouns, if we use them at all, in speaking of Greece than we do when speaking of Athens. And this seems to be the case with componential sortals such as 'animal' and 'mammal'. Here the mass nouns are largely terms like 'beef' and 'pork', which represent food made from the animals, not the organisms themselves.

Many of the data on cognitive psychology have to do with how children acquire languages and on the reports of grownups as to what words are supposed to mean. There is a problem here, because how people use words does not necessarily correspond to how they think, and it is all too easy to let assumptions about the relationship between thought and language slip in uncriticized. There is no reason to believe that people simply must use the same classification system when thinking about something and when talking about it. It also seems rather questionable to use the acquisition of a natural language by a child as a model for inductive reasoning, especially when it is so difficult to figure out what is supposed to be induced, all the more so when the experimenter presupposes a naive and simplistic model of induction.

Evidently children use adults as sources of authority when learning how to use terms. For better or for worse, there is no monolithic source of authority in such cases. Among other things, the vernacular language stands in a very complex relationship to the scientific. Children are supposed to shift to more sophisticated criteria as they learn the rules. These days everybody supposedly knows that whales are not fish, and no doubt most educated adults know enough to tell an inquisitive child that they have "warm blood" and breathe air just like the rest of us mammals. But what proportion even of college graduates, cognitive psychologists not excepted, really understands that the basis of scientific classification is genealogy?

Furthermore, it seems very likely that the students of cognitive psychology and folk taxonomy may have underestimated the intellectual sophistication of their subjects in some respects, perhaps overestimated it in others. Often in experiments the subjects are asked to "classify" pictures of organisms that are quite unlike anything that exists, or is at all likely to exist, in the real world. For example, mammals with extra pairs of legs attached to the belly in no position to be able to function as locomotory organs. Must we presuppose that human beings have no idea what constitutes a real, functioning organism, with plausible anatomy? Children have a lot of experience manipulating their own bodies and interacting with other children and pets during play. One wonders just how much background information is brought to bear upon language learning.

It has seemed a puzzle that such "atypical" birds as cassowaries and ostriches are sometimes, though not always, treated as perfectly good birds in folk classification (Bulmer, 1967). And yet, given the facts about growth and reproduction, this makes a lot more sense than one might think. Baby birds all hatch out of eggs, and when they do, they may be able to run around quite easily, but they do not fly. And one does not need to know about Von Baer's Law to apprehend the resemblances among baby birds in general. Organisms change as they get bigger.

Among zoologists the tradition of drawing parallels between folk and scientific classification goes way back. Mayr (e.g., 1963a) noted the close parallel between the species that he studied in the field and the groups that were known to the natives. Diamond (1965) reached very similar conclusions. Johannes (1981), a marine biologist, studied the fishermen in Palau (West Caroline Islands). He was profoundly impressed by their detailed and accurate knowledge of the natural history of the animals, and by their ability to evaluate scientifically meaningful questions. Their knowledge of the fact that color phases often represent males and females might have tipped anthropologists off to the possibility that some of the groups in question were not natural kinds. But for both zoologists and anthropologists, the interest of this sort of parallelism has lain elsewhere. Rather, it is taken to provide reasons for taking a "realistic" view of the systems in question. Supposedly we are equipped by nature to perceive a natural order that is really there (Brown, 1984; Berlin, 1992). One wonders what to make of the fact that so many people all over the world believe that the world is flat. Nonetheless there does seem to be much in common among folk classification systems, and the differences between those and the ones that scientists use should not be exaggerated.

Indeed, some of the findings of cognitive psychology and anthropology can be of great value in helping us to understand why scientists classify the way they do. For example, it has been found that folk systems recognize a quite limited number of categorical levels and treat some of them as more basic than others. The reason would seem to be a matter of what Rosch (1978) has called "cognitive economy." We find it burdensome to the memory and to our ability to process data if we have to deal with systems that have more than about five to seven categories (Miller, 1956; Wallace, 1961). So what was said earlier about having a few categories, the members of which are treated as stable and otherwise useful as reference points, makes a certain amount of sense. Also, people prefer to have taxa that are "salient" in the sense that they are readily apparent. The genus, in particular, has tended to be a fundamental category, the members of which are readily and conveniently identified without recourse to detailed examination. Leaving aside the merits of such notions, it makes a lot of sense to take account of the psychology of classification in deciding matters of policy, and to avoid creating a system that is incompatible with the properties of the human brain. How to do this without sacrificing scientific truth is a difficult issue and deserves further study.

The study of folk taxonomy has had to struggle mightily to overcome the notion that classification has to be artificial rather than natural, and phenomenal rather than etiological. Berlin (1973), for example, renounced the position that he and his botanist collaborators had taken during the heyday of phenetics (Berlin, Breedlove, and Raven, 1966). He now takes a strongly realistic position, at least insofar as the folk taxa are supposed to correspond rather well with those that are recognized by scientists, and also rejects the notion that classification is culturally determined (Berlin, 1992). Social relativism retains a strong following among anthropologists. Its most extreme formulations (Durkheim & Mauss, 1903, translation 1963), which would have it that classification is nothing more than a projection of social needs and values, are generally taken as going too far (Douglas, 1986). Whatever terms we apply to colors, they seem to have a basis in physiological genetics (Berlin & Kay, 1969). Natural selection, not society, has done the classifying. And organisms, including systematists, organize their own knowledge. On the other hand, we ought to recognize that classification does have social importance, and that supraorganismal individuals, such as the International Commission on Zoological Nomenclature, play an important role in deciding how we classify. Of course legislatures and bureaucracies decide what groups are recognized and deemed important, what their social status is, and all sorts of other things. And people want to be free and unencumbered by such constraints, and to be treated like individuals rather than mere instances of classes (Starr, 1992). Yet the sort of pluralism that denies the objective reality and scientific importance of natural kinds and individuals sacrifices truth upon the altar of a political agenda. It would seem more reasonable, if one wants to justify such liberal attitudes, to invoke the individuality thesis. For if the world is full of individuals rather than natural kinds, if such groups have no essences, if their characteristic features are incommensurable in principle, and if human nature is nothing more than a metaphysical delusion, then the pillars of traditional society are rotten, and we had better come up with some alternative.

It is abundantly clear that both the cognitive psychologists and the comparative anthropologists have missed the boat when it comes to individuality. Their literature says virtually nothing about how we deal with individuals, even the sort of individuals that everybody recognizes as such. About all we find are occasional references to proper names and the like. Time and again, it is simply taken for granted that biolgical taxa, whether folk or scientific, are natural kinds. The alternative is virtually never mentioned, and if so it is taken to mean that the scientific classifications have become fundamentally different from those of folk biology. This view has been ably explored, but to me unconvincingly argued, by Atran (1990). It seems to me that he is trying to make us into more of the sort of "hard-wired" Aristotelians than the data really justify.

One of the major themes in the historiography of classification has indeed been that our technical classification is basically a refinement of the vernacular. Our forbears supposedly came into this world equipped to deal, if perhaps somewhat crudely, with the classes of objects that surrounded them. And supposedly, again, with increasing experience and intellectual sophistication, we were able to develop ever subtler class inclusion hierarchies. Accordingly, any shift to individuals and incorporative hierarchies must have been a much later accomplishment. Perhaps it happened in 1975, perhaps in 1859, but not as early as the time of Linnaeus and Buffon.

Such a position, one that of course has never been articulated explicitly, makes no biological sense. The ability to recognize individuals is at least as old and as fundamental as the ability to recognize classes. Chromosomes recognize what (individual) lineage they pair up with during mitosis and meiosis. The phenomena of immunology depend upon an ability to recognize an individual— one's self. And recognizing one's parents, one's offspring, one's siblings, one's conspecifics, and the like are all of fundamental importance to living organisms. Learning who such persons are and coming to know their proper names are among the earliest tasks of education.

As to supraorganismal wholes, the nuclear family is not the sort of thing that a human being, however young and however primitive, can readily overlook. Every normal human being from time immemorial must have been familiar with extended families and such larger individuals as tribes, clans, and nations. The literature of ancient society is full of references to such organizations as armies and churches. So too with farms, hunting grounds, and territories—in short, the world abounds in a vast range of individuals that are important enough that we often attach proper names to them. There is nothing new about that.

Considering biological species, does it really seem likely that sex and procreation have only been discovered within our own lifetimes? Or considered important enough to think and talk about? Primitives and agriculturists have long been aware that "kinds" of organisms generally consist of both males and females, and that these may be quite dissimilar in appearance. Nor has the idea of genealogy been a modern development. Therefore the idea of a sexually re-productive organization, a supra-organismal individual with a proper name, is hardly out of line with pre-scientific experience.

In point of fact, one of the most striking features of discourse between students of folk taxonomy and their informants all over the world is the native speakers' use of terms like "brother" and "family" when explaining relation-ships within their systems. Naturally the folk taxonomists interpret this as "metaphor," and there may be some truth in what they say. But how do we know that such expressions are nothing more than figures of speech? The human mind can easily pass from a metaphor to an analogy to the subsumption of a particular under a more general case.

Furthermore, it is more than just likely that in striving to improve our classification systems, biologists have been more or less aware of such possi-bilities. The categories of empire and kingdom suggest such individuals as the Roman Empire and France. Curiously, the original "Strickland Code" of zoo-logical nomenclature, to which Darwin was a signatory, actually refers to the names of taxa as proper names (Strickland et al, 1843). And if the framers of the Code had something like the highland clans in the backs of their minds, then it makes perfectly good sense to call the names of taxa proper, for they are not the names of classes, but of individuals.

There was nothing about the ontological capabilities of our scientific pre-decessors, nor is there about the everyday metaphysics of our less tutored brethren and sisters, that means they must treat everything in the cosmos as if it were a natural kind. They didn't and they don't. Wholes and parts with ge-netic relations among them are perfectly familiar ground to those who inhabit the real world of everyday life. We can follow the path to truth without having to walk upon the water.

13

CHARACTERS AND HOMOLOGIES

"Homology clears away the mist from such terms as the scheme of nature, ideal types, archetypal patterns or ideas, &c.; for these terms come to express real facts. The naturalist thus guided sees that all homologous parts or organs, however much they may be diversified, are modifications of one and the same ancestral organ; in tracing existing gradations he gains a clue in tracing, as far as possible, the probable course of modification through which beings have passed during a long line of generation."

—Darwin, *The Various Contrivances by which Orchids are Fertilised by Insects,* second edition, p. 233

According to the old philosophy of taxonomy, taxa are classes, and these classes have certain "characters" that serve as defining properties of those taxa. According to the individuality thesis, this cannot be. Taxa are individuals, and these have no defining properties at all. They can be described, but not defined. In some ways this does not make a great deal of difference. The properties in terms of which an individual can be described are often sufficiently constant and otherwise reliable that, when our goal is merely keeping track of the objects around us and clearly distinguishing between one thing and another, it doesn't really matter that things could have been otherwise. Although it is not a law of nature, I can be pretty confident that an organism with hair and mammary glands also has a backbone.

When systematists diagnose a taxon such as Mammalia, they do so by listing a series of "characters", such as the possession of hair and of mammary glands. This seems pretty straightforward, and one would think that the notion of a character is unproblematic. Such is by no means the case, and this deceptively simple term turns out to be highly equivocal and the source of much confusion. Inglis (1991) has asserted quite rightly that taxonomists have no clear idea of what 'character' is supposed to mean; he goes so far as to call characters "the central mystery" of systematics.

Falling back upon the system of categories suggested in Chapter Two, we might suggest that 'character' is just a synonym of 'property' or 'attribute'. But usage can be very broad indeed, to the point that "characters" become virtually equivalent to the empirical data upon which classification systems are constructed (Fristrup, 1992), and, as we shall see, it often is used for things that properly fall under other categories. In practice or by habit, usage of the term tends to be restricted to the intrinsic properties of individual organisms, so that the color or the shape of an animal would be a taxonomic character, its geographical range or genealogical relationships would not. In theoretical discourse, it is sometimes asked how many characters an organism possesses, and in this context what probably is meant is how many properties may be truly predicated of it. But it might mean all sorts of other things, including, as we shall see, how many parts, or parts of a given kind, it possesses.

Yet in the context of diagnosis and identification, characters are by no means the whole range of predicables, but a much more restricted one, which is rather crudely referred to as the "distinguishing marks" of a thing which tell us that it is indeed *that* or *one of those* rather than something else. Individual organisms have such distinctive properties, although monozygotic twins sometimes create substantial challenges. When the individuals in question are groups of organisms, then the problem is to find properties that allow one to tell of which group an organism is a component, and usually the properties of the group as a whole are not what gets taken into consideration. Rather we proceed much as we would if such groups were classes, perhaps unaware of the distinction.

So the logic here is close to that involved when dealing with class membership. The characters are differentiae: they differentiate the groups. Properties held in common by all elements of both of two groups that are being distinguished are of no use for telling them apart. The presence of hair allows one to tell a mammal from a bird, but not a kangaroo from a bear. Which characters are interesting here depends upon the hierarchical level. Nonetheless the characters that are diagnostic at a higher level are of course present at all the lower levels; so when a property is said to be a character, we mean that it is diagnostic at one level or another.

Merely saying that a character is diagnostic at some level is not very informative. We need a terminology that will allow us to specify at precisely which level a character plays its crucial diagnostic role. Hennig (1950) coined the terms 'synapomorphy' and 'symplesiomorphy' to make this distinction. Both mean shared characters, but there is an important difference. A synapomorphy at a given level is a character that is diagnostic of a taxon at that level and no higher; a symplesiomorphy at a given level is one that is diagnostic above that level. Recurring to our examples, the hair of mammals is a synapomorphy at the level of the class but a symplesiomorphy at the level of the subclass—a synapomorphy of Mammalia, a symplesiomorphy of Theria and its components, such as Rodentia and *Mus musculus*.

Synapomorphies, or shared, derived characters, are important because their presence in the common ancestor of a clade provides a criterion for distinguishing that clade from its ancestors and collateral relatives. Were taxa classes, and were the synapomorphies defining properties, it would stand to reason that all the "members" of each taxon would possess those properties. Such reasoning would seem to have underlain the claim of Platnick (1979) that snakes have legs. Leggedness, after all, is a synapomorphy of Tetrapoda, and snakes are tetrapods. "Once legged, always legged" would be perfectly reasonable were leggedness something that really cannot be changed. Once a native of a place, always a native of a place, but the same is by no means true of being a resident. Platnick unwittingly devised a *reductio ad absurdum* of his own metaphysical position.

Another argument for the leggedness of snakes is as follows. We must distinguish between characters, on the one hand, and character states, on the other. The distinction between characters and character states was originally made in the context of coding schemes for descriptive data (Colless, 1985). For instance, jargon would have it that eye color is the character, and red, blue, and yellow are its "states," rather than predicating red-eyed, blue-eyed, or yellow-eyed of the organism. Now, for an amusing bit of sophistry, one might suggest that the characters are legs, the states long, short, and absent, and argue that snakes have the character, legs, but these are in the state called absent. So again, snakes have legs.

Often 'character' is conflated with a class of characters. In many cases this is quite innocuous, as in a paper by Bock (1964) on bill shape as a (kind of) taxonomic character in avian systematics. Whitten likewise (1959) refers to the tracheal system as a taxonomic character among insects, and again this is obviously just a harmless but technically improper way of referring to a class of properties of the respiratory system. But it suggests our next sense of 'character' which has proven very pernicious indeed, that of a part.

One would think that the distinction between an organism or its parts, on the one hand, and attributes predicated of an organism or its parts, on the other, is sufficiently obvious that category mistakes would be no real problem. One dissects a bilateral organism, not its bilaterality, its digestive system, not its herbivorousness. But there are serious problems here (Ghiselin, 1969c, 1984b; Colless, 1985; Rodrigues, 1986; Fristrup, 1992). A major reason for this confusion is that many of the words that refer to parts are used "attributively"; animals are said to be legged or winged, for example, and this makes it easier to confuse the leggedness with the legs or the wingedness with the wings. That the two are conceptually distinct in some people's thinking is apparent from their use of the term 'feature' when speaking of both attributes and parts.

The part-attribute ambiguity first came to my attention in the context of discussions on the relative utility of "adaptive" and "non-adaptive" characters.

This has been a most crucial issue in discussions on the role of function in phylogenetics, and impinges upon a whole range of other epistemological issues (see Chapter 18). At any rate, when one says that a "character" is non-adaptive, it makes all the difference in the world whether one refers to a part or to an attribute of that part. It may be pretty much indifferent, from a functional point of view, whether an organism's major nerve trunks lie ventral to the gut, as in annelids, or dorsal to the gut, as in chordates. But it makes all the difference in the world whether or not an animal has a nervous system at all. For an analogy, consider the advantages of having everybody drive on the same side of the road; whether they all drive on the left or the right is a very different question, and deserves a very different answer.

And it turns out that a lot of very useful diagnostic characters are like driving on the left side rather than the right. It doesn't make much difference one way or another; which gets adopted is basically a matter of historical accident, and it is not easy to change from one condition to the other via adaptively intermediate steps. Does this mean, as some have suggested, that only physiologically unimportant changes should be considered reliable? Uncritical thinking often leads people to jump to just this conclusion, perhaps on the assumption that natural selection will produce the same adaptation repeatedly. But experience and sound principles alike lead us to reject such a simplistic view of things. When physiological advance is hard to evolve, that in and of itself reduces the probability of such convergences, and when the part is very complicated, the resemblances tend to be superficial.

Along a similar vein, we often encounter claims that a classification is based upon a "single character." What this usually means is not that a single trait diagnoses a taxon, but rather that the properties studied are all those of a single part, perhaps a single organ or perhaps a single organ system. In phylogenetics, as in the sciences generally, the greater the amount and diversity of evidence that one brings to bear upon a question, the more likely one is to come up with the right answer. So if one uses properties of all the organ systems of the body, one is apt to have more evidence. Furthermore, the parts within a single organ system tend to evolve as a functional whole, many of the changes being causally interconnected and highly correlated, as is sometimes recognized by the rather unfortunate term 'character-complex' in referring to such parts of the body as the vertebrate skull (Schaffer, 1952:104).

Such correlation among anatomical parts can be very strong indeed. For example, it is remarkable how many properties of the "higher" tetrapods are adaptations that allow them to breathe while they run (Kluge, Gauthier & Rowe, 1988). If one tries to carry the notion of a part to its logical conclusion, it is hard to say where to stop when dismissing a classification as based upon a "single character" in this sense, for one might with perfect legitimacy call a complete map of the genome a single character.

Which brings us to the question of how many characters a given organism might have. In framing such a question, one might have all sorts of perhaps tacit reservations as to what sort of properties are to be counted. Should one include things like geographical range or aesthetic appeal? It should be apparent that there is no definite number of properties that might be predicated of an organism. From time to time it has been suggested that the number of genes in the genome corresponds to the number of phenes in the phenome, or something of the sort. Even if there were a straightforward correspondence between the structure of the chromosomes and that of the soma, and there is not, a little analysis shows that one faces insuperable obstacles in trying to decide how many "characters" lie before one. To take a very simple example, consider two strings of nucleotides:

ACGT

AGCT

Do they differ in two characters, so that, for example, the one on top can be transformed into the one on the bottom, by changing a C to a G and a G to a C? Or is it just one character, as might result from inverting the GC segment? If one answers that it all depends on what did in fact happen, one has made a subtle shift from describing the properties of the nucleotides themselves to describing their history.

It would seem that the effort to "reduce" phenotypic characters to genotypic elements, when attempting to answer the question of how many characters an organism might be said to possess, boils down to the same basic problem of confusing parts with properties. Organisms are indeed composed of parts, and simply by counting them one can indeed determine the number of components of a given kind within a given organism. Such a number may vary with age, and it may not be constant within a species, but it can provide useful data for systematics and much else besides. The chromosome number is a very good example, the number of cells not a bad one either. But the properties here are the numbers of parts of a given kind, and the question of how many parts there are is not thereby answered.

Another manifestation of the same basic category mistake occurs when it is said that evolution is a process of character transformation. This seems an odd way to put it. What gets transformed when a species evolves is a population, or a lineage, much as what gets transformed in an ontogeny is an organism. It is not the property that changes, but that of which the property is predicated. (When a boy becomes a man, an immaturity does not become a maturity.) If anything ceases to exist when change happens, it is an instantiation of one property ceasing to exist and an instantiation of another property coming into being. (Adolescence is a transitional period when a childhood ends and an adulthood begins.) I am not sure how much real mischief this manner of speaking has done, perhaps very little.

A common metaphor has it that evidence should be weighed, and not just counted, meaning that some arguments are more compelling than others. In systematics we often encounter the expression "character weighting" when one would hope that what weight is being attached to is arguments. Again, the elementary category mistake here would be quite innocuous, were it not for the tacit misrepresentation of what counts as evidence and how it is evaluated. The term 'character weighting' presupposes that the evidence used in systematics consists of properties of organisms, and that nothing else is germane. Calling it that automatically leaves out of consideration both descriptions of historical events and such higher level generalizations as laws of nature, as one can see from the methodological literature (Neff; 1986, Wheeler, 1986).

But perhaps the most serious consequence of confounding attributes and parts is to be found in the literature on 'homology'. The controversies surrounding this comparative anatomical term provide a good example of how a failure to appreciate the individuality thesis can lead to gross confusion, as can be seen from a recent symposium volume that nonetheless contains some valuable contributions (Hall, 1994). Perceptively, Müller and Wagner (1996) point out that interest in 'homology' has increased with efforts to shift the focus of attention in evolutionary studies away from the gene and toward the organism.

In the light of the individuality thesis, virtually all of the vast literature on 'homology' pales into triviality, and to review it in detail here would be tedious and uninstructive (see, however, Ghiselin, 1976; Donoghue, 1992). Nonetheless a few comments about received opinion would seem to be in order. It is common knowledge that the term goes back to pre-Darwinian times and that it was redefined in evolutionary terms. However, the redefinitions were not always satisfactory. Often 'homology' and its contrast-term 'analogy' were defined as kinds of similarity. Among Synthetic Theorists we have the example of Haas and Simpson (1946:323), who define 'homology' as "a similarity between parts, organs, or structures of different organisms, attributable to common ancestry." Treating homology as a kind of similarity due to common ancestry remains a common practice in textbooks and dictionaries, as well as in the scholarly literature. To this we should add those who consider it similarity but prefer not to define it in evolutionary terms (Riedl, 1979; Haszprunar, 1992; Rieppel, 1992, 1993, 1994). As to 'analogy', it continues to be defined as similarity due to shared function, in spite of my protestations to the contrary (Ghiselin, 1976).

Some such persons hearken back to pre-evolutionary days and endorse the definitions of Owen (1843), who defines (p. 379) 'homolog' as "The same organ in different animals under every variety of form and function" and (p. 374) 'analog' as "A part or organ in one animal which has the same function as another part or organ in a different animal." Among such neo-Owenites we may mention Boyden (1943, 1947) and Hall (1994).

Because Owen has become a sort of culture-hero among contemporary anti-evolutionists (in the broad sense), including historians such as Desmond (1982) and Rupke (1994), it seems appropriate to say a little bit more about him at this point. Owen deserves high marks for treating homology as a relation of correspondence between parts, but when it comes to the basis for determining what is meant by "the same" he embarked upon an exercise in unbridled mysticism and occult metaphysics. He explained things by reference to what he and other idealistic morphologists call an "archetype" (Owen, 1848). Owen (1849:1) explicates the metaphysical rationale of the archetype as follows:

> The 'Bedeutung,' or signification of a part in an animal body, may be explained as the essential nature of such part—as being that essentiality which it retains under every modification of size and form, and for whatever office such modifications may adapt it. I have used therefore the word 'Nature' in the sense of the German 'Bedeutung,' as signifying that essential character of a part which belongs to it in its relation to a predetermined pattern, answering to the 'idea' of the Archetypal World in the Platonic cosmogony, which archetype or primal pattern is the basis supporting all the modifications of such part for specific powers and actions in all animals possessing it, and to which archetypal form we come, in the course of our comparison of those modifications, finally to reduce their subject.

The term 'Bedeutung' is evidently a literary allusion to the essay on the vertebral theory of the skull by Lorenz Oken (1807). As is well-known, Oken and his contemporary, the German poet Goethe, were rivals for priority with respect to the vertebral theory of the skull, of which Owen was a zealous advocate. It is less well-known, but very well documented, that Goethe, who coined the term 'morphology', was literally a practicing alchemist, and that his so-called biological publications were actually exercises in that pseudoscience (Gray, 1952). For alchemists, the world was a projection of God's mind, which they attempted to understand through intuition. That Oken was an alchemist is abundantly obvious from his publications (Oken, 1805). As to Owen, it is hard to say, but he did arrange for an alchemical textbook by Oken to be translated into English and published (Oken, 1847). We shall say a little more about the idealistic tradition in morphology at the end of this chapter. Here we shall merely explicate the scientific, in contradistinction to the pseudoscientific, terminology.

Homologies are relations; more particularly, they are relations of correspondence between parts of wholes which are in turn parts of larger wholes (Ghiselin, 1981a; McKitrick, 1994). The parts which stand to each other as terms in this relation are called "homologues," and, as with other relations, all of the terms must be supplied if the word is to mean anything. In its most important evolutionary signification, the parts in question are parts of organisms, which are in turn parts of lineages (more precisely, but more obscurely, of genealogical nexus).

For a simple example of a lineage, let us imagine a sequence of nucleotides produced by replication, such that the ancestor (I) of the rest is above its immediate descendant (II), and so on to the most remote descendant (VI).

AAAA   I
AAAC   II
AAGC   III
ATGC   IV
ATGCC   V
GTGCC   VI

The first A in I corresponds, or is homologous, to the A just below it in II, III, IV, V and to the G in VI. The second A in I is likewise the homologue of all the As and Ts beneath it in the other nucleotides. For the third A, we would have to say that it is homologous to the third A in II, to G in III, IV, and V, and to the G in the middle of VI, but not to the one on the end. If we assume that the fourth A in I has given rise to C in II, and that the C in IV has duplicated, then the fourth A in I is homologous to C in II, III, and IV, and to CC in V and VI.

Note that the relationship is symmetrical. If the second A in I is homologous to the T in V, then the T in V is homologous to the second A in I. (Unlike the parent-offspring relation, which is asymmetrical: I is the parent of II does not imply that II is the parent of I.) Like the relationship of ancestor and descendant, and unlike the relationship between parent and offspring, homology is transitive. And this makes a great deal of sense, insofar as it is legitimate to say that the parts are the ancestors of other parts. (Being an ancestor or a descendant is to stand in an asymmetrical relation, although there is also of course a symmetrical relation between direct lineal relatives.)

We might change things around a bit and make III the common ancestor of two lineages: with III giving rise to II and II to I, on the one hand, and III also begetting IV, IV in turn V, and V finally VI, on the other. Exactly the same relations of correspondence would obtain. In other words, one can trace the homologous parts either up or down the temporal sequence, and off onto separate lineages via the common ancestor.

Now consider two very important points. In the first place, the relationship of correspondence between the parts of the whole nucleotides is very different from the relationship of similarity between the whole nucleotides, and the relationship, though symmetrical, is not transitive. I is similar to II, II is similar to III, and so forth, but with each generation the wholes become increasingly different. This demonstrates that homology is not a relationship of similarity. It also provides a convenient opportunity to explain what is wrong with the notion of "percent homology" that so often appears in the writings of molecular biologists (e.g., Neurath, Walsh, and Winter, 1967; for further criticism see Hillis, 1994). In the above example, such persons would say that AAAA is seventy-five percent homologous to AAAC and fifty percent homologous to AAGC. What they are really talking about is not percent homology, but percent homologue identity.

In the second place, whether the parts are homologous has no necessary connection with whether they are an A, C, G, or T. They can change, but even though there has been a kind of innovation, it does not affect their homology, for they still correspond. On the other hand, when we get the duplication of C into CC we now have two distinct nucleotides, in other words an evolutionary innovation in which a new part has been added, and there are separate relationships of correspondence between the terminal ones, on the one hand, and those adjacent to the G in the middle, on the other. Indeed, a second C might originate as an "insertion," or in other words, arise *de novo*. When any such changes occur, even though homology is unaffected, we have shared, derived properties that can be used to infer relationships, or synapomorphies. Propinquity of descent is a function of homologue identity. But homology is neither identity nor synapomorphy; it is correspondence.

Why, then, is homology so important? Because the relative positions of parts within the whole helps us to identify them as individuals, in the sense of parts of the entire lineage. Again, consider VI above, which we also depict a few lines below: it has G at two different positions, but these are parts of different individuals (individual nucleotide lineages). Does this mean that position is defining of the homology relation? No, for we might have an inversion, for example, of four of the nucleotides as follows:

GTGCC   VI
GCCGT   VII

In this case the homologue of the first G in VI is the first G in VII, and the other homologies are not (left to right) T:C, G:C, C:G, and C:T, but T:T, G:G, C:C, and C:C.

One might want to expand the term 'homology' so that any property, as well as any component, in the sense of an organ or other part, might be spoken of as homologous. And in fact this is commonly done, so that this is why in defining 'homology' we often say that "features" rather than "organs" or "parts" are homologous, if and only if they can be traced back to the same ancestral precursor. This creates an ambiguity, such that we must "stipulate" which is meant, the usual example being that wings of birds and bats are homologous as limbs but not as wings; the corresponding parts were there in the common ancestor, but they lacked the properties that would have been necessary had the common ancestor been able to fly. This usage now seems much less coherent to me than it did when I began to discuss such matters in print many years ago. For one thing, the term loses its anatomical precision; for another, it does not do justice either to its other senses or to its contrary, 'analogy', and that they are indeed contraries we shall now proceed to show.

Analogy, like homology, is a relation of correspondence between parts of wholes. But in an analogy, unlike an homology, those wholes are not parts of some larger individual; instead they are members of a class. And this makes all the difference in the universe, for analogies are by no means spatio-temporally restricted, and analogous organs are analogous irrespective of their locations in time and space. Perhaps the best way to conceive of this is to imagine some distant planet in which life has arisen and undergone an evolution such as ours, but quite independently. We can be reasonably certain that the inhabitants of such a world would be quite different from ourselves. Unlike the bogus extraterrestrials portrayed on television, for example, they would not speak anything even remotely resembling English. Nonetheless, we would definitely expect to find certain features in common among unrelated extraterrestrials, given the fact that life exists and evolves in conformity with the laws of nature. The ecosystems might be peculiar, but under circumstances that are quite plausible, one would expect there to be producers, consumers, and reducers. And among these we might well find photosynthetic organisms being fed upon by herbivores, and these in turn fed upon by first-level carnivores, and so on. We might have sessile organisms and free-living ones, creatures that float in the water, that fly through the atmosphere—indeed just about anything physiologically workable that can survive and reproduce in the face of stiff competition in a real environment.

Now, if we imagine such an animal (in the sense of an occupant of a higher trophic level rather like a tellurian metazoan) and suppose that it made its living by hunting for smaller organisms and feeding upon them, then we might expect that it would have certain kinds of properties adapting it to that way of life. It would have a means of locomotion and, since it is generally advantageous to emphasize movement in one direction, a front end and a rear end. And, given that a gut with two distinct ends has distinct advantages, it would have a "mouth" at one end, an "anus" at the other, where, respectively, food is taken up and waste materials are discharged. Putting the mouth up front has various advantages, not the least of which is the possibility of using it in subduing prey. Also located in the leading direction, near to the mouth, might be other organs such as teeth and jaws, used in subduing prey, breaking it down, and passing it to the mouth—also organs of sensation, helping to locate food, and a kind of control center, or brain. This would give us something that might reasonably be called a head, set off a bit from a body that specializes somewhat in locomotion, with reproductive organs tucked out of harm's way more or less toward the rear.

One doesn't have to be much of a zoologist to realize that there are all sorts of alternatives to such a hypothetical creature. Sea stars, for example, are superficially radial, with the mouth located at the middle, and many of them lack an anus. The only point is that sometimes, through a combination of historical accident and physiological and other necessity, quite unrelated organisms have similar parts arranged in more or less the same way within their bodies, and these parts often, though not necessarily, discharge the same function. So that the heads of our hypothetical extraterrestrial creatures would correspond to our own, and we would say that they are analogous parts. Because homologues must, by definition, have a common historical origin, and because analogues must, by definition, have a separate historical origin, the two terms are mutually exclusive; in other words, they are indeed contraries.

Two crucial points need to be stressed at this stage in the exposition. First, since analogy is a relation between parts of members of classes, the members themselves must be similar in one respect or another, whereas there need be no such relationship of similarity between the whole organisms the organs of which are homologous. But this by no means tells us that analogy is a kind of similarity; it is a relationship of correspondence between parts of wholes that are similar by definition.

Second, although common function is one cause of the similarity between the wholes, it is not a defining property of the relation of analogy. It is neither a necessary nor a sufficient condition for two parts to be analogous. It is not necessary because the similarity can result from purely fortuitous circumstances, a kind of coincidence if you will. An opening at the front end of the body need not be the functional equivalent of a mouth; for example it may house a part that is used in subduing prey instead of leading into the gut. Neither is it sufficient, for there are all sorts of parts that have the same function, and yet nobody ever refers to them as analogues: a mushroom's poison and a clam's shell both reduce the organism's likelihood of getting eaten. Although I drew attention to this point some years ago (Ghiselin, 1976), few commentators seem to have paid any attention. It would seem that they have failed to make the connection here with a more general sense of "analogy" of which the anatomical one is a special case. If we consider the analogy between, say, the nucleus of an atom and the nucleus of a cell, we can appreciate how even the most far-fetched parallels between the parts of things can be expressed in terms of notions of correspondence. On the one hand, such correspondences, whether far-fetched or not, are often the source of valuable new ideas in the sciences. On the other hand, they are notorious for the role that they have played in mythology and pseudoscience.

'Nucleus' is a Latin diminutive which means a little nut, and the metaphorical use of that term is easy to appreciate. When the term 'homology' was introduced into comparative anatomy, it was obviously a new coinage and not, as is occasionally asserted in dictionaries and textbooks, something to do with a Greek word for "agreement." Its roots are *homo*, same, and *logos*, word, so it means a namesake; homologues are the organs that deserve to have the same name. 'Analogy' begins with a root that means "up," as in 'anatomy', which literally means "cutting up" (cf. "anabolism" and "catabolism"). So an analogy is a namesake, too, but a different kind of namesake.

In what at least seems to be a very different sense, anatomists often speak of homologous parts within the same organism. So do other kinds of biologists, as when cytologists refer to the homologous chromosomes which pair up during meiosis. Here the correspondence is between individual chromosomes that are parts of a larger individual, which is not just a single cell's complement made up of two genomes, but the chromosome complement of the species as a whole as well. The fact that a pair of homologous chromosomes exists in the majority of cells of most of us diploid organisms nicely illustrates the point that we have a certain modularity to our organization, such that the parts stand to each other as elements of some causal nexus. They are namesakes because of a common control system that underlies the development and functioning of the whole, and that also has a certain historical aspect. Some years ago I coined the term "iterative homology" for such correspondence, and it has gradually been adopted (Ghiselin, 1976; V. L. Roth, 1984, 1994). It means the correspondence between repeated parts and has the same root as 'reiterate'.

If you place your hands together, palm to palm, as if in prayer, the iterative homologues between the two so-called "antimeres" will line up: for example, your left thumb with its iterative homologue, your right thumb. A similar comparison can be made between your left hand and your left foot, in which case you are comparing so-called "metameres," and the iterative homologues are your left thumb and your left toe. The human body, unlike that of a centipede, is not altogether obviously made up of serial units that are, so to speak, variations on the same theme. However, if one takes a look at the appropriate anatomical drawings, it is remarkable how many such units, such as vertebrae, and vertebrae together with ribs, are in evidence. It is also quite striking how many parts have corresponding elements (antimeres) on the two sides, though we are not quite so bilaterally symmetrical as one might think. Another kind of iterative homology is that between the two sexes. In mammals in particular, the male and female reproductive organs again seem to be built upon the same plan, and one organ after another of one sex has a corresponding element in the other, sometimes evidently without a function in one of the sexes, and sometimes apparent only in the developing embryo.

Nor is iterative homology limited to anatomy. We see it in machines with their standardized parts, for example, among the cars in a railway train, something that one might fancy rather like a centipede or millipede. In inflected languages, we may make out a strong pattern of resemblance among the various declensions and conjugations. Social wholes are also quite generally organized with the components organized in much the same way, and with equivalent functionaries in each of the repeated units—universities have their colleges with deans, departments with heads or the like, for example. And it is noteworthy that when the whole changes, there are coordinated modifications among the parts. Components within a whole seem to have something in common that maintains the common pattern and maintains the equivalence of the corresponding elements. Whatever that something may be, it makes the assemblage of similar parts function as a single individual. The parts are integrated, coordinated, and compatible one with another. Organisms are organized, and so are supra-organismal wholes.

Having such antimeres as one's left and right arm function in much the same way and share a common control system has straightforward advantages in coordinating their activities. Similar advantages can readily be imagined for the metameres, including an arm and a leg. In both, however, the common pattern seems also to depend upon their sharing mechanisms that control their embryological development. This is clear from the fact that genes "for" limb length commonly affect all of the limbs; short-legged dogs and sheep don't have separate leg-length genes for each appendage. Just how all this apparatus is controlled is a matter under active investigation and is yielding fascinating results at the time of writing. But we needn't wait for the results of this research to see how such phenomena might come to be misinterpreted. So having explained what homology is, let us now proceed to examine what it is not.

Quite a number of authors have tried, with varying degrees of success, to champion a "biological" homology concept based upon shared developmental processes (Wagner, 1989, 1994). When this is a special case of iterative homology, there would seem to be little objection beyond the not too obvious point that the causes need not be purely developmental. Be this as it may, when one attempts to mix this "biological" notion with that of evolutionary homology, serious errors often result. The idea that homology can be defined in terms of parts having a similar mode of embryological development, and not just diagnosed by it, goes back at least as far as the nineteenth century. On the one hand, de Beer (1938) suggested that homology is problematic because homologous parts are not necessarily under the control of homologous genes. On the other hand, Hodos (1976) and later, but only for a while, Roth (1984, cf. Roth 1988, 1994) tried to make a case for homologues being parts that are under the control of homologous genes.

The difficulty here is again relatively straightforward. It is not just the organs that evolve, but the developmental processes that give rise to them as well. Development is quite generally under polygenic control, and it is not difficult for one assemblage of genes that control the development of a given part to be replaced by a quite different assemblage. This is exactly how one would expect evolution to take place, given that species are individuals with individuals as parts. But if one is in the habit of treating taxa as classes, then it stands to reason that one will seek for some stable determinant of morphology, and a kind of "essence" in the genome may seem an attractive possibility. That a lot of authors try to invoke such "essences" makes one wonder about how deeply committed they are to evolutionary thinking.

Alternatively, someone who believes that taxa are classes might want to explain the regularities in embryological development and the correspondences between the parts of its various products on a very different basis—as the consequences of laws of nature. This again is a very old idea, and has traditionally been argued on the basis of the analogy between the development of an embryo and the formation of mineral crystals. In recent years such an interpretation of morphology has been vociferously championed by quite a number of so-called "structuralist" authors (Michaux, 1989). Some of these authors are quite forthright in acknowledging that the individuality thesis poses a serious difficulty for their views (Webster, 1989). The reason is that if morphogenetic processes are a matter of contingency, such that the how and why of embryology is a matter of historical accident, the nomothetic wind is taken out of their sails. All they can do is blow hot air from on deck. And having taken the stance that history doesn't explain anything, they find themselves cast adrift on a sea of unfathomable mystery.

Now, to be consistent with what gets said elsewhere in the present work, it must be acknowledged that embryological development is indeed governed by laws of nature in a sense not altogether different from that in which the growth of a crystal is said to be so governed. However, any such laws that we may one day discover will have to be of a very general and abstract character, and they will have to apply to classes of developing organisms irrespective of time and place, and making no reference to any particular taxon. So Wagner (1996) was quite correct in questioning the notion that homologies are natural kinds; they definitely are not natural kinds. For that very reason we cannot accept the assertion of Fortey and Jefferies (1982) that homology means "fundamental structural identity between species, which means a recurrence of the same law content."

In evolutionary comparative anatomy, there is a straightforward criterion for what makes homologues of two organisms "the same" organ, that is to say, namesakes. This is the actual succession of ancestors and descendants in real historical lineages—community of descent. Were it not for evolution, and were taxa classes, there would be a serious question of what determines sameness, and it would be hard to get along without some kind of abstraction—a set of defining properties. In the case of analogy, the arrangement of parts within organisms that bear some specified kind of similarity to one another provides this criterion, and one might specify what kind of similarity it is by means of a morphological diagram. This would give us a so-called "body plan" that is abstracted from all the members of the class. And pre-evolutionary morphology supplied just such abstractions for their non-evolutionary taxa (which, of course, were classes), and their diagrams could often be interpreted as "pictorial definitions" or "pictorial diagnoses" of such taxa. Essences, in other words. Such diagrams often had Platonic connotations, either implicit or explicit, but other kinds of essentialism would do just as well. The "archetype," or *Bauplan*, would then be, if not as it was for Oken and Owen, an Idea in the Mind of God, one kind of metaphysical delusion or another.

Some "modern" authors have continued to insist that such abstractions are somehow necessary for the conduct of comparative anatomical research (Jacobshagen, 1925; Kälin, 1935; Young, 1993; Brady, 1994). Woodger (1945) tried to define 'homology' by reference to such *Baupläne*. That this is quite unnecessary is clear from the way in which we defined the term above, by depicting a series of replicating and evolving nucleotides. The attempt by Gould (1983) to say that a little essentialism is really a good thing, and to rationalize this claim by proposing that essences can be redefined into something material, was grossly misleading from the point of view of physics and metaphysics alike. If the physical properties of material bodies, including economic and genetical systems, are such that they tend to resist change, then all one needs are concepts like inertia and friction to deal with them. A mutable essence, like a round square, is not a logical possibility and is not a viable metaphysical option.

Sometimes it is alleged that evolution "does not explain" homology (e.g., Panchen, 1992). If by an explanation we mean relating something to its underlying causal basis, this makes a certain amount of sense for iterative homologues, for which the correspondence is between parts that share a common developmental mechanism as well as a history of interconnectedness. On the other hand, since the existence of a common ancestor is a defining property of evolutionary homology, there is obviously something wrong with such a claim. It is like saying that reproductive isolation does not explain species.

What really seems to have been meant by such claims is that evolution does not explain the existence of taxonomic groups between the elements of which relationships of homology appear to exist. Pre-evolutionary systematists were perfectly capable of arranging their materials into "natural" groups within which the anatomical parts had manifest correspondences. The locus classicus for such comparison is a sixteenth-century diagram by Belon, in which the homologous bones of a bird and a man are shown in some detail (published in 1555, reproduced in Cole, 1949:8). So what is really explained by evolution is the fact that groups with that sort of correspondence between the parts of the elements could be recognized, and not the correspondence itself. And clearly the claim that evolution fails to explain that situation is wrong. Indeed it is not the only possible explanation, although as a matter of fact it is the correct explanation. It might have been that the underlying order was a matter of law, rather than history, so that the homologies were really analogies. In this case laws of nature would have explained the system with its correspondences. Yet another possibility would be that it pleased the Creator to construct organisms as variations upon some ideal pattern. And in point of historical fact, such alternatives have had their vociferous advocates. From a purely logical point of view, all of these explanations are perfectly legitimate.

The claim that evolution fails to explain homology rests upon a straightforward and quite elementary logical blunder, namely confusing what follows temporally in the sequence of scientific investigation, on the one hand, with what follows logically in the sense of following from the premises, on the other. It is a mistake to which I and others have drawn attention in the past, but which continues to be made. Let us consider a couple of examples. The periodic table of the elements was discovered before the "anatomy" of atoms was worked out. Indeed, many of the properties of the elements were discovered upon an empirical basis by artisans and alchemists long before chemistry existed as a modern science. One might argue how far atomic physics "explains" the properties of chemicals, but what the alchemists knew has nothing to do with that. In like manner, pre-literate peoples were perfectly well aware of correlations between the apparent movements of celestial bodies and such phenomena as the tides and seasons, and a great deal of empirical knowledge about what they conceived of as the heavens above them was possessed by mediaeval and renaissance navigators and astrologers. Now, it would seem utter folly to claim that the heliocentric theory of the solar system and the physics of Newton and Einstein do not explain such appearances. The only obvious reason for making such a claim, beyond having an intellect that is very muddled indeed, is desperation in the effort to rationalize some alternative. And if the only legitimate explanation is the "wisdom of the ancients," the entire scientific enterprise becomes nothing more than a pseudoscientific belief system that is grounded in mysticism and occult metaphysics.

Another very common mistake about 'homology' is the claim that defining it in terms of phylogeny amounts to a vicious circle. Remane (1952) attributed this notion to Oskar Hertwig. Whoever deserves the blame for originating this notion, it has been repeated time and again by persons who ought to have known better (Inglis, 1966; Jardine, 1967, 1969; Schoch, 1986) in spite of having been time and again refuted (Hennig, 1966; Ghiselin, 1966a; Hull, 1967; Patterson, 1982). The claims in question are somewhat diverse, so we have to rebut more than one argument. In the first place, there is no way in which the definition of 'homology' in terms of phylogeny has to be logically circular (Ghiselin, 1966a). A circular definition is one in which the definiendum (the word defined) is contained in the definiens (which provides the defining property), for example, "A circular definition is one that has the property of circularity." The trouble with circular definitions is that one cannot understand them without already knowing what the term means. But in the case of 'homology' there is no question of that. One can define the word 'phylogeny' without using the word 'homology': phylogeny is, by definition, the history of lineages.

The other, more common form that the allegation of circularity takes invokes an epistemological circle rather than a definitional one (see Hull, 1967). In a circular argument, the truth of the conclusion is presupposed when justifying the premises. In phylogenetic reasoning, it is perfectly possible to come up with circular arguments, especially if one is not careful to identify one's premises and make them explicit. If, for example, one wants to justify the proposition that the notochord of chordates arose only once in the history of life, it would be legitimate to point out that all chordates and no other organisms have them; but this particular argument would be worthless if the only reason for thinking that the chordates are a monophyletic group is that all of them possess a notochord. One might of course invoke some other line of evidence, perhaps the improbability of evolving so complicated a structure more than once, or less controversially some other feature held in common by all organisms with a notochord.

From such a legitimate possibility of circular reasoning, it is easy to fall into a muddle and maintain that there is an inevitable circularity, because the only way that we can know what the homologies are is by inferring a phylogeny on the basis of homologies. For several reasons this does not follow. For one thing, what are in fact homologies are not the same thing as what we suspect, believe, or opine are homologies. The data give conflicting results, and because certain features do evolve more than once, we do not base our conclusions on posits about homology; in other words, given the weight of evidence, we accept that some possible homologies are real, whereas others are merely apparent, the consequence of evolutionary convergence.

More importantly, the evidence by which phylogenies are induced is by no means exhausted by the list of putative homologies. The synapomorphies, or shared, derived characters, that serve as evidence can be something other than just shared parts of organisms or attributes of such parts. They can be properties of entire organisms, such as their size or shape. Or they can be properties of populations, such as the sex ratio. The evidence, furthermore, need not be "characters" in the usual sense, nor need it be anything that can be homologized in any ordinary sense either. For example, one may legitimately invoke the position of specimens in time and space. And finally, the hypothesis that a particular genealogical relationship is true can be tested by reference to laws of nature.

In suggesting that more is involved in phylogenetics than just characters, I am rejecting a very old, pre-Darwinian tradition in comparative anatomy. By this I mean the notion that classifications should be purely "morphological" in the original sense of that term. Unfortunately, the language has become somewhat corrupted, so that most biologists are only vaguely aware of what they are talking about when they refer to "morphology." It should be clear from the etymology what it means, but these days, alas, most people don't know a Greek root from a fig leaf. In its strict and original sense, morphology is the study of form, and physiology is the study of function, but the term has lost its precision and come to be a vague catch-all for anatomy, embryology, and all sorts of other things.

So when Marvalee Wake (1992) defined morphology as "the study of form and function," it was not quite like defining 'gynecology' as the study of men and women, but it comes close. Her criticisms of my view that morphology contributed essentially nothing to the Synthetic Theory make sense only if one reads her definition into my text (Ghiselin, 1980b). During the years that led up to the Synthesis, the kind of evolutionary functional anatomy that Wake and I have long advocated was largely restricted to the Russian school, notably Sewertzoff, Schmalhausen, and their followers. The anti-evolutionary morphology that was fashionable in those days is well explained by its advocates, such as Agnes Arber (1937, 1950). These days anti-evolutionary morphologists often refer to themselves as "structuralists."

Wake (1992) adopts the arguments of Waisbren (1988) in attempting to discredit my claim that morphology did not contribute to the Synthesis. They maintain that Edwin S. Goodrich and his students were indeed important contributors. As to Goodrich himself, I explicitly cited him as an example of a Darwinian who perhaps might have contributed to the synthesis, but as a matter of fact did not. Of course his teaching must have exerted a favorable influence, given the fact that his students included Gavin de Beer and Julian Huxley. But the contributions of the students of a morphologist are by no means the same thing as the contributions of morphology itself. In the thirty-five pages of references at the end of his foundational document, Julian Huxley (1942) cites not a single publication by Goodrich.

Pure morphology has nothing whatever to do with function, and that is one reason why I prefer to call myself a functional anatomist, rather than a functional morphologist. 'Function' in this context may mean either or both of two things. On the one hand it may refer to the functions of organs, or what they do in the economy of the body, as when we refer to the respiratory function of the lungs. On the other hand, it may refer to the functioning of organs, or how they work, as when we say that a pair of lungs functions like a bellows. We shall consider the term 'function' in greater depth in Chapter Eighteen.

Morphology is by definition the study of what is called "form," and we ought to explain what this word means, too. Unfortunately, "form" is all too often treated as a kind of occult quality, and that is part of the problem. But it does have a fairly straightforward sense that can be explicated as follows (Ghiselin, 1991). Let us "do" the morphology of the sandwich. (Not a sandwich, but *the* sandwich, for this is an exercise in typology.) Let B designate bread, and F designate filling. We then get, with appropriate possibilities for elaboration, the following infinite series:

       B
       F
      B B
    F F F
    B B B . . .

The series of abstractions with its more particular cases, such as the open face and the double-decker, constitutes the morphology of the sandwich. It is a class of strictly "formal" properties. Note that in this simplified listing we have left out one logical possibility, namely:

    F
    B
    F

Although many a child would gladly prepare one if asked to do so, we rarely if ever encounter such an item in our daily lives. But, at least when doing morphology, we are not allowed to say why this is so, for morphology, again, has nothing to do with function.

　　Morphology, thus conceived as the science of form, must not be interpreted in a narrow sense either. The whole range of formal properties, including shapes and symmetries of the entire body as well as organs, are important aspects of morphology. Contrary to some loose talk about "molecules versus morphology" in the study of phylogenetics, most of the evidence in molecular phylogenetics thus far has been based upon the morphology of the molecules in question, their physiology being largely ignored. Another false disjunction is apparent in the habit of contrasting behavioral characters with morphological ones. Activities and movements have form. So do social organizations. Poems have form: consider the sonnet with its fourteen lines in iambic pentameter and its rhyme schemes. And of course so do all sorts of other things. Come to think of it, one is hard pressed to come up with an example of something that does not have form.

　　Classifying on the basis of pure morphology means arranging the materials in such a manner that those items with the largest number of strictly formal properties are placed closest together. In comparative anatomy, it means an arrangement in which the closer relatives have a larger number of putative homologues in common, and a larger proportion of these are identical. This would be a very straightforward process were the data complete and unambiguous. If we had a complete chromosome map of representatives of all branches within any group of organisms that we want to arrange in this fashion, then "all" that we would have to do would be line up all the homologous sites in a way that minimizes the number of differences between putative relatives and get the answer—or at least something very close. Not being able to do that, we have to attempt an approximation based upon much more fragmentary evidence. And the evidence frequently points in more than one direction. Although any difference that might correspond to a change in the structure of the genetic material is evidence for relationship of common ancestry, it often happens that the "same" change has occurred more than once. So the trick is to find the arrangement which is least likely to be false, and this is generally done by means of what is called "parsimony," which simply stated means that the best tree, *ceteris paribus*, is the one which entails the least number of steps.

In coming up with a purely formalistic arrangement, one might want to leave evolution out of consideration altogether and simply say that what one is doing is minimizing the differences among abstract schemata. After all, pre-Darwinian morphologists got what seemed to be meaningful results without any evolutionary perspective. And given the fact that formalistic schemes can be constructed for groups of minerals and other things that do not evolve, clearly one can get along, more or less, without conceptualizing what one does in terms of history and process. And in recent years this has been precisely the goal of such anti-evolutionary movements as "pattern cladism" and "structuralism," as it was for older generations of "idealistic morphologists" whose ideals were mediaeval, not modern.

Nobody seems seriously to contest the proposition that one can get something of an approximation to the correct genealogy using strictly formal properties and even a quite simplistic notion of parsimony. Therefore it seems justified not to elaborate upon how one proceeds using such methods, especially since this area has been treated ably and in detail by other authors. On the other hand, the alternatives to such restrictive approaches have received far less attention than they deserve. But to explain them, we need to approach such matters from a fresh perspective. It will be easier for the reader to understand the arguments after we have considered what role the laws of nature might play in our thinking, and in the context of a very general discussion on the principles of historical inference that apply to all branches of knowledge.

14

LAWS OF NATURE

"The Grand Question, which every naturalist ought to have before him, when dissecting a whale, or classifying a mite, a fungus, or an infusorian, is 'What are the laws of life'[?]"

—Darwin, Notebook B

Because the individuality thesis makes a clear distinction between the nomological order and the historical order, it enriches our understanding of both. On the one hand, it helps us better to understand what truths, if any, are legitimate laws of nature, and consequently how better to distinguish such laws from certain kinds of generalizations that are often mistaken for them. On the other hand, it provides valuable insights as to the appropriate role of those laws in discovering and explaining history writ large and, conversely, as to how the historical sciences contribute to our understanding of the laws.

In discussing such matters, one might feel obliged to mention that philosophers of late have been quite critical of the very idea of a law of nature (Cartwright, 1983, 1989; Van Fraassen, 1989). Indeed philosophers in general seem hard pressed to come up with a satisfactory explanation of what this term 'law' is supposed to mean (Armstrong, 1983). Not that they haven't tried, of course. For the present discussion, this is just background information. We really don't have to address the issues that seem to be vexing philosophers as they grapple with these problems. We need only acknowledge that the individuality thesis makes a legitimate distinction between laws of nature and certain other kinds of generalizations. To develop that insight, it doesn't really matter either whether Hempel's version of the scientific method, with its nomological deductive approach and all that, contains more than a mere hint of the truth (Hempel, 1965). It is worth noting, nonetheless, that the individuality thesis has made some of the older ideas about laws and explanation seem more reasonable than had been thought. And we needn't really concern ourselves with the merits of the "semantic" view of the scientific method or its competitors (see for exposition Beatty, 1980; Lloyd, 1988; Thompson, 1989; and for critique Ereshefsky, 1991a; Hodge, 1991). These are matters of logic and epistemology that are only indirectly related to the ontological task at hand, and are mentioned here because the individuality thesis may perhaps be germane to them.

What does matter is the fundamental distinction, implicit in the individuality thesis, between two kinds of natural order. Science as we know it could not exist without some sort of commitment to the uniformity of nature and to its particular uniformities, however problematic and inscrutable these may seem. And what scientists call laws of nature are precisely such regularities. Indeed, it has often been said that they are nothing more than those regularities. Beyond this ontological point, however, we also must add that only regularities of a truly universal character qualify as laws, and their instantiations are something else again. (So an individual law of nature would be metaphysically impossible, and an instantiation of a law might be thought of as an individual lawfulness.) The regularities that qualify as legitimate laws of nature are therefore those which are true irrespective of time and place. In other words, they are spatio-temporally unrestricted, as we have already emphasized. And finally they have a certain physical necessity, suggested by the saying that they are "true of everything to which they apply" (Reed, 1981).

In contrast, there are "regularities" that have a contingent and localized character, even though for practical purposes it may not make a great deal of difference. A person's temperament or personality, for example, is fairly uniform on a day to day basis, and for the most part changes only gradually with the passage of time. But it does change, which is why our ability to predict a child's behavior becomes far less reliable when extrapolated to the mature adult. And if species are in fact individuals, we can say pretty much the same thing about them, too. Our descriptions of them are useful in forecasting, but more so in the short term than in the long.

At this point, two questions raise their ugly heads. First, are there other such regularities in biology, and perhaps in other sciences, that are not laws but have been mistaken for them? And second, having struck some generalizations off our list, do any remain? (We may recollect that the individuality thesis itself was propounded partly in response to allegations that there are no laws in biology.)

We might want to press the investigation still further and argue, through exclusion, for the nonexistence of laws of nature in all the sciences. Here, however, we shall attempt a more modest task of defending the lawfulness of biology, arguing that if one accepts that there are laws of nature in other sciences, denying that they exist in biology creates some very unpleasant dilemmas. Furthermore, we shall give some examples of laws in biology which, although not exactly what everybody had expected, seem to be perfectly legitimate. To make things worse, however, we shall then proceed to show that the historical aspect of biology, and of other sciences as well, plays a far more important role in our understanding of the universe than has been generally recognized.

From the outset we must emphasize the point that discussions on whether there are laws of nature in biology have involved confusion about quite a range of issues. A major theme has been whether the biological sciences differ in some fundamental way from the physical sciences. There are several logical possibilities here. First, there might be laws in both the physical and the biological sciences. Second, there might be laws in the physical sciences but not in the biological sciences. Third, there might be laws in the biological sciences but not in the physical sciences. And finally, there might be no laws in either.

Accepting the last alternative (no laws at all) might please certain philosophers who don't like the notion of a law; but that would merely sweep the problem under the rug, because the apparent orderliness of nature would not go away. It is important, however, insofar as one is driven to it, if certain arguments are taken to their logical conclusion. The third possibility (that biology has laws but physics does not) is an amusing one, and if shown to be true, it would make a lot of physics-minded philosophers look silly, but otherwise there is no obvious reason why anybody would wish to entertain it except as a mere logical possibility. So that leaves us with evaluating the relative merits of the first and second possibilities, but deciding between them is not a straightforward matter.

The ulterior motives of those who wish to maintain that there are no laws of nature in biology would seem to be two in number. First, there are those who want to have it believed that physics is a science, whereas biology is not. Second, there are those who wish to have it believed that although biology is a science, it is somehow different, and thereby to establish its so-called "autonomy." Or speaking pejoratively, one might wish to say that the aim is to establish its "provinciality" (Munson, 1975; M.B. Williams, 1981). On the one hand we have those who maintain that all one needs to know to understand biology is the laws of physics and perhaps some history. In that case, there would be laws of physics and chemistry in biology but no "biological" laws. A lot of "reductionism" takes this guise. On the other hand, one might want to establish that there is something about living matter that exempts it from the physical or chemical laws that apply to atoms and molecules. That certainly smacks of vitalism.

Now, somebody who truly attempts to maintain that there are no laws in biology has to account for all the uniformities among living organisms. Having eliminated laws of nature, there would seem to be two, and only two, possibilities left: contingency and miracles. The first of these means that everything is purely a matter of historical accident, or of what is so often and so misleadingly referred to as "chance." If we prefer the second, we in effect opt out of science altogether, and for some persons this has been an attractive alternative.

It seems unlikely that any biologist in command of his senses would really disbelieve in the applicability of the laws of chemistry and physics, whatever these laws may be, to living systems. Even when we are not working in the laboratory, we presuppose their applicability to ourselves and all the creatures around us. As a marine biologist, I have spent a lot of time studying animals under water. When using SCUBA equipment, it is very important that one not attempt to defy the perfect gas laws. Unless one really wants to commit suicide, it is absolutely crucial not to hold one's breath when heading toward the surface, but rather let the expanding air flow out of the lungs. Irrespective of how others may choose to conduct their lives, nobody is going to persuade me that air embolism is no real problem, on the grounds that there is no logical justification for induction, or that these days laws of nature are out of fashion among philosophers!

So we can pretty well narrow the legitimate controversy down to a matter of whether any laws of nature exist that are strictly biological, in the sense that they apply to living entities and nothing else. Here two different claims might be at issue. First, one might deny that there are any such biological laws. Second, one might admit that such laws may exist, and indeed even claim that they do exist, but nonetheless deny that they are known to us or even that they are knowable. These two issues often get confused, and the confusion is compounded when the argument gets translated into a weaker form, namely that in biology laws of nature have relatively less importance in biology than they do in physics or chemistry. If that were the sole issue, then there would be very little to argue about once some basic points had been clarified. In the first place, it makes a lot of sense that the objects studied by physicists and chemists, being simpler and often more readily manipulated experimentally than those studied by biologists, might have more readily discoverable laws. In the second place, physics and chemistry are non-historical sciences, whereas biology deals with both laws and history, so that the comparison is not very accurate. If we compare biology with physics plus astronomy, or with chemistry plus certain of the earth sciences, the contrast becomes far less pronounced.

Making all of these distinctions should save us a considerable amount of work, because they simplify much of what was said in earlier discussions, especially those between Mayr and me (Mayr, 1987a; Ghiselin, 1987e, 1987f). We do need to emphasize how the individuality thesis helps us to understand what the laws of biology are, and how they function in the thinking of biologists. We have repeatedly urged that the laws of nature are strictly universal. There are no laws "for" or "about" or "making necessary reference to" any individual. All of them are "about" classes of individuals. In physics + astronomy this seems pretty straightforward. The laws of nature here are all "about" celestial bodies in general, about planets and galaxies in the abstract. They are not "about" the concrete, and therefore individual, instances of such classes, such as the Earth or the Milky Way. For that matter, there are no laws for the Universe or for its Origin, the one being, as it were, the most incorporative individual in the class of everything that exists, the other an individual event.

Indeed, cosmological speculation during the last few years has suggested that many of the features of matter as we know it are somehow the consequence of historical accident. Other universes that might have existed, or even that may now exist outside of our own, could perhaps have elementary particles with quite different properties, so physical necessity might be quite different. With respect to the issues under discussion here, I see this as only a complication, implying only that the most fundamental regularities have to be conceived of at a higher level of abstraction than that to which we are accustomed.

Switching now to biology, it seems intuitively obvious that there are no laws for such individuals as you and me, or for any other organism that has ever lived or ever will. What has not seemed intuitively obvious to just about anybody is that there are no laws for individual species, or, for that matter, for any taxon in the Linnaean hierarchy that is a natural group in the sense of a population or a clade. And this of course is because people have treated such entities as if they were classes rather than individuals. Because artificial groups, such as Invertebrata, have individuals as their basic components, with just some of the parts abstracted, there are no laws of nature for them either. The mere fact that there are laws for some classes does not imply that there are laws for all of them.

Furthermore, there should be no laws for all sorts of other things, such as an individual extinction event or the origin of tellurian life. It is easy to forget that individuals need not be substances. All this creates considerable difficulty for sciences that perhaps have not fully appreciated the historicity of their subject matter: various kinds of economics and anthropology for example. And cosmology.

But let us get back to the topic of the biological laws themselves. Although there cannot be laws about individual organisms, there can be laws about classes of organisms, and the classes in question might play an important role in physiology, ecology, or functional anatomy. Such laws would of course have to be applicable to all organisms past, present, and future, and indeed throughout the universe. One might, for example, want to compare big organisms with small organisms, and evolutionary biologists in particular have identified a lot of trends that occur with increasing body size (Rensch, 1959). The graviportal limbs that have evolved convergently in large mammals and in large archosaurs are a good example. But there is something of a question as to whether we should treat these as straightforward applications of physicochemical laws, rather than as biological laws per se. Likewise all sorts of laws may apply at a higher level of abstraction to organisms and other organized beings, such as machines. Birds, bats, insects, and airplanes have features that provide for balancing and lift.

We might better look for laws of nature that govern all "organized be-ings" that make use of resources—that is to say, for "economics" in the broad-est sense of that term (Radnitzky and Bernholz, 1986; Radnitzky, 1992). All such systems, whether they are organisms or economic units composed of or-ganisms, or for that matter machines and groups of machines or mixtures of or-ganisms and machines, display all sorts of regularities that seem not to be simply derivative of physico-chemical laws, although of course they are con-sistent with them. For instance, lowering the cost of carriage increases the ex-tent of the market, or, in more laymanesque terms, when transportation gets cheaper there are more potential customers. And, as Adam Smith (1776) made clear, the more extensive the market, the greater the potential for a division of labor. Larger organizations can have greater return on investment when they have a greater number of specialized parts, and it doesn't matter whether the organizations in question are organisms, insect societies, or universities.

One should also be able to find laws of nature that are about classes of species, and about other classes of populations, as well as for the kinds of processes in which such entities participate. Also there should be laws for other kinds of supraorganismal wholes, such as classes of languages. Indeed there should be laws that treat both species and languages at a higher level of ab-straction. As a straightforward example, let us consider what I have called "Mayr's Law" both to honor its leading proponent and to present him with something of a dilemma (Ghiselin, 1989). The thesis that Mayr's law is indeed a legitimate law of nature has been endorsed by both philosophers of biology (Ereshefsky, 1992) and philosophical biologists (Eldredge, 1993). Simply stated, it asserts that under ordinary conditions, speciation cannot occur in sym-patry. We have already explained the rationale for believing that this general-ization is true. And it would seem that the scientific community accepts it. Even such ingenious advocates of sympatric speciation as Bush (1975, 1981) do not contest the law; rather, they seek to show that the extraordinary conditions are frequently realized. And in the case of speciation by polyploidy, there are plenty of well-substantiated examples.

One might want to argue that Mayr's law is different from those in the physical sciences, because all it does is say that certain kinds of events can never happen in the real world. And yet the first and second laws of thermody-namics tell us that energy can neither be created nor destroyed and that nobody can possibly construct a perpetual motion device. If they told us nothing more, they would still be laws.

Sometimes the opposite complaint is heard: that laws in biology only assign relative probabilities to certain kinds of events. Certainly some of them do at least that. Consider, for example, the laws that govern genetic drift and other kinds of "sampling effect" in population genetics. I long ago analogized the fluctuations that occur in small populations to the Brownian movement, a sort of bouncing around of small particles such as soot or pollen grains as the result of being buffeted by molecules, which is easily seen under the light microscope (Ghiselin, 1969d). As genes get transmitted from generation to generation, their frequencies are partly a matter of chance, and they fluctuate from generation to generation, producing changes in addition to those that result from mutation and selection. The fluctuations are particularly pronounced in very small populations, and in the limiting case of a species of self-fertilizing hermaphrodite reduced to a single diploid organism, the number of alleles per locus can be no larger than two. To cast this principle in the form of laws of nature, we need only relate the fluctuations and the fixation of alleles to the effective population size. Now, although one cannot predict which allele will happen to be fixed, nonetheless for any locus the probability of the number of alleles decreasing becomes greater as the effective population size is reduced. This is a perfectly good law of nature, and it functions as such in modern population genetics irrespective of whether biologists call it a law or not. Does its probabilistic character set it off from the physical sciences? By no means. The mere fact that population genetics is more like statistical mechanics than celestial mechanics does not make it any less like physics. Or any less mechanical.

If one wants biological laws that predict what will indeed happen, and happen of (physical) necessity, if certain conditions are met, these too are by no means far to seek. In the absence of selection pressure, evolution is inevitable, and it occurs in the direction of physiological degeneration. This because of mutation pressure. Were things otherwise, the second law of thermodynamics would be false, for replication cannot possibly be perfect, especially in the long run. One might want to ponder whether this law is biological, physical, or both, but at least we can rule out the possibility that it is neither biological nor physical. I would say that it is biological, for it refers to populations in the genetical sense of that term.

There is some question about classical genetics, but the thesis that it contains legitimate laws of nature has found some able supporters (Ruse, 1970; Rosenberg, 1989). Population genetics might furnish some good examples of laws, but once one gets into the details everything becomes very tricky and technical, even for a population geneticist. The genetical laws that interest us may be the sort of thing that need to be formulated at a higher level of abstraction than geneticists commonly deal with. But certainly, the evolutionary population genetics of Haldane (1932), for example, contains a lot of what look like laws to me. Van Valen (1973) authored a well-known paper on what he called "A new evolutionary law." Among philosophers of biology, Reed (1978) has argued for the lawfulness of evolutionary biology, as has Mary Williams (1992), who gives a fine example. She compares the laws of biology to the perfect gas laws (Williams, 1973).

Some questionable areas notwithstanding, it should be obvious that providing a massive list of biological laws would be a mere chore. Rather than do that, it will no doubt prove more enlightening to consider some well-chosen examples and to expand the scope of the discussion. Therefore I have chosen to consider some of the laws that govern evolution in sexually reproducing populations, but without treating them at the level of the genotype, and ones that are familiar to me from having worked with them.

It is hard to find anything even remotely resembling sex in nonliving systems, so we have less of a problem with respect to whether the laws in question are truly biological ones. Moreover, Dupré (1993:83) has suggested that "the way in which the basic sexual categories—male, female, neuter, hermaphrodite—divide the natural world tells us nothing about either the extent to which such categories will give rise to general laws, or, more importantly, what will be the range of application of whatever interesting laws do involve those categories." It seems to me that in his zeal to debunk anything that stands in the way of his sexual egalitarianism, he overlooks what some of us biologists have already accomplished.

I should emphasize that my own work on reproductive strategies was the direct consequence of reading all of Darwin's books and then writing one of my own that explained their epistemological rationale (Ghiselin, 1969d). In the course of that study, I was most impressed with Darwin's theory of sexual selection, and my own work no doubt helped to rekindle interest in that long-neglected topic. Sexual selection soon became quite popular, as can be seen from an outstanding review by Andersson (1994) which provides approximately 2,500 references. Even so, the strategic importance of Darwin's theory of sexual selection as an integral part of a more general theory has not received the attention that it deserves. The fact that it is so often treated as a particular kind of natural selection is compelling evidence that Darwin's theory has yet to be fully appreciated by the scientific community.

Darwin divided selection into three modes: artificial, natural, and sexual. These terms derive from whether differential reproduction results from the choice of the breeder, the struggle for existence, or from the organisms' efforts to attain a monopoly with respect to mates. Sexual selection, whether with males fighting over the females or competing with respect to their attractiveness, has more in common with artificial selection than it does with natural selection. And this is born out by the fact that in Darwin's theory the ornamental poultry produced by the breeder parallel the brilliant plumage of birds of Paradise. Sexual ornaments and weapons are mere instruments of reproductive monopoly: unlike, say, genitalia that function in transferring sperm, they have no real utility in the struggle for existence as such. And indeed they can be downright maladaptive from the point of view of the species, which is precisely the point: in Darwin's theory, nothing is done because it is advantageous to the species.

So, far from being an ad hoc hypothesis that was supposedly devised to explain away facts that cannot be explained in terms of natural selection, Darwin's theory of sexual selection is a crucial element in a more general theory of selection. It accounts for three distinct classes of phenomena on the basis of differential reproduction, but specifies that they will occur under different circumstances and predicts different outcomes for each of them. And the data of experience provide a beautiful confirmation from the point of view of unexpected and otherwise inexplicable predictions that derive from what are really quite simple propositions. One could hardly wish for a better example of theoretical elegance in science.

Another beautiful example of such reasoning is to be seen in theories of the sex ratio. Under ordinary circumstances, but ones that make no sense whatsoever except in terms of the theory of natural selection, the ratio of males to females tends to be around one to one. Fisher (1930) explained this on the grounds that fathers contribute just as much to the ancestry of future generations as mothers do, and scarcity of the rarer sex leads to a kind of economic balance between son-producers and daughter-producers. Hamilton (1976) developed this line of reasoning by considering the circumstances that are out of the ordinary. Namely, when there is mating among siblings, the females contribute proportionately more, and the ratio of females increases with the amount of inbreeding. Sex-ratio theory is a particularly difficult and technical area, and rather than try to explain it, we may merely refer interested readers to the general literature (Charnov, 1982 and references therein). The philosophical point to be made here has rather to do with the logic of selection theory. The detailed predictions in deviation of the sex ratio from unity stand to the general theory of selection, as the detailed predictions in the deflection of photons as they pass by the sun stand to the theory of relativity. I make this comparison precisely because Popper (1962) used such phenomena as an example of the hypothetico-deductive scientific method at its best, to underscore the irony with respect to that great philosopher's unfortunate difficulties with the theory of natural selection. Indeed, when Popper (1977) recanted, he said that sexual selection had changed his mind about the testability of the theory. Darwin's theory is every bit as well-established as Einstein's, and if either of them lacks laws, we had better say the same of both.

Getting back to the particulars of reproduction, my own research began at a time when the laws that govern the evolution of hermaphroditism were beginning to receive some attention. According to the traditional "low-density model," organisms which are simultaneously male and female are at a selective advantage if opportunities for mating are few. In the case of organisms that encounter a conspecific only once in a lifetime, and in which the sex ratio is 1:1, it should be obvious that half the encounters will be between males and males or between females and females, rather than between males and females, so that half the organisms will be unable to mate. The probabilities can of course be calculated for different population densities and for different sex ratios, though such technicalities need not be considered here (see Tomlinson, 1966). All we need to say is that in a population affected by selection for simultaneous hermaphroditism, the frequency of hermaphrodites will increase as population density decreases, to justify treating the low-density model as a law of nature. Of course, as with other laws of nature, one can present it in more elaborate form and with all sorts of qualifications. Sex-allocation theory includes ways of explaining, for instance, why simultaneous hermaphroditism is disadvantageous. Having fewer reproductive organs economizes on resources by reducing what economists call "fixed costs" (Heath, 1977).

The "size-advantage model" was evoked by my realizing that although the low-density model explains simultaneous hermaphroditism, in which the organisms are both male and female at the same time, it fails with respect to sequential hermaphroditism, which means switching sex, generally either from male to female (protandrous hermaphroditism) or from female to male (protogynous hermaphroditism), although there are other possibilities such as male to both male and female. In attempting to test an hypothesis that turned out to have little if any value, I came across a study on sex-switching fishes, in which it was incidentally observed that the males were brightly colored. From my work on Darwin I immediately recognized this as a likely sign that sexual selection had been operative. During my research on phylogenetics, I had familiarized myself with the work of Bernhard Rensch, already mentioned above, on the laws of nature that have to do with body size. Realizing that larger animals are at an advantage when fighting, it occurred to me that the younger fishes were reproducing as females until they became large enough to defeat males in sexual combat.

At the time I discovered the size advantage model, I was very much aware of the laws of nature that govern such investigations, and my whole research program at the time was consciously guided by my understanding of those laws. But that is a story in itself, and should not divert our attention from the basic theme. The size-advantage model is about as simple an hypothesis as one can ask for, and that is a very good thing. It has been developed so as to explain protandrous, as well as protogynous, hermaphroditism. It has proven highly tractable to mathematical treatment and generates remarkably precise predictions. It has been extensively tested both in the laboratory and in the field and has taken its place along with the sex ratio in modern "sex-allocation theory" (Charnov, 1982). However, its nomothetic character, including the extent to which it fits in with a larger body of theory, seems not to have been widely appreciated. This in spite of clear and explicit statements to this effect by Charnov (1982, 1993). So we need to consider some connections at a higher level of generalization.

Both the low-density and the size-advantage models relate the mode of reproduction to the frequency of encounters between reproducing conspecifics. In the case of the low-density model, the difficulty of finding a mate becomes an increasingly overriding influence as effective population density goes down. I say "effective" density because highly motile organisms like ourselves have less of a problem than slow-moving organisms like slugs and especially attached ones like barnacles. In the case of the size-advantage model, on the other hand, with increasing effective population density the frequency of encounter goes up, and this increases the amount of competitive interactions, especially the sort of male combat that leads to the evolution of great bodily size and weapons such as tusks and horns. Darwin fully recognized this, and argued that such secondary sexual characters as are serviceable in male combat are best developed in animals that breed in large herds. Elephant seals, which breed in dense aggregations, are a fine example. They are much more dimorphic than, say, Hawaiian monk seals, which breed at much lower density.

Darwin seems not to have realized that the same principle of diminished opportunity for sexual combat and analogous forms of reproductive competition can be extrapolated in the opposite direction, and explain the kind of "dwarf males" that he discovered in barnacles (but see Darwin, 1873). With decreasing effective population density the males not only cease to be larger than the females but actually become smaller. This because there are fewer and fewer opportunities to monopolize the females by fighting, and the trick is just to find the females as soon as possible. So the males mature younger and smaller in proportion as the effective population density goes down. There are beautifully graded series of such creatures among parasites and inhabitants of the deep sea, precisely where one might expect them. In some deep sea anglerfishes, both sexes descend from the shallower waters where they spend the earliest part of their lives, but only the females continue to feed and grow, the males immediately becoming sexually mature, mating with an older generation of females, and perishing. In others, the females actually keep the little males alive, and their bloodstreams are actually connected. It would seem that the problems of finding a mate have become so accentuated as to make the males a sufficiently valuable resource that it pays the females to support them. In some parasitic marine gastropods and cirripedes, a similar trend in size reduction occurs, but the end result is even more spectacular: the males are reduced to mere gonads within the bodies of their mates. (For further details see Ghiselin, 1974a, 1987a.)

Thus from quite simple and straightforward principles or mechanisms, we can derive some very general statements that make no reference to any individual taxon or to any particular moment or place, and are necessarily true of everything to which they apply. Whether such statements are enunciated formally or informally, whether they are qualitative or quantitative, and irrespective of which quantities are intensive and which are extensive, we have to recognize that they function as laws of nature.

One of the most important functions of laws of nature in all sciences is that they allow us to envision a whole range of otherwise disconnected phenomena from a unitary point of view. That is what monism and synthesis are all about. Such unification is not just ontological, for it has powerful epistemological implications. If the laws of nature tell us what can and cannot happen, irrespective of time and place, then they ought to be very useful in working out the history of life on earth, and of the earth itself, and of everything else. In the chapters that follow we shall examine how this is accomplished.

15

The Principles of Historical Inference

"If man had not been his own classifier, he would never have thought of forming a separate order for his own reception."

—Darwin, *Descent of Man*, first ed., Vol. 1, p. 191

When philosophers discuss the nomothetic disciplines, they continue to emphasize physics. If this practice has its disadvantages, at least they can draw upon some straightforward examples of what is generally agreed to constitute good science. In discussing the historical sciences, on the other hand, the tradition has been to discuss the opinions and practices of historians in the narrow (academic) sense (Dray, 1957; White, 1965; Walsh, 1967; Danto, 1985). Although this literature is by no means uninteresting, the examples it provides are neither clear nor particularly illuminating, let alone representative of the natural sciences that interest us. Herein we spurn that tradition and do the exact opposite, drawing upon the earth sciences, with their well-earned prestige, and their many and straightforward examples of historical reasoning (Laudan, 1992).

Geology, including paleontology, provides a sort of bridge that connects astronomy to such biological sciences as comparative anatomy. But in drawing attention to that very connection, we need to emphasize that there is nothing either ontologically or epistemologically privileged about the status of geological and paleontological materials. The notion that geology and paleontology provide "direct" evidence for the history of life on earth has been mindlessly repeated in countless textbooks. But such claims have nothing to do with the actual procedures that have been applied in the course of research. Earth scientists have had to reconstruct the history of continents, strata, oceans, and extinct creatures from very subtle, and often quite fragmentary, evidence.

In this process of historical reconstruction, certain principles are applied and certain assumptions presupposed that are not always explicit. When dealing with history, we tacitly assume that the objects that interest us are individuals, rather than classes, whether or not we give it much thought. And that makes a great deal of difference. Simply because individuals are spatio-temporally restricted, they must occupy one position or another in space and time, and they must do so continuously throughout their existence. While they exist, they must be somewhere, for they cannot be everywhere, nor can they be nowhere. Such ontological necessity definitely limits the range of logically possible histories. The "living fossil" mollusk *Neopilina* represented a taxon that was known only from Lower Paleozoic rocks until it was dredged up by the Swedish Deep Sea Expedition in the 1950s (Lemche, 1957; Lemche and Wingstrand, 1959). The taxon (individual) must have been in existence during the intervening time span.

It is not just the individual organisms and lineages that are spatio-temporally restricted, but also what they do, such as feed, reproduce, and speciate. Although many species are said to be "cosmopolitan," these are widespread rather than omnipresent. To my knowledge not a single metazoan or metaphyte species has a normal range of distribution that extends all the way from the deepest trenches in the ocean to the highest mountain peaks. And the activities in which they and their parts participate have to be quite local. Simply because speciation events are a relatively local matter, lineages are restricted to definite geographical areas. Were taxa classes rather than individuals, we would not have such characteristic faunas as that of Australia which obviously have evolved *in situ*. The most compelling argument for evolution is the biogeographical one, and this argument makes sense if and only if species and monophyletic higher taxa are individuals. Likewise the pattern we get in the coevolution of parasites with their hosts, with the genealogy of host and parasite showing remarkable correspondences, makes sense if and only if the association itself forms an individual at a supra-organismal level (see Szidat, 1956; Brooks, 1988; Brooks and Brandon, 1988; Brooks and McLennan, 1991).

Individuality again provides the rationale for rejecting "anachronisms," or putting an individual into a temporal position in a reconstructed history either before or after it really existed. From another point of view, the ability to identify an individual means that its presence at a given time or in a given place allows one to identify that time or place, and to associate any other individuals that occur together with it as contemporaries. This is why certain fossil taxa that had fairly short periods of existence, but have characteristic features that allow them to be identified with a high level of reliability, have been so useful in stratigraphic correlation. The presence of such a species in two widespread fossiliferous deposits clearly shows that the deposits were formed at approximately the same time. Deposits of mineral "species" are obviously less useful for the very reason that they need not have this individualistic character.

Much of the success of modern geology is rightly attributed to the principle of uniformitarianism which, again rightly, is largely attributed to the contribution of Charles Lyell (1830–1832). Here is not the place to go into the history or the subtleties of that principle; I only mean to show how it makes much more sense in the light of the individuality thesis. Crude notions of uniformitarianism—and Lyell's views were rather primitive—tended to presuppose that the earth exists in a "steady state" with change being only local. And indeed it was influenced by Aristotelian thinking, which envisioned the world changing cyclically with materials being eroded in one place and deposited in another, much as water in the hydrological cycle is seen to evaporate in one place and be precipitated in another. (See Aristotle's *Meterologia*.) It also got associated with the notion of gradualism, or the de-emphasis, at least, of sudden and cataclysmic upheavals.

In more modern interpretation, however, uniformitarianism is taken to mean that the laws of nature are uniformly the same irrespective of time and place, so that what is necessarily true today had to be necessarily true even in the most distant past (see Gould, 1965). So if one is going to reconstruct the past, the reconstructions must not contradict any legitimate law of nature. Given that stricture, it would be perfectly legitimate to develop an historical account of the history of the earth in which during earlier times temperatures were warmer, tidal amplitude was greater, and the composition of the atmosphere was quite different, all of which are in fact widely maintained by paleobiologists. Our planet, being an individual, can change. And if one wants to reconstruct its history properly, one should not treat mere historical contingencies as if they were legitimate laws of nature.

The extent of such confusion at the present time is hard to assess, and I only want to draw attention to one contemporary example. This is what is sometimes called the "principle of taxic uniformitarianism" as invoked by some paleoecologists. If one wants to reconstruct a fossil animal and perhaps thereby infer its relationships to other organisms in its environment, one source of information might be its closest living relatives. By experiment and careful field observation, one might find out a great deal, say about what the organisms eat and how they find their prey, and one might reasonably ask whether the extinct relatives did likewise. But from the mere fact that all the extant "members" of a group have a particular kind of food, what could we conclude about the extinct forms? Nothing, for taxonomic generalizations are not laws of nature, and groups of organisms are perfectly capable of evolving new diets. The panda, for example, belongs to the mammalian order Carnivora but is a strict vegetarian. Likewise, although all the extant Cephalopoda of which I am aware are predators, there seems to be no reason why all the fossil ones must have been predators. A factitious appearance of uniformity in diagnostic features results from systematists screening out the characters that vary; this sort of artifact will be discussed in Chapter Seventeen.

If one wants to infer the mode of life and other aspects of the biology of fossil organisms, one had better have something more reliable to go upon than taxonomic diagnoses. And indeed we can do much better. There is a substantial literature on the techniques whereby one infers function from structure in fossils (e.g., Huxley, 1856; Falconer, 1856; Rudwick, 1964; Bruton, Jensen & Jacquet, 1985; Fortey, 1985; Hickman, 1988). However, the principles employed are not fundamentally different from those used by functional anatomists working with extant organisms, except that certain kinds of evidence are not available. One formulates an hypothesis and tests it by finding out whether or not the organisms under study have the sort of properties that it implies. And this means their fitting in with generally accepted engineering principles and laws of nature. If, for example, one wants to reconstruct the alimentary biology of a fossil animal, one looks for things like teeth and asks whether they have the appropriate properties for dealing with various kinds of food items. Teeth used in crushing, for example, are stout and flat, whereas those used in cutting and shearing are sharp and bladelike. Furthermore the position and arrangement of teeth correlates very closely with the mechanics of the jaw, and one can tell a great deal by calculating the forces that could be exerted on the food or the prey.

Someone once asked me how it is that we know something to be a tooth, except by homologizing it with a tooth in some particular taxon. The answer is quite straightforward. Organisms, irrespective of taxonomic position, have arrangements of parts that agree with the laws of physics and biological organization. Teeth do not occur at random in the bodies of animals, but rather where they are most effectual in subduing prey and chewing food. For that reason they occur in the mouth, rather than the anus—sea-cucumbers with somewhat tooth-like anal plates notwithstanding. We need not belabor the nonissue of how we tell the mouth from the anus. The point here is that our inferences from structure to function are derived from laws of nature, which again are true irrespective of time and place, and not from the sort of taxonomic information that is purely historical and contingent.

But before we do leave the topic of the alimentary canal, we might dwell a bit on the "directionality" of bowel movements. There is a definite tendency for materials to move in one direction through the digestive tract, but as the example of vomiting makes clear, this process is not absolutely irreversible. Indeed organisms such as sea-anemones, in which an anus has never evolved, and some brachiopods, in which the anus has secondarily been lost, take in food and eject waste through the same orifice. Yet a paleontologist would never reconstruct a fossil so as to have it take in food via the anus and defecate it via a buccal cavity armed with strong jaws and teeth.

A lot of natural processes involve change that occurs only in a particular direction, and some of these indeed have a definite and constant rate. Both are important in establishing chronologies, although only the latter gives absolute dates. In dating rocks, geologists take advantage of the fact that radioactive decay takes place at a constant rate. If a rock is formed with a known amount of a given isotope that changes at a particular rate into a second isotope, then the ratio of the two can be used to calculate the time that has elapsed since the rock was formed. Here the laws of nature are such that we can rely upon the constant rate of change from one state to another as a means of calculating absolute dates.

On the other hand, not every change from one state to another has rates that are set by laws of nature. The so-called "molecular clock" resulted from an overly optimistic hope that branching sequences in evolution could be calculated that way. DNA x DNA hybridization is a technique that gives an index of "similarity" for two strands of chromosomal material. It works by the tendency of such materials to pair up more closely the more that they have in common, but that does not really matter for this discussion. What does matter is that as evolutionary time passes, the chromosomes evolve, beginning with a state of virtual identity at common ancestry. So as time passes, the percentage of homologous sites that are identical decreases, and hybridization provides a crude index of elapsed time. It is therefore very much like isotopic dating insofar as it depends upon a change in the configuration of matter from one state to another. However, the change in question does not occur at a constant rate, which is in large measure a matter of contingent fact. The "clock speed" itself evolves. That people were a bit overly optimistic about DNA x DNA hybridization and other molecular techniques may have resulted in part from the tacit assumption that the clocks are classes rather than individuals.

Stratigraphic geology has relatively straightforward ways of establishing a relative chronology. If one is attempting to establish which among layers of rock was deposited first, the usual interpretation, based upon the principle of superposition, is that the older rocks are always at the bottom. And indeed this is the relationship of the vast beds that erosion has exposed in the Grand Canyon of the Colorado. (Geologists often speak of a "law" of superposition, but "principle" is preferable because it seems to be less a law than a kind of methodological guideline for applying laws.) The principle of superposition finds its rationale in the manner in which deposits are understood to be formed, something that can readily be justified in terms of commonly accepted laws of nature. The strata are formed from materials that are carried about by wind and water, then settle and get consolidated. It is not physically possible for layers of sedimentary rock to be formed either under or between other layers of rock. So they have to be formed, at least, on top.

It is physically possible, however, for layers of rock to be displaced in various directions, so that under some conditions the older rocks might overlie younger ones. Here, criteria in addition to the principle of superposition may have to be applied. For example, entire beds are sometimes overturned, so that the sequence is reversed and the rocks become younger, not older, with increasing depth. In some cases one can find evidence that this has happened by tracing strata from place to place as if they were folded rugs. A more instructive kind of evidence, however, is of a more intrinsic character, and an entire book has been devoted to this topic (Shrock, 1948). One can look at a stratum and see clear evidence that it has been overturned. The laws of nature impress a distinctive pattern in relation to the direction of gravitational attraction and other formative influences. For instance, materials tend to sort out according to their relative size and density, the larger pebbles settling out before the smaller ones, sand before silt, and so-forth. Often one can see clear signs of orientation in the form of "varves" or layers that were deposited at a single time. Sometimes lake beds dry up, and the mud that is exposed to the air forms characteristic cracks; when the lake refills, and the sediments fill the cracks, they can impress upon the rock a clear indication of which way was up at the time when it was formed.

We should emphasize that in such inference we are not trying to show that a particular interpretation of the strata is true because it seems consistent with a given law of nature, but rather to show that some alternative is not consistent with such a law. Furthermore, such laws are not like the a priori deductions of mathematics. They are induced from experience, they might turn out not to be true, and for any number of reasons one might go wrong in trying to apply them. Let us briefly consider two very famous examples, first the effort of Lord Kelvin to estimate the age of the earth (Burchfield, 1975). His estimate depended upon the assumption that the sun's heat depends upon the gravitational attraction of the matter that it contains, and the figure he got was much lower than advocates of evolution, especially Darwin, felt comfortable with. Kelvin of course did not know that the sun's heat is produced by nuclear fusion. The other notorious example is the falsification of continental drift on the grounds that it seemed most implausible for continents to plow their way through sheets of rock. The architects of the modern theory of plate tectonics were able to show that this was by no means what happens. The sea floor spreads and gets subducted, and the continents move together with it, rather than plowing through it. (Hallam, 1973; Glen, 1982; Menard, 1986.)

In both of these cases, biological theory seemed to be out of line with theory that was accepted in the physical sciences as then understood, and the lesson that ought to be drawn here is that when refuting hypotheses, we should draw no rash conclusions with respect to what it is that has been refuted. But that by no means negates the legitimacy of the approach. The premises in such reasoning are not "assumptions" but hypotheses and, like all hypotheses, are tentative and subject to reconsideration.

In attempting to work out the history of life, biologists have invoked quite a variety of laws and principles. Some of these have indeed been discarded as investigation has proceeded, but others have long remained in service. It will be unnecessary to give more than a few illustrative examples here to explain the rationale of such reasoning. On the other hand, getting biologists to face up to reality here may prove a daunting task. So heavy is the weight of tradition with respect to such possibilities, and so entrenched the opposition to even considering them, that they have yet to receive a fair hearing.

To begin, we may consider a law or principle that is generally attributed to Étienne Geoffroy Saint-Hilaire (1818; 1830) and played a major role in his celebrated debate with Cuvier, as it has in comparative anatomy ever since (see Isidore Geoffroy Saint-Hilaire, 1847; Cahn, 1962; Appel, 1950). This is the proposition that although organisms may change their proportions and organs their functions, the anatomical connections or relative positions of parts are immutable. As a categorical generalization that asserts that no rearrangements ever occur, this would make a fine law, but it is not true. Parts do change their connections. One might nonetheless treat it as a legitimate law if equipped with appropriate *ceteris paribus* qualifications, statements of boundary conditions, and the like. Changes of connections are improbable, partly because they tend to abolish function, an important point that we shall discuss in due time. And particular kinds of connections, especially those between nerves and the organs that they innervate, do seem to be particularly stable. So it is at least possible to formulate criteria of justification for minimizing such changes in terms of laws of nature.

And indeed we do have such arguments. They form the conceptual underpinnings of some of the most elegant phylogenetic research that has ever been done. This is the phylogenetics of *Drosophila* based upon the study of chromosomal inversions. (Classical literature includes Sturtevant and Dobzhansky, 1936.) The mapping of chromosomes in these animals revealed that sections of chromosomes had come to have an "inverted" order, and by comparing the maps it was possible to find included and overlapping stretches of chromosomes, each of which could be interpreted as unique innovations, or what have come to be called synapomorphies. There are compelling reasons to believe that each of these inversions is so improbable that the underlying event that produced them can only have happened once. To produce an identical inversion, the chromosome would have to break and then re-attach at two absolutely identical points, and this is most improbable. Furthermore the conditions for the fixation of an inversion are very stringent, for selection disfavors them, and small population size is necessary. The reliability of the method is attested to by the lack of conflicts between different parts of the chromosome complement; in other words, because convergencies and reversions never happen, one does not have to resort to mere parsimony and simply minimize the number of such events.

The interesting thing here, from an epistemological point of view, is that we have gone from a law which tells us that certain events never happen, and which is used as a means of refuting certain possible historical hypotheses, to a law that leads us to reject all hypotheses save one, because the law predicts that certain hypotheses are so improbable that for practical purposes we can assume that we are dealing with unique events. Before proceeding, we should note that reasoning of this form has sometimes proved misleading. Some remarkably detailed cases of evolutionary convergence are known, so it behooves the phylogeneticist to have more than just a subjective criterion for believing that something can only have happened once.

It seems almost a truism that any hypothesized series of ancestors and descendants must have corresponded to viable organisms in real environments. Organisms must survive and make a living if they are to reproduce, and this is true of every generation. The sequence of ancestors and descendants cannot be broken by even a single generation that does not meet this criterion of continuity. So any clearly nonviable or grossly maladaptive hypothetical intermediate has to be ruled out. This canon has frequently been mentioned in the literature, for example by Butler (1982). Perhaps it would be invoked more often were its obvious violations not considered so preposterous as not to warrant explicit consideration.

How one decides what is and is not apt to have been viable depends upon various criteria, one of which was mentioned in an earlier chapter. Namely, the various parts of the body are functionally interdependent, so that if the result is to be a viable organism, any change in one part of the body has to be accompanied by adjustments in other parts of the body. Under ordinary circumstances this means that major changes have to take place over a long series of generations during which the parts are mutually adjusted, or, in other words, there can be no "macromutations" or "saltatory" evolution. We may observe that the exclusion of saltations on this basis follows from the invocation of a law of nature and not, as is sometimes claimed, from a metaphysical assumption that "nature makes no jumps." The physical (rather than metaphysical) support for excluding saltations is based upon the study of artificial breeding, under which it is indeed possible to produce "hopeful monsters," but these are only maintained with great difficulty by breeders, and their analogues in nature seem not to have produced any detectable lineages (Darwin, 1868). In order to apply such laws and principles, one must really understand them, and a great deal of contingent fact may have to be known if one is to do the job properly. For instance, new species can be formed in a single generation and with some impressive differences between parent and offspring by allopolyploidy. The descendants have complete sets of chromosomes from two different ancestral stocks. Such doubling of the chromosomes is accomplished without abolishing the harmonious functioning of the whole, so the law in question has not been violated. Rather, it does not apply.

Natural selection can only produce an organism that is adapted to past conditions or that happens to be adapted to present ones. The theory provides no mechanism for anticipating the conditions that will be encountered by future generations. And this generalization, which is readily understood as a law of nature, rules out all sorts of logically possible transformations in structure and in function. It is not sufficient that a change from one condition to another produce some advantage for it to evolve; all the intermediate stages must be adaptive as well. A vestigial part, such as a tooth in the jaw of a fetal baleen whale that never even erupts, must not be interpreted as preparing the way for whales to have teeth, but rather as a reminiscence of a past condition when the ancestors of baleen whales still had teeth.

Other laws of nature used in working out the history of life have to do with the narrow range of circumstances under which certain kinds of change are possible. Parts do not originate absolutely *de novo* and *e nihilo*. They arise from pre-existing parts. Therefore any reconstructed history for the evolution of an organ must allow for the origin of that organ from some plausible precursor, something that already existed in the right place and at the right time. And since the transformation of parts has a physiological aspect, any hypothesized precursor must have the appropriate functional properties as well.

The logic here was initially developed by Darwin, and especially by his follower Anton Dohrn, who developed a "principle of the succession of functions" whereby new organs evolve through the modification of pre-existing ones by change of function and consequent further modification (Dohrn, 1875; also translation and commentary in Ghiselin, 1994b). Unfortunately, the examples given by Darwin and Dohrn were often dubious and sometimes obscure, but then again there are plenty of others. Perhaps the stock example is the transformation of some of the bones that were originally elements in the jaw of vertebrates into ear ossicles (Rowe, 1996). The bones in question had to be in just the right place and get reduced at just the right time to allow them to shift from what was originally a supportive role to that of transmitting sound to the rest of the auditory apparatus. The very improbability of such a chain of events provides compelling evidence that the relationships among the various vertebrates have been correctly established. Nor should we overlook the point that here we have a sequence of transformations that would make no sense if it were read backward, with parts moving out of the inner ear and gradually getting incorporated into the jaw, for there would be no advantage to the intermediate forms in having their physiology changed in that direction.

At the molecular level, it has become a commonplace that evolution of new parts occurs by duplication of a pre-existing molecule followed by the functional modification of one of the two "paralogues." Here the stock example is respiratory pigments. Haemoglobin is thought to be modified cytochrome *c,* and the various haemoglobins, such as foetal haemoglobin and such modified haemoglobins as myoglobin, nicely fit a pattern of continuity of function between the original and the derived molecules and of nothing originating that is altogether new. In working out the history of such molecules and the organisms that contain them, it is not just our ability to trace them through lineages by homologizing the various sites in the molecule that compels our assent to the reconstruction, but the very lawfulness of the transformations.

The respiratory pigment haemocyanin, found in a wide variety of mollusks and arthropods, provides a less familiar but equally impressive picture of functional change with plausible precursors. These respiratory pigments evidently arose from the enzyme tyrosinase. As a student of invertebrate phylogenetics, I had hoped that perhaps the evolution of haemocyanin would provide an argument for a close relationship between mollusks and arthropods (Ghiselin, 1988b), but those who work on haemocyanins have tended to rule this out because the sequences are quite different (Van Holde, Miller, & Lang, 1992). On the other hand, the haemocyanins do support the monophyly of Mollusca and of Arthropoda, at least in part.

The idea that our phylogenetic trees should be based upon shared derived characters finds its ultimate justification in the improbability of innovations. And yet some kinds of changes are known to have occurred time and again. So the fewer the better, and that at least goes a long way toward justifying parsimony in evaluating the merits of alternative trees. However, any criterion that would tell us that some change is particularly improbable should lead us to prefer a tree that has it happen once and only once. The trouble is, how does one decide how much "weight" to assign to any such probability argument? There are various criteria here, but perhaps the most important one is that the more complicated the structure, the fewer times something like it will originate. And in spite of the protestations of some authorities, there is good evidence that complexity is what makes for a low probability of undetectable convergence, as Donoghue and Sanderson (1994) rightly point out. I say undetectable convergence, for repeated origins of complex structures from distant sources, such as the eyes of cephalopods and fishes, are generally accompanied by all sorts of minor details that betray separate origin. In this connection I should point out that the claims for extensive polyphyly among metazoan eyes made by Salvini-Plawen and Mayr (1977) were based upon the senior author's flawed premises about the animal tree (see Ghiselin, 1988b). Evidence from molecular genetics strongly supports an homology of cephalic eyes among a wide range of metazoan families (Quiring et al., 1994). Of course there has been a lot of parallelism in eyes and all sorts of other things, but this particular case has been grossly exaggerated, and to make things worse it is being taken up by textbook writers and vulgarizers (Gould, 1994).

More importantly, it is far easier to lose a complex part than to gain one. The truth of this proposition in many aspects of our lives is apparent from how much easier it is for someone who uses a computer to lose a file than to create one. If one thinks about such complex organ systems as the gastropod buccal mass, with its radula and immensely complicated musculature, or the lensatic eyes of vertebrates or cephalopods, it seems more than just intuitively obvious that such an organ is more readily lost than gained. So having such an organ evolve only once, but be secondarily lost, provides a more likely account of what actually happened than having it evolve twice, or having it be lost and then re-evolve, in any reconstructed history. Strathmann (1978, 1993) provides cogent evidence that the rather complex larvae of marine invertebrates have repeatedly and irreversibly been lost rather than being convergent, at least to the extent that some authors have been wont to maintain. In this connection it is worth emphasizing that when larvae have in fact re-evolved, as in the case of the tadpole larva in urochordates, they are quite different in structure from those of related lineages.

In many cases, parts can be lost without leaving a trace of their former existence. And yet, even where there are evident traces of a formerly existing part, there seems to be a great deal of resistance to invoking such an event. I say this after having considered the evidence for the former existence of segmentation in mollusks and several other groups of animals (Ghiselin, 1988b). One wonders to what extent such persons as will not acknowledge such loss are presupposing a metaphysical posit, namely that the groups in question are classes, and the features in question are defining. Time and again I have been asked, "Are mollusks segmented?" and have had to restrain myself from giving the correct response: "Don't ask silly questions!" The appropriate question is not whether they are segmented, but whether their ancestors were.

The fact that a part can be gained and then lost implies that evolution can go in either direction, even though evolution in one of these directions is more probable. Even if this is just a trend, it provides a criterion for weighting in cladistic analysis that is relatively uncontroversial (Harper, 1976; Farris, 1977; Wagner, 1982; Sanderson, 1993). On the other hand, there are certain kinds of changes that seem to be altogether irreversible. But the mere fact that changes seem always to go in one direction may reflect mere contingent fact, rather than a legitimate law of nature, and it behooves us to understand the causal basis of such apparent directionality before we invoke nomic necessity.

According to "Cope's Law" or rule, organisms always get bigger in the course of evolution; therefore any phylogenetic sequence has to be "polarized" according to the "states": body size small $\rightarrow$ body size medium $\rightarrow$ body size large. Although there are plenty of well-documented examples of phyletic size increase (Newell, 1949), there are also well-documented examples of groups that become smaller, especially those that have invaded habitats such as the interstices between sand grains where minification has some advantages (Swedmark, 1964; see also Hanken and Wake, 1993). Stanley (1973) pointed out that the common ancestors of new groups as a matter of historical fact turn out to be animals closer to the physiologically minimal size, so that the only direction in which their descendants can evolve is toward larger body size. So there is no law here that precludes minification, but instead the trend is a matter of contingent fact which may be explicable in terms of some other law. The enlargement might be explicable in terms of the laws that govern interactions among populations that diversify in an adaptive radiation, such that competitors evolve away from their closest relatives, and when, as a matter of contingent fact their competitors are small, they are driven in the direction of size increase.

However, there are other directional changes that would seem to be much better candidates for lawfulness, in particular the shift from external fertilization to internal fertilization (Remane, 1963). The phylogeny of Metazoa as currently understood admits of no exception to the rule that once fertilization has come to occur inside the body, it never reverts to happening outside the body. There may be some borderline cases that require some qualification if we are to have a good law of nature, but these would have to be limited to organisms such as sponges that are largely open to the seawater and do not really violate the principles involved. (It is rather like the "exceptions" to the perfect gas laws in which molecules combine with increasing temperature or pressure.) Certain other irreversible phenomena might be treated together with this one as special cases of a more general law of nature. The most obvious example is internal parasites, which are generally looked upon as having undergone a kind of "specialization" from which they cannot escape; although they may switch hosts, they cannot re-evolve the ability to live outside of one host or another.

When fertilization becomes internal, spermatozoa are transferred from the body of one animal to that of another, so that they no longer come into contact with the external medium. On the one hand, they develop new adaptations to life inside the body of the animal in which fertilization takes place, undergoing considerable morphological modification and also evolving such physiological adaptations as anaerobic metabolism. On the other hand, they lose the sort of shape and physiology that allowed them to function outside of the parental body. Adjustment to the conditions inside the body is by no means a major change; for the sperm are produced within the testicles. Not much really has to be gained. But a great deal may be lost, including the appropriate locomotory apparatus, the ability to maintain proper osmotic pressure, various metabolic pathways, and all sorts of other things. The origin of these features depends upon unusual historical contingencies, and the probability of all of them re-evolving is the product of each of them separately. To this we may add that moving the reproductive processes ever closer to, and then inside, the maternal body is always something that can be produced by selection, provided that there is some advantage to it. However, once the process goes on well within the body, a reversal would have to entail an advantage to each and every intermediate stage in the transformation. So in principle, at least, it would seem possible to recast the empirical rule in terms of laws, though admittedly I can only give a very preliminary sketch at present. On the other hand, some good theoretical work has been done in this area (Muller, 1939; Bull & Charnov, 1985).

The notion that evolution is always irreversible is commonly referred to as Dollo's Law, though it appeared in conjunction with other laws, one of which said that evolution is discontinuous and another of which that said it is limited (Dollo, 1893). Dollo's Law seems rather disappointing as a legitimate law of nature (see Oudemans, 1920; Cuénot, 1951; Gould, 1970). If we take it as a general principle that any evolutionary change is improbable, then the more changes that occur when a population is transformed in one direction, the less the probability that all of them will go back to their original state when the general course of evolution gets reversed. This statistical law is not very predictive, especially since it does not tell us which features will get reversed and which will not. And it says nothing in and of itself about any particular change. Furthermore the reasons why a particular reversal does or does not happen are quite diverse (lack of suitable genetical variants, and the lack of selection pressures, for example).

So if we consider the various movements that organisms have undergone from the sea to the land and fresh water, we find among other things that these have occurred a fair number of times, but almost exclusively in a direction away from the sea. Frequent claims that a particular group of freshwater or terrestrial animals has secondarily gone back to the sea have virtually all turned out to be flawed. Particularly remarkable is the complete failure of the insects to produce lineages that are truly marine. The putative exceptions about which one might want to quibble all turn out to be aerial creatures that have penetrated into the intertidal zone or moved out onto the surface of the water. Moreover, the secondarily marine fishes retain metabolic traces of their earlier freshwater existence. And if the whales and pinnipeds look sort of fishy, the resemblance is quite superficial, and they breathe air and suckle their young just like the rest of us mammals.

Generalizing, then, with respect to the place of laws of nature in working out the history of the earth and its inhabitants, we may confidently draw the conclusion that such laws do indeed exist, and also that they may indeed be used to infer the truth of historical hypotheses. However, in reconstructing history on this basis, there are many serious difficulties, not the least of which is the problem of distinguishing between legitimate laws of nature and historical contingencies. On top of that, we have to know what the laws are, and also how to apply them. And finally, when we do apply them, we find that we cannot do so in the absence of a vast quantity of empirical data about individuals. The laws of nature may tell us what is possible, but that only limits the number of acceptable hypotheses. An ideal evolutionary biology would present the entire history of life, in a manner that made it clear what was historical accident and what was nomologically necessary, and of course what laws applied and why, but that ideal is only beginning to be realized.

16

EMBRYOLOGY AS HISTORY AND AS LAW

"Thinking of effects of my theory, laws probably will be discovered of correlation of parts, from the laws of variation of one part affecting another."

—Darwin, Notebook E, page 51, November 1, 1838

In the last chapter we discussed how the particulars and the laws of nature can be fitted together to formulate historical narratives. But one might well ask how it is that the laws are to be discovered. If we say, *simpliciter*, that we use evolutionary history to discover the laws, and the laws to discover evolutionary history, one might suspect that the reasoning is circular. And in fact a lot of circularity, or at least begging of questions, has entered into phylogenetical reasoning. But that does not make the difficulty more than just an annoyance, as may be illustrated by a few examples.

Experience teaches that hypotheses about the laws that govern evolutionary processes can in fact be tested in the context of genealogical reasoning. This was precisely the rationale that underlay much of my own efforts to test hypotheses about the adaptive significance of various kinds of hermaphroditism and other "reproductive strategies." Given even a rough idea about the branching sequences, one can plot out the various phenomena of interest on a tree-like diagram and see if one can find correlations that are implied by the hypothesis that one wishes to test. For instance, there seems to be a relationship between hermaphroditism and brooding of the developing young, and hermaphroditism seems to be an adaptation that has evolved after brooding. One nice piece of evidence is that among echinoderms brooding and hermaphroditism both occur scattered about in quite a number of groups that may reasonably be assumed to be natural (monophyletic), but within these groups there is a rule that although all hermaphrodites are brooders, not all brooders are hermaphrodites. So it looks like brooding came first, much as the chicken was preceded by the egg. On the other hand, not all the evidence is phylogenetic; we can show that the brooders can raise more young if they are hermaphrodites by comparing the manner in which they care for the young across taxa and by making appropriate measurements. The genealogical and the physiological studies are mutually supportive and to a considerable extent logically independent.

By the time I did my own research on hermaphroditism and on the phylogenetics of hermaphroditic gastropods, knowledgeable zoologists had pretty well abandoned the notion that hermaphroditism is a "primitive" trait. There were plenty of well-documented examples of hermaphrodites evolving into gonochorists, as well as of gonochorists evolving into hermaphrodites. For earlier generations of phylogeneticists, however, the assumption that hermaphroditism is primitive was a very serious source of error, and the error has tended to persist in spite of the facts. Yet this "law" also coexisted throughout much of the nineteenth century with various other "laws," among which was that evolution always proceeds from the simple to the complex, except in the case of parasites, which always degenerate. Another major criterion of polarity was that evolution always goes from lower to higher, and this for the most part meant becoming increasingly more like Man. It was a hangover from preevolutionary times, and we still haven't altogether recovered from it. Since people have separate sexes, gonochorism has to be the derived condition. This "law" was widely accepted, to the point that it was generally believed that our own lineage had passed through an hermaphroditic stage.

Fortunately a Belgian malacologist, Paul Pelseneer (1895), worked out a reasonably good genealogy of the mollusks, and there was no way to reconcile his trees with the primitiveness of hermaphroditism in that group. Although hermaphroditism could be used as an incorrectly "polarized" character, a wide variety of evidence was applied, and therefore it was possible to correct this error. As the adaptive significance of hermaphroditism has become increasingly well-understood, we have been put in a far better position to fit it into our phylogenetic narratives. Furthermore, these examples clearly illustrate the advantage of having investigations of phylogenetic history, on the one hand, and the processual laws in terms of which it is explained, on the other, be carried out as an integral part of the same research program. And preferably by the same investigators. The way to get the right results is to understand both sides of the problem, not to ignore one or the other.

By way of further illustration, we may devote the rest of the present chapter to another area that might be clarified in the light of the individuality thesis: the relationship between ontogeny and phylogeny. This is an ancient and venerable topic, to which a vast body of literature has been devoted, but attempting a general overview would merely distract attention from the task at hand. The reader may wish to consult some of the general literature on the topic (Kryzanowski, 1939; Rensch, 1959; Smitt, 1961; Gould, 1977; Humphries, ed., 1988; Mayr, 1994b). Also of considerable interest have been efforts to deal with ontogeny and phylogeny from the point of view of cladistic methodology (e.g., Rieppel, 1979, 1993; Fink, 1982; De Queiroz, 1985; Kluge, 1985; Kluge & Strauss, 1985; Kraus, 1988; Mabee, 1989a, 1989b, 1993). This discussion will rather emphasize the theme of this book and attempt to show how the individuality thesis helps to clarify what can only be described as a very muddy picture. But we do need to consider how different ways of viewing the problem have affected thinking upon such matters, and some historical perspective helps.

The basic phenomenon to be understood is the apparent correspondence between the elements of certain systems of classification and the stages of life-history. The history of the topic begins far earlier than the writings of Ernst Haeckel (e.g., 1866), who coined the terms 'phylogeny' for the history of the lineage and 'ontogeny' for the history of the individual organism. The parallels had a broader context. Ancient and mediaeval thought was full of notions about the mystical correspondences among the various parts of the cosmos. In the macrocosm/microcosm analogy, the properties of the universe as a whole were mirrored in the properties of the human body. Traditional thinking often involved analogies between a human being's progression through the stages of infancy, childhood, youth, maturity, and old age and all sorts of other changes. We find these persisting into modern thinking in the guise of notions about the maturing and senescence of cultures, nations, and even taxonomic groups such as species.

One could go on and on. But for our purposes the most important mystical notion is that of the *scala naturae* in which the various inhabitants were ranked from higher to lower, with each of us having his place, whether among angels, animals, minerals, or whatever (Lovejoy, 1936). The developing organism was seen as passing from a lower to a higher condition in a sequence that corresponded to that same hierarchical arrangement. When the *scala* became temporalized so that there emerged a kind of proto-evolutionary view of things, there would be an historical succession with increasingly higher organisms putting in an appearance in the course of geological time. And there would be parallels between the two. The situation was rendered somewhat more complicated by the fact that Linnaeus had erected a hierarchy of groups within groups, and it became increasingly difficult to place everything in a single, linear scale.

Nonetheless the picture that began to take shape at the beginning of the nineteenth century was that of some kind of correspondence between the classificatory hierarchy, the succession of fossil organisms, and the stages of embryological development in different groups. There was no particular reason to link the various groups together as genealogical units with common ancestors, even if, as did Lamarck, one believed in lineages. Each lineage could go back independently of all others to the primordial slime. The parallels could be treated, as they were by Louis Agassiz, as manifestations of a Divine Mind that thought consistently in working out His Ideas. It was a long succession of miracles (Agassiz, 1857, 1859, see re-issue, ed. Lurie, 1962).

As discussed in Chapter One, quasi-evolutionists had two main ways of dealing with change. First, they could think of it as the result of something that was already laid down from the beginning, so that the developing embryo could be thought of as going from a stage of potentiality to a stage of actuality. For ontogeny this was by no means unreasonable, but when extrapolated to phylogeny, it becomes a kind of orthogenesis. The ontogenetic order would not be seen as the results of a series of historical contingencies but as the remote consequence of some initial miracle. The other possibility was to explain everything in terms of laws of nature, and here the analogy of crystallization played an important metaphysical role. The development of the organism and of the lineage alike was the consequence of unknown laws of nature, analogous to those of chemistry. God might have helped by ordaining the appropriate laws, and given the immense complexity of developmental processes, His participation lent a bit of plausibility to such views. Whatever, it should be emphasized that such ways of viewing the phenomena presupposed that taxa are classes.

We should also emphasize that throughout the long history of comparative and descriptive embryology, which goes all the way back to Aristotle's empirical researches, there were efforts to document the stages of development and to explain the changes in causal terms. By the end of the eighteenth century, it was generally agreed that as development proceeds, passage through stages of increasing complexity actually does occur, and is not just a matter of appearances. Such embryologists as Caspar Friedrich Wolff (original 1759, see translation 1896) attempted to explain development in terms of mechanical or quasi-mechanical processes. On the one hand, it seemed quite reasonable to look for some parallels between ontogeny and the classification schemes of systematists. On the other hand, it also seemed reasonable to try to explain development as a law-like process.

In response to the recapitulationist musings of Meckel and of Serres, Karl Ernst von Baer produced a statement that set the tone for much of the discussion ever afterward. His pronouncement is quite commonly referred to as von Baer's "four" laws, but if we turn to the original publication, we find not four laws, but one law with four parts. On top of that, the "law" has been so often mistranslated that it would seem a good idea to quote it (deleting some irrelevant parenthetical material) in the original German (von Baer, 1828:224) and then provide what seems to me a reasonably accurate English equivalent:

> Wenn wir oben bemerkten, dass man, um die Uebereinstimmung zweier Thierformen zu finden, in der Entwickelung um so weiter zurückgehen muss, je verschiedener diese Thierformen später sind, so erkennen wir daraus als Gesetz der individuellen Entwickelung,

1. Dass das Gemeinsame einer grössern Thiergruppe sich früher im Embryo bildet, als das Besondere.
2. Aus dem Allgemeinsten der Formverhältnisse bildet sich das weinger Allgemeine und so fort, bis endlich das Speciellste auftritt.
3. Jeder Embryo einer bestimmten Thierform, anstatt die andern bestimmten Formen zu durchlaufen, scheidet sich vielmehr von ihnen.
4. Im Grunde ist also nie der Embryo einer höhern Thierform einer andern Thierform gleich, sondern nur seinem Embryo.

Since we noted above that if one is to find a correspondence between two animal forms, one must go back increasingly far in development the more different these animal forms are later on, from this we make out as a law of individual development,
1. That that which is general to a major animal group is formed earlier in the embryo than that which is particular.
2. Out of that which is most general among the relationships of form arises that which is less general, and so forth, until finally the most special comes forth.
3. Each embryo of a definitive animal form, instead of passing through the other definitive forms, departs increasingly from them.
4. Fundamentally, therefore, the embryo of a higher animal form is never equivalent to another animal form, but only to its embryo.

Thus von Baer conceived of the taxonomic hierarchy in basically a Scholastic fashion, that is to say, as a class-inclusion hierarchy, with the categories indicating the degree of generality of the taxa, and the taxa being more general classes that include more particular classes. As development proceeds, he tells us, the embryos pass through stages, each of which in turn manifests the defining properties of the taxa to which they belong, following a sequence which is precisely the order of higher to lower categorical rank. So the defining properties of the phylum would develop first, then those of the class, and so on down to the species. The developmental stages of different taxa would remain identical so long as the defining properties of mutually exclusive taxa to which they belonged had not yet begun to manifest themselves. So were one to compare two animals of different orders but the same class at a time when only the class-level differentiae had put in their appearance, one would see no differences between them.

Von Baer's four major taxonomic groups based upon embryology agreed neatly with Cuvier's four "types" based upon comparative anatomy. And their concordant results tended to reinforce the prevalent anti-evolutionism, though by no means unequivocally, for certain adjustments were possible. Be this as it may, it is a most interesting question whether von Baer's law, or for that matter any part of it, constitutes a legitimate law of nature or whether, perhaps, it is pure description and maybe not very accurate description either. However accurately it depicts reality, the formula that has just been quoted might be thought of as just like the taxonomic hierarchy as now understood. That is to say, it presents an arrangement of historical contingencies, something that could have been otherwise, and not something that simply must be true. On the other hand, it is cast in terms of taxa in general, and if we take it as meaning that whenever there is a taxonomic hierarchy the ontogenies must necessarily display this pattern of correspondence, then it would indeed qualify as a legitimate law of nature.

If we treat von Baer's "law" as a legitimate law of nature, then, given that it must apply to each and every part of all taxonomic hierarchies, we can easily refute it. The gill clefts that occur in Chordata and Hemichordata, and perhaps originally in Echinodermata as well, appear after the notochord in the Chordata, although the presence of this structure is diagnostic of Chordata. Further examples are provided by de Beer (1958).

One way to salvage such a "law" is to restrict its scope of applicability, and this has been attempted by having it apply only to "single characters," a view that Osche (1982) rightly attributes to Weismann (1904) and Naef (1917) (see also Remane, 1960). As was explained in the preceding chapter, the term 'character' is equivocal, so that it could mean an attribute, a class of attributes, or a part or group of parts. At any rate, the gill clefts and the notochord are parts of different organ systems. As such qualifications are added, the law of course comes to tell us increasingly less, and one wonders if any proposed exception could be rationalized by choosing the appropriate definition of 'character'. As usual the authors who have discussed these matters are most unclear as to what the term 'character' is supposed to mean (De Quieroz, 1985). As to Naef (1917), it is abundantly clear from the text that he meant parts, not attributes, and he attempted to ground his theory in the developmental mechanics of his day. Different parts are linked up to different developmental pathways, and therefore their ontogenetic sequences are somewhat less constrained. We shall soon say a bit more about his views.

Another way to salvage such a "law" is to make it a tautology: the general features are defined as the ones that appear first, and exceptions are dismissed because secondary changes do not reflect the real order of things. Nelson (1978), who takes such a tack, seems to have put his colleague Platnick's "snakes have legs" argument to a new use.

Rather than fuss with the verbiage or attempt to rebut the sophistry, we might consider the possibility that here we may indeed have a situation in which there are legitimate laws of nature, but we do not understand them very well or have not properly taken the causal basis of development in hand. The situation in the embryology of von Baer's day might be compared to that in some early period in the history of astronomy, when regularities in the movements of celestial bodies could be made out, but nobody understood why. Biology awaited its Copernicus and its Newton.

It was Darwin who provided the basic solution to the problem of the relationship between ontogeny and phylogeny. This was to treat the evolution of morphology as the consequence of the evolution of the morphogenetic mechanisms that give rise to that morphology. According to his theory, morphological evolution occurs as the consequence of selection among variant organisms. And variation was the product of changes in developmental processes. So a series of diversifying ontogenies would be modified along various lines, depending upon what environments they found themselves in. In our days we tend to conceptualize such matters in terms of transmission genetics, but Darwin conceptualized them in terms of developmental genetics, and we can do that too.

Darwin owed a great deal to a treatise on teratology by Isidore Geoffroy Saint-Hilaire (1832–37), who in turn derived a great deal from his father, Étienne Geoffroy Saint-Hilaire (see I. Geoffroy Saint-Hilaire, 1847; Cahn, 1962). The elder Geoffroy's basic idea was that abnormal development is a variant upon normal development; both have a law-like character, but something may go wrong and produce "birth defects" or monstrosities. Darwin believed that an understanding of the laws of variation would provide very valuable clues to evolutionary history as well as evolutionary mechanisms. He elaborated upon this theme in great detail in *The Variation of Animals and Plants under Domestication* (Darwin, 1868, second edition 1875).

In *The Origin of Species* Darwin provided a sort of model for how developmental processes might interact with natural selection and produce the sort of correspondence between ontogeny and the taxonomic hierarchy that his predecessors had documented. As he saw it, variation can occur at any stage in the life cycle of the organism, and change can occur due to selection among the variants thus produced, but it will affect only the later, post-variant, stages of the life cycle. (We would say that the genes are subject to selection only from the time that they have been phenotypically expressed.) From this virtual truism we get a kind of statistical law, which tells us that the more nearly terminal stages of the life cycle will be more labile in evolutionary time; it generates something like von Baer's law, but it is only statistical and requires a *ceteris paribus* clause—all else has to be equal. There is another way of justifying something like von Baer's law, which is that the earlier that modifications occur in developmental processes, the higher the probability that they will have some lethal or semilethal effect by disorganizing the whole system. This is in itself a law of nature, for it applies to classes of organized systems, is spatio-temporally unrestricted, and does not refer to any individual in particular. Naef (1917), as already mentioned, thought that he could justify a law of terminal lability on somewhat similar grounds. But for him the reason was somewhat different: the developmental constraints are more restrictive earlier in ontogeny.

However, Darwin also believed, as do we, that selection can modify the early stages of ontogeny, so the amount of change that occurs in the young stages will depend in part upon the selection pressures that affect them. So he thought that free-living young organisms would tend to diverge more than those which live, say, inside egg capsules or within the body of the mother, where they would be protected from the environment. And these might retain ancestral traits. We should make it abundantly clear that Darwin had special expertise with respect to the relationships between ontogeny and systematics. His first book on evolution, *A Monograph on the Sub-Class Cirripedia* (Darwin, 1851, 1854), was also the first book ever written on phylogenetics, and as it happens he dealt with a group in which embryology had lately given quite important results. The researches of J. Vaughan Thompson (1828–1834, 1835) had shown unexpectedly that the larvae of barnacles were very much like the larvae of less aberrant marine crustaceans, providing conclusive evidence to a relationship that was still questioned in some quarters. In the years just preceding the publication of *The Origin of Species,* systematists were taking an increasing interest in embryology, not out of any belief in evolution but because similarities in early development sometimes provided useful characters.

In his research on barnacle phylogenetics, Darwin studied the developmental stages in great detail, evidently in order to investigate the relationships between embryology and systematics as well as to work out the genealogical relationships. In some cases he was able to document how ontogeny has been modified in the course of evolutionary history. A particularly good example can be seen in his research on the "dwarf males" of barnacles mentioned earlier. They are "degenerate" little creatures that live on much larger hermaphroditic or female barnacles. This condition has evolved, in response to selection of course, largely by the elimination of terminal stages in the life history; the animals stop growing much earlier, and they mature at a much younger age, and only as males. Such elimination of the late stages of development is now called "progenesis." The progenetic males obviously do not "recapitulate" the entire phylogeny of their species, irrespective of whether their mates do.

From this example we can see why Darwin believed that he could explain evolutionary changes as the consequence of modification in developmental processes. The elimination of later stages, coupled with earlier sexual maturation, is one possible way in which this can occur. It fits in quite well with the law that later stages are more evolutionarily labile than early ones; that this kind of modification is not impossible may be considered a law of nature, albeit one that doesn't allow us to predict very much. Darwin also had an example of larval stages in barnacles being eliminated. Again, we can find a sort of law here, for as we have mentioned, losses are more easily produced than gains. And yet neither of these laws really specifies enough to do more than narrow down the range of possible scenarios that might be proposed as historical narratives. They tell us what is possible, but leave it an open question how to chose among the variety of possibilities.

Darwin's ideas about the relationship between ontogeny and phylogeny were almost immediately seized upon by one of his most gifted supporters, Fritz Müller (1864, English translation, 1869). Müller, whose name is immortalized by his discovery of Müllerian mimicry, was one of the few nineteenth-century biologists who really understood natural selection. And since he worked on the systematics and embryology of crustaceans, he was in a particularly good position to understand Darwin's barnacle monograph. It was he, not Haeckel, who first applied the principle of recapitulation in a phylogenetic context, at least explicitly. Following up on Darwin's insights, Müller realized that if the Crustacea had evolved by the addition of new stages at the end of development, then each stage through which they passed would represent the sequence of adult ancestors. He suggested that such terminal addition seemed to have happened in certain cases. Due to his influence, the idea that the nauplius and zoea larvae of crustaceans correspond to adult ancestors that swam about in the plankton was widely accepted and had quite a number of distinguished supporters, such as Anton Dohrn (1870, 1871). But within a few years it became clear that the ancestral crustaceans, like the worms from which they were descended, had passed through their early developmental stages while swimming around in the plankton and then completed their life cycles on the bottom. In other words, the larvae were not recapitulating adult stages.

Fritz Müller definitely knew that terminal addition, which is the only mode of transmutation that produces clear-cut recapitulation of the history of the lineage, is by no means the only possibility. He was perfectly aware, from reading Darwin's barnacle monograph as well as from his own research, that stages can be deleted either early or late in ontogeny. And he also discussed how the descendants could deviate from the original path, as well as the possibility of all sorts of adaptive specializations to be secondarily acquired in larvae and elsewhere. The following principle, or rule, or perhaps we should say law of nature, is not the sort of thing that one would expect from a naive recapitulationist (Müller, 1864:81):

> Die Urgeschichte der Art wird in ihrer Entwicklungsgeschichte um so vollständiger erhalten sein, je länger die Reihe der Jungendzustände ist, die sich gleichmässigen Schrittes durchläuft, und um so treuer, je weniger sich die Lebensweise der Jungen von den Alten entfernt, und je weniger die Eigenthümlichkeiten der einzelnen Jugendzustände als aus späteren in frühere Lebensabschnitte zurückverlegt oder als selbständig erworben sich auffassen lassen.

I translate:

> The primordial history of the species gets more completely main-
> tained in its developmental history the longer is the series of young stages
> which they pass through by regular steps, and more accurately the less
> the way of life of the young departs from that of their elders, and the less
> the peculiarities of the particular young stages can be interpreted as hav-
> ing been pushed back from a later stage in life or independently acquired.

In his *Generelle Morphologie* Haeckel (1866) first propounded his "biogenetic law" (*biogenetisches Grundgesetz*), but he did not call it that until later.
Given that both he and his critics have provided somewhat edited versions,
again it seems best to quote the original verbatim and then provide a translation. This is all the more important because it is difficult to decide wherein the
law itself consists, having been presented as a series of theses on page 300 of
the second volume:

VI. Thesen von dem Causalnexus der biontishchen und der phyletischen
Entwicklung.
    40. Die Ontogenesis oder die Entwickelung der organischen Indi-
viduen, als die Reihe von Formveränderungen, welche jeder individuelle
Organismus während der gesammten Zeit seiner individuellen Existenz
durchläuft, ist unmittelbar bedingt durch die Phylogenesis oder die Entwickelung des organishchen Stammes (Phylon), zu welchem derselbe
gehört.
    41. Die Ontogenesis ist die kurze und schnelle Recapitulation der
Phylogenesis, bedingt durch die physiologischen Functionen der
Vererbung (Fortpflanzung) und Anpassung (Ernährung).
    42. Das organishche Individuum (als morphologisches Indi-
viduum erster bis sechster Ordnung) wiederholt während des raschen und
kurzen Laufes seiner individuellen Entwicklung die wichtisten von
denjenigen Formveränderungen, welche seine Voreltern während des
langsamen und langen Laufes ihrer paläontologischen Entwickelungen
nach den Gestezen der Vererbung und Anpassung durchlaufen haben.
    43. Die vollständige und getreue Wiederholung der phyletischen
durch die biontische Entwickelung wird verwischt und abkekürzt durch
secundäre Zusamenziehung, indem die Ontogenese einen immer ger-
aderen Weg einschlägt; daher ist die Wiederholung um so vollstandiger,
je länger die Reihe der successiv durchlaufenen Jungendzustände ist.
    44. Die vollständige und getreue Wiederholung der phyletischen
durch die biontische Entwickelung wird gefälscht und abgeändert durch
secundäre Anpassung, indem das Bion während seiner individuellen

Entwickelung neuen Verhältnissen anpasst; daher ist die Wiederholung um so getreuer, je gleichartiger die Existenzbedingungen sind, unter sich das Bion und seine Vorfahren entwickelt haben."

I translate:

VI. Theses about the causal nexus of biontic and of phyletic development.

40. Ontogenesis, or the development of the organic individuals, as the series of changes in form that each individual organism passes through during the entirety of its individual existence, is directly conditioned by phylogenesis, or the development of the organic stem (phylon) to which it belongs.

41. Ontogenesis is the short and rapid recapitulation of phylogenesis, conditioned through the physiological functions of heredity (reproduction) and adaptation (nutrition).

42. The organic individual (as a morphological individual of the first through sixth order) repeats, during the hurried and brief course of its individual development, the most important of those changes in form which its ancestors have gone through, during the slow and protracted course of their paleontological development, according to the laws of heredity and adaptation.

43. The complete and reliable repetition of phyletic development by the biontic becomes obliterated and shortened through secondary regression, as ontogenesis strikes out along an ever more direct route; therefore the longer the series of young stages passed through, the more complete the repetition.

44. The complete and reliable repetition of phyletic development by the biontic gets falsified and modified by secondary adaptation as the bion adapts to new circumstances during its individual development; therefore the repetition is the more reliable, the more similar are the conditions of existence under which the bion and its ancestors have developed.

Thus, although Haeckel tells us that the relationship between ontogeny and phylogeny is reasonably straightforward, he also emphasizes the point that the ontogenetic traces get obliterated with the passage of time, especially when there is a change in the conditions of life. Furthermore he specifies the conditions under which the ancestral conditions are most likely to be retained, namely where the young stages retain a complex pattern of development and where the organisms have continued to live in more or less the same way that their ancestors did.

The obvious procedure for a phylogeneticist would be to use the developmental sequences as approximations to the evolutionary ones, but to bring in additional criteria to screen out the secondary modifications. It was particularly obvious to Haeckel, for example, that certain features were secondary adaptations in the younger stages due to the evolution of parental care. A placenta is obviously not an ancestral reminiscence of a period in which the adult animals lived solely upon resources provided by their mothers. Likewise where the eggs are provided with a large amount of yolk, the developing embryos are highly modified, and could hardly have existed as free-living creatures (Haeckel, 1875). Thus, because it was nomothetically necessary that the reconstructed ancestral forms be viable as free-living adults, he could eliminate some of the possible sequences of ancestors and descendants.

Later in his career, Haeckel clashed with Elias Metschnikoff with respect to the common ancestor of the Metazoa. Haeckel had proposed that all multicellular animals are descended from a common ancestor called a "gastraea," a creature with a mouth and two cell layers that supposedly swam around in the Precambrian ocean. He proposed a sequence of phylogenetic stages roughly corresponding to stages that had been found in many developing animals. Thus the uncleaved egg (cytula) would correspond to a unicellular adult animal (cytaea). Then the result of the first few cellular divisions in ontogeny, a solid ball of cells or "morula," would correspond in phylogeny to a hypothetical "moraea" with the same structure. Next there would come a hollow ball of cells, called the "blastula" in the ontogenetic series, the "blastaea" in the phylogenetic series. Finally the surface of the body would form a depression that would become the gut, with an opening called the blastopore; the "gastrula" of ontogeny would correspond to a "gastraea" in phylogeny. (The terms and the stages varied somewhat in different versions of the theory, but this need not concern us here; see Haeckel, 1872, 1873, 1875, 1877.)

Metschnikoff (1876, 1879, 1880, 1882, 1883, 1886) proposed that instead of being a gastrula-like organism, the common ancestral metazoan was something quite different, more like what he had found in certain other animals. Rather it would be a solid creature, with a surface layer surrounding a mass of cells in the interior, but without a mouth or gut cavity. This "parenchymella" fed by taking up small particles at the surface and digesting them inside of the solid mass of cells. Metschnikoff argued that this hypothetical ancestor provided for a more plausible physiological scenario. He had good evidence that in the earliest Metazoa, digestion did not take place within an extracellular cavity, but inside the digestive cells; so it would have had no use for a gut. This controversy led to something of a standoff, and the same theme continues to be debated. The extant animals do not provide us with the kind of intermediates that would be needed to give a very satisfactory resolution. For our purposes, however, what matters is the kind of evidence that was being put forth in the controversy. It was not simply a matter of fitting ontogenies to phylogenies, but trying to come up with a scenario that was physiologically plausible.

In other words, the ontogenetic criteria were not simply a matter of formalistic comparison. Haeckel was quite clear in insisting that a wide range of evidence ought to be applied to phylogenetic research. He treated paleontology, comparative anatomy, and embryology as mutually supportive. He did, to be sure, overemphasize the ontogenetic criteria and treated terminal addition as a rule with but rare exceptions. In his popular writings especially, he strongly deemphasized progenesis and other kinds of heterochrony, but nonetheless he acknowledged their existence. It is unfortunate that so much of our impression about how science is done is derived from vulgarizations.

As Gould (1973, 1977) has so emphatically and so rightly stressed, much of the literature on the "biogenetic law" deals with a travesty of the views of Haeckel and others who have attempted to apply the principle (see also Lovejoy, 1959; Ghiselin, 1969d). Beyond just plain bad scholarship, there are several reasons for such misrepresentation, some of them having to do with politics, both academic and otherwise, especially on the part of experimental embryologists. To explain development in terms of laws rather than history had two main advantages. First it aided experimental embryologists in their struggle to obtain academic chairs; the writings of Jane Oppenheimer (1967) provide good examples of this kind of anti-evolutionary propaganda. Second, it enabled professors in Berlin and elsewhere to support the Established Order; a good example is to be found in the writings produced in his later years by Oskar Hertwig (1922). De Beer (1958) may have been overreacting to a kind of evolutionism that he felt was tainted with Lamarckism, as indeed it was to some degree.

The context of early discourse on the principle of recapitulation has all too often been ignored. Many of the earlier writers, including Darwin, Müller, and Haeckel, were presenting arguments to the effect that evolution has in fact occurred. Instances of recapitulation would be hard to explain otherwise, even if recapitulation was by no means considered an invariable rule. Furthermore, if one's goal is to reconstruct the history of life, the mere fact of secondary modification is no more grounds for giving up the search for historical vestiges in ontogeny than the incompleteness of the fossil record should discourage the efforts of paleontologists to fill in their series.

Be this as it may, it is somewhat of a puzzle as to whether the biogenetic law is a legitimate law of nature, and whether or not it is true, given the formulation of it that has been quoted above. If all that is asserted is that, as a matter of (contingent) historical fact, ontogeny recapitulates phylogeny, then this is hardly a law, for even if true, things might have been otherwise. But taken as a connection between ontogeny and phylogeny that is necessarily true irrespective of time and place, it would be a law, for things could not have been otherwise. Yet, given the fact that Haeckel says that the parallel is imperfect and that his formulation provides ways of accounting for apparent exceptions, we cannot simply adduce examples of non-recapitulatory ontogenies as refuting it, in spite of what so many authors have maintained. We might say that the law applies only under ordinary conditions, namely where there has been terminal addition. Then additional laws, principles, and of course some historical contingencies would have to be invoked to account for these deviations from what is considered the norm.

Such an analysis of the principle of recapitulation could probably be defended quite effectively, at least from a logical point of view. From an ontological point of view, however, it would seem that the fundamental process to be explained in terms of laws of nature is a succession of ontogenies being gradually modified in an orderly fashion such that vestiges of earlier conditions may or may not be retained, depending upon what laws apply and what conditions are realized. What we need are some very general laws of nature that account for the basic patterns of lability and stability, and more particular laws of nature that account for the details. Then the historical details can be integrated into an explanatory historical narrative. And this seems to have been the goal of many historically-minded students of evolutionary mechanisms, beginning, as usual, with Darwin.

We may recollect that Darwin had some notion of reasons for the ontogenetic "program" being rather stable, some of which were extrinsic, others intrinsic. A purely statistical effect of earlier changes being less frequent but having a larger phenotypic effect can be treated as a law of nature. But in and of itself this seems inadequate. A more basic cause for such stability is that the ontogenetic processes that give rise to an adult organism are causally interdependent, so that changing one part without affecting many others is difficult. This may be treated as a more particular instance of Cuvier's principle of the correlation of parts, discussed in an earlier chapter. So the stability in question is a manifestation of a law that governs all complex, organized beings (Simon, 1962). But such a very general truth does not get us very far, because the interconnectedness among parts is neither simple nor independent of history. The more particular laws might tell us that when two parts are components of a single functional complex, they will evolve together or develop together, but being a component of such a complex is a matter of contingent fact. Darwin (1868, 1875) gave the example of selection for white coat-color in domesticated cats producing a breed that was also blue-eyed and deaf, to illustrate his principle of correlated variability. I am not sure why this particular phenotype showed this particular pleiotropic effect, but it seems to be due to a combination of fortuitous circumstances restricted to a particular lineage.

Quite a number of authors have attempted to analyze and classify the various ways in which ontogeny might be altered so as to produce a modified kind of descendant (Sewertzoff, 1924, 1927, 1931; Gould, 1977; Alberch et al.,

1979). The terms and classifications depend upon various combinations of modified growth rates, changes in proportions, addition or deletion of stages, earlier or later formation of "characters" (parts or attributes) and the like. We should mention a few of these terms just to give an idea of what the terminology is like. The general term 'heterochrony' means a change in the timing of developmental events. Paedomorphosis is the shifting of properties of the juveniles into the older stages; gerontomorphosis is the opposite, with properties that evolved in the later stages coming to appear earlier. Deletions of stages or properties can be terminal (progenesis) or subterminal. Deviations can be earlier or later. Coming up with a good nomenclature and classification has proved more difficult than one might think. One point rightly stressed by Alberch et al. (1979) is that we need to keep the modification of the underlying developmental process conceptually and terminologically distinct from the evolutionary outcome. But even their terminology seems to me not altogether satisfactory. For instance, they refer to progenesis as a kind of "reverse recapitulation." But what really happens in progenesis is not that the stages of development get reversed; the later stages of ontogeny are eliminated altogether and the organisms become sexually mature at a much earlier age.

Rather than attribute the terminological confusion to an undesirable combination of pattern and process (D. Wake, 1996), I would argue that the problem lies more with a failure to distinguish properly between the nomothetic and the historical aspects of both ontogeny and phylogeny and to put them together into a coherent and unified whole. So the appropriate question is how ontogeny might get modified in order to reveal how it actually has been modified, though by no means neglecting other lines of evidence. This way of thinking seems to have been at least implicit in the work of Darwin on barnacles and many other researchers too numerous to enumerate here, though a few outstanding examples of this approach should be mentioned (de Beer, 1956; Davis, 1964; Delsol, 1965, 1977). Therefore the only innovation here is to show how the individuality thesis fits in (see also Morss, 1991).

The list of known ways in which evolution can occur by the modification of ontogeny is by no means coextensive with the list of logically possible ones. Although the order of appearance of certain traits may get reversed, there is nothing really comparable to an insect, say, instead of going egg → larva → pupa → adult, proceeding in the opposite direction: adult → pupa → larva → egg, even though there are apparent partial exceptions. So the phylogeneticist might reasonably reject certain possible relationships or hypothesized transformations on the grounds that these would be contrary to the laws of nature. But given so many possibilities, quite a variety of scenarios would by no means be ruled out, and to chose among them a great deal of additional information would be necessary. Whether the developmental process would be extended or shortened, or whatever, would depend upon all sorts of contingent facts and historical accidents, such as what predators happened to be in the environment at a given time.

Furthermore, terms like "terminal deletion" and "post-displacement" are formalistic and descriptive. They do not provide any principles, laws, or matters of contingent historical fact that might help us to predict or to explain what has happened. To do that we need to understand developmental mechanics, but that in itself is a very big order. For although there must be quite a number of very general laws of nature that govern the development of all organisms irrespective of time and place, these have to be formulated at a high level of abstraction, and the particular way in which an organism is formed is largely contingent upon its history. The manner in which an organism develops is another of those many things that evolve: it could have been otherwise and maybe it will not always be so in its descendants.

If we find, for example, that the presence of a notochord is a necessary condition for the development of the central nervous system in vertebrates, we are by no means dealing with a law of nature, but rather with a matter of historical contingency. If we want to find a law of nature here, it will have to be something like the following: as long as a structure is indispensable in the formation of another structure that is retained in the adult organisms, it will not be lost in the young ones. This stricture cuts the ground out from under those who want to explain evolution as the consequence of developmental laws, for what they have been calling laws turn out to be accidents of history. It should not cause the rest of us to despair, however. On the contrary, it tells us that we can indeed have a phylogenetics of development that is both nomothetic and historical, provided that comparative biologists are willing to do the empirical research.

If we are going to have an historical embryology, then we want to have a narrative account of the history of ontogenies—a series of "biographies" in which one kind of biography replaces another within populations that form evolving lineages. The laws of nature that govern classes of ontogenies and classes of populations of lineages will help us to construct that narrative history. And because the process of development stands in such intimate relation to the process of evolution, the narrative will explain those processes in terms of each other. What I have in mind is a kind of tree-like diagram, in which the branches are, as usual, species and lineages. However, instead of a tree annotated with diagnostic "characters," there would be a running account of the morphogenetic processes getting modified through time in conformity with law, with the relevant contingencies supplied. This seems to be what is emerging these days from efforts to reconstruct the history of metameres and body regions by studying the genes that control their development. The genes themselves can be homologized across phyla and changes in their function documented (McGinnis, 1994; Akam, 1995; G. Müller and Wagner, 1996). It is even possible to produce a kind of stain that allows one to identify homologous parts on the basis of the genes that affect their development (Holland, 1996).

Conceiving of the problems of ontogeny and evolution from such a phylogenetic perspective might help to correct some serious misconceptions about epistemology. In particular we have the notion that systematists can only proceed by grouping objects together in terms of shared characters. For example, we might be able to detect instances of progenesis (terminal deletion) by finding that one lineage well within a group lacks stages present in related ones. Darwin's barnacle phylogeny with its dwarf males provides an excellent example of that, especially since the hermaphrodites and females within the same species are not progenetic.

But there is another way to test that kind of hypothesis: by experiment. Quite a number of lineages of amphibians have ceased to undergo metamorphosis, in effect becoming sexually-mature larvae. However, they can sometimes be induced to undergo metamorphosis by the injection of hormones. The physiological basis of such "experimental atavism" is reasonably well understood. By laboratory manipulation we can re-establish phenotypes that are no longer penetrant enough to appear in natural populations. There is a substantial literature on this subject, with long-lost feather patterns, femurs, gill arches, and even avian teeth being evoked (Hampé, 1959; Steiner, 1966; Raikow, 1975; Garcia-Bello, 1977; Kollar and Fischer, 1980; Sander, 1982; Hall, 1984; Lauder, 1988; Mishler, 1988). It would be as tedious and silly to explain why these "characters" have to be vestiges rather than "prophetic types" as it would to explain why the earth is not flat. But we should emphasize that the reconstitution of long-lost phenotypes could be as much a part of routine practice in systematics as is counting chromosomes.

More importantly, the execution of such an experiment provides a fine example of a systematist doing something more, and something far more profound, than just putting things together on the basis of shared characters. It is a test of an hypothesis. As such it allows us not only to test the empirical hypothesis that evolution has occurred by suppression of terminal stages in ontogeny, it also provides a test of the methodological proposition that phylogenies are constructed solely on the basis of shared characters. Neither the empirical theories nor the methodological assumptions of systematists have to be a priori in the only legitimate philosophical sense of not being justified by experience and perhaps needing no such justification. Some authors have in fact suggested that methodological assumptions are indeed a priori, not capable of refutation, and "metaphysical" rather than physical or biological (Bonde, 1984; Weston, 1988). If they really believe that, then one may seriously question whether what they are up to deserves to be called science.

17

THE ARTIFACTUAL BASIS OF MACROEVOLUTION

"Authors sometimes argue in a circle when they state that important organs never vary; for these same authors practically rank that character as important (as some few naturalists have honestly confessed) which does not vary; and, under this point of view, no instance of an important part varying will ever be found: but under any other point of view many instances assuredly can be given."

—Darwin, *Origin of Species*, first edition, page 46

In previous chapters attention was drawn to the anti-evolutionary character of much of the literature about embryology and evolution—or what purports to be evolution. The notion of orthogenesis makes sense only if phylogeny has the same sort of pre-established sequence of changes that a developing embryo does. In ontogeny, the structure of the zygote causes a particular range of adult structure to be produced, in spite of externally-produced perturbations, a certain amount of flexibility and resultant diversity notwithstanding. If we reject orthogenesis and assume that the germ itself evolves, then morphogenesis does not cause evolution, but merely restricts the range of variants that can appear, limiting or constraining rather than ordaining its course. So too with genes and mutations. Mutations are, of course, responsible for meeting one of the necessary conditions for evolution by natural selection to occur by providing a source of inherited, phenotypically expressed variation within populations. But except in that trivial sense, the notion (so widely expressed by anti-evolutionists) that it is mutations that cause evolution has to be dismissed as reversing the real order of things. The existence of cotton and wool is not what causes us to wear garments of the one in warm weather and of the other in cold, or indeed to wear clothing at all.

"Mutationism" attempted to account for evolution, at least the origin of species, purportedly on the basis of mutation alone (De Vries, 1919). However, a pure form of mutationism, without anything like natural selection to impose directionality, seems very much like the "fortuitous concourse of atoms" invoked by such Epicureans as the Roman poet Lucretius in his *De Rerum Natura*. It stands to reason that orthogenesis has had a much larger following, especially since its advocates have been rather vague about what they had in mind. I will not here discuss the history of orthogenesis in any detail, the topic having been ably discussed by several authors (Bowler, 1979, 1983; Mayr, 1983a; Richardson and Kane, 1988; Rainger, 1989, 1991; Pinna, 1995). However it deserves emphasis that throughout the first half of the present century orthogenesis was widely advocated as an alternative to Darwinism (Osborn, 1902, 1917; Vavilov, 1922; Berg, 1926; Lotsy, 1931; Gregory, 1935, 1936; Schindewolf, 1950). Its advocates were taken seriously enough to evoke rebuttals from such Darwinians as Sewertzoff (1924) and by the architects and advocates of the Synthesis (Rensch, 1939, 1959; Simpson, 1944, 1953; Jepsen, 1949; Romer, 1949; Dobzhansky, 1979). As Simpson points out, the term 'orthogenesis' has been so loosely defined, if defined at all, that it is readily conflated with orthoselective trends and the results of developmental constraints, both of which are quite consistent with Darwinism old and new.

The rebuttal of orthogenesis was so successful that one is surprised to find anyone taking it seriously at the present time, but it continues to be advocated in some quarters, especially by cladists and structuralists (Baroni-Urbani, 1979; Grehan and Ainsworth, 1985; Van der Hammen, 1989). And a fair number of authors invoke mysterious unidirectional trends or drives of the sort that makes one wonder (Salvini-Plawen, 1980). The "empirical" basis for such claims usually consists in evidence that certain characters have evolved more than once within a lineage. In other words, parallel evolution has occurred. But the claim that what goes on has some kind of (mysterious) intrinsic cause, rather than some (perhaps straightforward) extrinsic basis, has always been gratuitous. In other words, why orthogenesis, when orthoselection fits the facts? Brooks and McClennan (1991:75, 84) lately invoked orthogenesis in commenting on the fact that brooding has repeatedly evolved in frogs. However, the anatomical and behavioral features that facilitate such brooding are quite heterogeneous within the Anura. Sometimes the male cares for the young within a brood sac, sometimes they are cared for in pouches on the dorsal surface of the parent, and so forth. The selection pressures and environmental circumstances that favor parental care are still under investigation, but the environmental correlations are evident enough.

In every case where we can argue from knowledge rather than from ignorance, the facts contradict rather than support orthogeneticist interpretation. Again I draw upon my own experience in the laboratory. The opisthobranch gastropods are a notorious case of parallel evolution, with group after group independently undergoing reduction of the shell, so that snails have repeatedly evolved into slugs (Gosliner and Ghiselin, 1984). The change from snail to slug can reasonably be interpreted as the consequence of modifications of developmental processes. However, the group as a whole is not characterized by a disposition to undergo a particular kind of change in developmental processes, for in getting rid of the shell, different lineages have modified the developmental processes in quite different ways. Often the body proportions have been altered by differential growth, yet it is clear at a glance that in different lineages different parts of the body have been reduced or expanded. As if this were not enough, we know what environmental circumstances have led to the reduction of the shell (Faulkner and Ghiselin, 1983). The animals began feeding upon various kinds of toxic and distasteful plants and animals, and then took to using the chemicals derived from food in their own defense. Enough is known about the genealogy of the group to show that such chemical defense was adopted while the shell was still present, so we have an interesting case of what has been called "pre-adaptation." Furthermore, there is very good evidence that at a later stage of evolution the animals have sometimes acquired the capacity to modify and to synthesize such chemicals, emancipating themselves from dependency upon the food source—but that is another story. The basic message ought to be clear. If we are to decide such matters, we need a real history of the group, with the genealogy known and the relevant causal influences properly taken into account.

The parallel evolution of sea-slugs and other groups is often referred to as a "macroevolutionary" phenomenon, in contradistinction to a "microevolutionary" one, the terms having been popularized by the architects of the Synthetic Theory, especially Rensch (1959). The basic distinction between the two is that microevolution has to do with what goes on within species, macroevolution with what happens after they have speciated. So under the rubric of macroevolution we would include such things as long-term trends, parallelism, adaptive radiation, and perhaps the larger patterns of extinction. The terms of course have been variously used, but this sort of thing would seem to be what most authors have had in mind.

The notion that the adaptive radiation of slugs demands an orthogenetic explanation illustrates the frequent claim that factors additional to natural selection have to be invoked for an adequate explanation of macroevolution. The rebuttal above suggests that such claims are not always well founded, although a lot of room is left open for other such possibilities. It also suggests that all too often such claims are the product of bad taxonomy, or at least of good taxonomy badly interpreted. This proposal is by no means new, and a lot of recent work will be cited in support of it. The point can perhaps be made better, more interestingly, and more forcefully by relating it to the individuality thesis, and this we shall now proceed to do.

Since so many of the claims about macroevolution have to do with the fossil record, we may as well begin with a look at the well-known geological time scale (see Harland et al., 1990). One doesn't have to examine it very

closely to observe that it is a straightforward hierarchical classification system, with smaller taxa falling under larger ones, and each taxon having a definite categorical rank. Just looking at some of the major units, we get:
Phanerozoic eon
   Cenozoic era
   Mesozoic era
    Cretaceous period
    Jurassic period
    Triassic period
   Paleozoic era
    Permian period
    Carboniferous period
     Pennsylvanian sub-period
     Mississippian sub-period
    Devonian period
    Silurian period
    Ordovician period
    Cambrian period
Cryptozoic eon

By now it should be obvious that the taxa in this system, such as the Permian, are individuals, and the categories, such as era and epoch, are classes of individuals. It should be almost as obvious that this is a whole-part hierarchy, so that, for example, the Mesozoic incorporates the Triassic, Jurassic, and Cretaceous. In this system, the sequence from lower to higher really represents an objective relation between the objects classified, that is, temporal succession. Any taxon lying below a taxon of the same rank is previous to it. One might want to argue that the sequence could equally well be represented by reversing the sequence, and in a purely formal sense this is quite correct, but it would be like writing somebody's biography from death to birth; in looking for older and older fossils in a given area, one generally proceeds downward, because deposition occurs on top. In effect we have a dual classification system, such that rocks or formations can be placed within the temporal order, according to the time at which they were formed, even though the individuals here are not the rocks or formations but stretches of time. There is, to be sure, a corresponding system of formations, which are also individuals.

But what is the basis for individuating the taxa here? There would seem to be nothing really equivalent to the species or the organism in this system; no kind of unit that plays some crucial role in geological processes. We can say that we are dealing with larger and smaller parts of the history of the earth, and perhaps compare that history to the tellurian phylogenetic tree in its totality. Reflecting on the history of the system, it is clear that it was built up largely on an empirical basis, using the fossils in the rocks as indicators of contemporaneity and of change. The geological taxa were differentiated by abrupt changes in the taxonomic composition of the fossilized biota and diagnosed in terms of the characteristic species and higher taxa within them. So taxa would be subdivided if and only if there occurred some break, whether apparent or real, in the succession of organized beings. It is important to stress that the limits and ranks have been adjusted so as to produce this very coincidence. Such a ranking procedure inevitably tends to accentuate the discontinuities. Furthermore, the taxa do not correspond to any particular quantity of time, and they have no particular number of contained subunits. Clearly, the succession of strata and species is a real phenomenon, but one seriously wonders whether rank and limits tell us much about nature.

Of course it is perfectly possible that the geological taxa correspond to distinct events and to discrete stages in the history of life. The various stages in the ontogeny of an organism reflect changing emphasis upon embryogenesis, growth, maturation, reproduction, and the like. And although Kluge (1988) is justified in being skeptical about breaking up ontogenies, and even in comparing them to chronospecies, such doubts seem groundless because we are dealing with events and their parts, which are legitimate individuals. There is no reason why the various geological taxa cannot somehow represent, if remotely, analogous differences in the kind of processes that have been going on.

On the other hand, the fossil record is notorious for the artifacts due to its incompleteness, and in recent years a great deal of effort has been devoted to estimating just how reliable it is. Between fossils getting destroyed, on the one hand, and not having been formed in the first place, on the other, there is plenty of good reason for thinking that a gap or a discontinuity in the fossil record may be artifactual. The so-called "Lazarus taxa" that disappear from the fossil record for substantial periods of time, only to reappear much later, can mean only that the organisms were surviving in spite of not being known as fossils, perhaps in some distant corner of the globe where subduction has destroyed the record (Flessa & Jablonski, 1983; Donovan, 1989; Maxwell, 1989).

A great deal of mileage has been made out of putative evidence for "explosive" evolution in the fossil record, especially at the beginning of the Cambrian. It turns out that a substantial variety of metazoans existed before that time (Fedonkin, 1996). The remarkably well-preserved early-Cambrian Burgess Shale fossils seemed to have appeared quite suddenly. However, it turns out that the Burgess Shale fossils represent part of a widespread biota, which, to judge from the level of endemism (which is indicative of local diversification), must have existed for quite a long period of time before the fossils themselves were formed (Fortey, Briggs and Wills, 1996). Thus, biogeography has once again been able to correct some of the deficiencies of the fossil record, and this is something that has been going on more or less continuously since 1859.

If there appears to be a discontinuity, it could be due to one or more of a variety of causes, and it may not be easy to distinguish between them. Darwin of course emphasized that fossils are formed only under unusual circumstances, and that once formed they are readily destroyed by erosion. The possibility suggested by Eldredge (1971), that apparent discontinuities are due to variable evolutionary rates, opened up a very fertile field for speculation, and not just with respect to the theory of punctuated equilibria (Eldredge & Gould, 1972). I must say that I was one of the participants in the symposium at which that theory was first proposed, and was puzzled at what seemed to be belaboring the obvious. The continued hype and distortions irked a lot of people (Mayr, 1992a). Nonetheless it was a good thing to draw attention to variable evolutionary rates as one possible cause for the well-known "gaps" in the fossil record. It was also an interesting possibility that Mayr's peripheral isolate model of speciation might be widely applicable and help to explain the general picture. So too, since according to Mayr's model a lot of change occurs at the time that species are formed, it was perfectly reasonable to ask just how much of evolutionary change occurs at, and how much between, speciational events.

This is not the appropriate place to review the evidence for and against the theory of punctuated equilibria, but it is well that we consider very briefly the nature of the purported evidence. So far as the legitimate evidence goes, it has to be a matter of finding samples of the fossil record that are rich enough to show us whether change occurs only at the time of speciation. Because it includes a universal negative proposition, the theory of punctuated equilibrium in an extreme formulation can be refuted by showing just one instance of evolutionary change between speciation events. These conditions would seem to have been met, and the theory has been refuted, though it may be appropriate to modify it rather than to reject it outright. Unless a lot of paleontologists are mistaken, populations with a good fossil record are readily shown to undergo transformation without splitting up into species (Chaline & Laurin, 1986; Chaline, 1987; Sheldon, 1987; Krishtalka, 1993). Putative examples of punctuated equilibria turn out not to bear critical examination (Brown, 1987). Part of the problem has had to do with equivocal empirical evidence, with alternative explanations fitting the data just as well. But part of it has had to do with taxonomic artifacts. Unlike the artifacts that result from accidents of deposition, preservation, erosion, and the like, taxonomic artifacts are the consequence of misinterpreting the data of systematics.

To a considerable extent, the apparent periods of stasis and discontinuity may be real, perhaps the result of long-continued stabilizing selection. On the other hand, there are all sorts of artifacts that result from the activities of those who produce and use the classifications. A "monographic high" in organic diversity is an apparent high level of diversity that reflects where a taxon has been extensively studied. Such peaks are apt to be located near specialists on the taxon, along roads, and in places that are pleasant to visit. Systematists also tend to specialize upon a stratigraphic or a geographic basis. They tend to assume that two forms are different simply because they are found on two sides of a temporal or spatial boundary. The manner in which stratigraphic ranges are scored may also produce artifacts. When the presence of a taxon has been recorded from anywhere within a stretch of time, the bar diagrams that are drawn to show this fact often depict the range as if it extended throughout the entire period, from the very beginning to the very end, when in actuality there is no real evidence one way or another that they began to exist or ceased to exist at those times. Indeed it is very difficult to determine whether a lineage that is approaching a stratigraphic boundary was really present, because the closer one gets to the boundary, the fewer the samples; this has been called the Signor-Lipps effect after the paleontologists who first drew attention to the problem (Signor and Lipps, 1982; see also Signor, 1985; Ager, 1988).

"Pseudoextinction" is a taxonomic artifact that results when a taxon that has evolved is treated as if it had ceased to exist (Van Valen, 1973; Fortey, 1989). The most notorious example has already been mentioned. This is, of course, the dinosaurs, one group of which, the birds, avoided the extinctions that wiped out a lot of their relatives at or around the end of the Mesozoic. The overestimation of extinction was studied in detail by Smith and Patterson (1988), who put their respective erudition about echinoderms and fishes to very good use. Many of the supposedly extinct taxa were paraphyletic assemblages from which the survivors had been removed. (For more on artifacts from the point of view of systematic theory see: Schopf, 1982; Padian & Clemens, 1985; Briggs et al., 1988; House, 1988, 1989; Donovan, 1989; Pinna, 1995; Foote, 1996.)

If one wants to understand biotic diversity, categorical rank is largely irrelevant, for it correlates very poorly with the structural and functional properties that are of ecological importance (Bambach, 1985). Paleontologists' awareness of this point has called into question the entire paleontological tradition of going through the *Treatise on Invertebrate Paleontology* and counting the ups and downs in the number of taxa that have been recorded in the fossil record. This enterprise has much in common with neontological work on species diversity, which has largely involved geographical comparisons, especially between higher and lower latitude biotas. Even the species is not a very satisfactory level at which to carry out such analyses. In my own discussions on such trends, I have mainly emphasized ecological patterns with respect to modes of reproduction, especially sex and dispersal (Ghiselin, 1974a). It seems to me that one gets less ambiguous results when one treats a single parameter, such as the amount of recombination, body size, or chemical or physical defense (Vermeij, 1978, 1987), and places the materials in an intensive ranking rather than try to produce a scale with extensive units. The accentuation of sexual dimorphism in the deep sea, with males becoming increasingly dwarfed in correlation with lowered food density, provides an example of a causal linkage that is more than just suggestive. By the same token it has been possible to correlate various aspects of larval biology with extinction events and other diversity phenomena (Gilinsky, 1981; Hansen, 1982, 1983; Valentine and Jablonski, 1983, 1986; Jablonski, 1986; Chatterton and Speyer, 1989). Of course the causal basis of such correlations is not so clear for the fossils as for the extant organisms, because the general biology and habitats are harder to determine.

On a crudely comparative basis, it certainly stands to reason that organisms will be more abundant and more diverse under optimal conditions of existence. So naturally we find more reef-forming corals, in terms of both organisms and species, in warm and shallow waters in the present-day ocean, and likewise their ups and downs in the fossil record can reasonably be explained by climatic changes. This in spite of the fact that the correct explanation is not a straightforward consequence of how temperature affects metabolic rates. And the numbers of higher taxa show the same kinds of patterns. However, when we try to come up with more than just a crude indication that the biota is expanding or contracting—perhaps some groups more than others—we face a serious problem with respect to the comparability of the units in question.

If we consider what it means to compare the numbers of biological species in various places and at various times, we do have a class of units with something objectively real and theoretically important in common. Alas, that something is merely the ability to evolve as a unit and to do so independently of other such units. There may be something that accounts for such a unit expanding or contracting in numbers, but in all probability it is not possessing the properties necessary or sufficient to make it a species. Whatever is responsible may include some emergent properties of populations, but probably it is a matter of the organisms' having certain contingent properties that allow the species in question to persist in the face of competition and perhaps other influences.

Now, of course, there is a general tendency for organisms that are more closely related to have more in common, so that competition between them is more perfect and more intense, so they are more apt competitively to exclude each other. But there is no particular relationship between the existence of such interactions and taxonomic rank; often very close relatives will coexist, whereas quite distant forms will not. So there should be some crude statistical correlation, such that the species within a genus will tend to compete more intensely than species of different genera but the same family, and so forth. But even that will disappear when we try to generalize about species or genera in different phyla.

In other words, above the species level, taxa of the same rank have nothing scientifically interesting in common. That they may appear to do so can readily be explained. In the first place, community of descent means that within each taxon the higher groups will have earlier common ancestry than the lower ones, and we get the misleading correlations just discussed. In the second place, systematists have felt that some kind of equivalence ought to exist and have attempted to provide one upon a subjective basis. This equilibration can readily be accomplished by raising or lowering the rank of the groups. The game tends to be given away by the tendency to inflate the taxonomic rank of groups that are deemed particularly interesting. Human beings in the broad sense provide plenty of bad examples; Mayr (1963b) in a seminal paper went a long way toward improving matters, but molecular work suggests that he did not go far enough (Cherry, Case, and Wilson, 1978). The birds, I am informed by one herpetologist, would make a good family of lizards, but ornithologists would probably dismiss such statements as hyperbole.

No doubt some entomologists feel that the orders of insects are somehow equivalent one to another, and some malacologists may say the same for the orders of gastropods. One wonders, however, how readily such persons would agree that an order of insects is equivalent to an order of slugs and snails. As a malacologist who has taken a few courses in entomology, all I can say is that the orders of gastropods are not equivalent to one another, and I would hardly know where to begin in deciding whether they are equivalent to those of insects. By explaining a lot of malacology, I could justify my conclusions. Indeed, Boss (1978) has done a fine job along such lines with the bivalves. But rather than burden the reader with a lot of arcane and technical information, let us try some examples that will be more widely intelligible, and consider whether the phyla of animals are in any sense equivalent units. In so doing we will reiterate some materials that are supposed to be presented in every introductory textbook of biology. It may help to grab an invertebrate zoology text and use the pictures to refresh one's memory while we proceed. There is a conventional order that smacks of the *scala naturae* in that it usually ends with ourselves, but that doesn't make any real difference, especially since we won't follow it except to begin with our remotest relatives.

We may pass over as really *incertae sedis* two minor groups that are phyla because nobody is sure where to put them: the very simple, free-living Placozoa and parasitic Mesozoa. The only thing we need add is that if the Mesozoa could be shown to be related to flatworms, as some workers have suspected, they would no longer be treated as a phylum and might be treated as a class or even an order with two families. The sponges are generally considered a very early offshoot of the lineage to which the rest of us belong. Even when treated as equal in rank to the rest of the Metazoa, they have generally been treated as a single phylum with three classes. One of these, Hexactinellida, is distinct from the rest and perhaps is their sister group, though older authors tended to put Calcarea in that position. In order to avoid changing the rank of one of the three traditional classes, one might reasonably treat the two clades as superclasses, or perhaps subphyla to make the relationship explicit. However,

Bergquist (1985) proposed that the Hexactinellida are so different from each other that the sponges ought to be treated as two distinct phyla. She provided no justification for that particular rank, however, and all the differences are technical matters of cytological ultrastructure, biochemistry, and the composition and geometry of the skeletal elements. If we add fossil sponges, we find that there was a long tradition of treating the Archaeocyatha as a separate phylum, even though they had the general configuration of sponges. There were two justifications for this ranking. First, it was possible that the resemblances are the product of convergent evolution. Second, they had some properties that are absent in extant sponges. In the first case, we have rank assignment based upon ignorance, in the second, without any "amount" of difference being invoked. In recent years with more fossils and cladistic interpretation, the Archaeocyatha have come to be treated more and more like a group of conventional sponges (Wood, 1990). Like the trilobites, they flourished in the Paleozoic but didn't make it to the Mesozoic.

The usual practice these days is to treat the Cnidaria and Ctenophora as two distinct phyla. The former includes the true jellyfish as well as corals and other nematocyst-bearing organisms, the latter only the "comb-jellies." A relationship between the two is a distinct possibility, but this has been denied, largely because the evidence is somewhat tenuous. Molecular evidence now supports the proposition that they are in fact closely related (Christen et al., 1991). If we accept that, for the sake of argument we could certainly revert to an older stage in the taxonomy of the group. Thus we could have a phylum Coelenterata with the Ctenophora as perhaps a class, equal in rank to the traditional Hydrozoa, Scyphozoa, and Anthozoa. (Treating the Cubozoa as a separate class is another issue.) The Ctenophora are a quite homogeneous, largely pelagic group, and making them a class would probably bother nobody. Indeed, the only reason for subdividing the Ctenophora into the classes Tentaculata and Nuda is that Ctenophora has been treated as a phylum; these groups, if they indeed deserve recognition, might just as well be subclasses or super-orders.

One might argue that although the Ctenophora are a small and homogeneous group, they differ in some sort of characters that are deemed "important" enough to justify distancing them from other coelenterates. Their possession of glue-cells (colloblasts) instead of stinging-capsules (cnidae) differentiates them quite neatly, but why is that sufficient to make them a phylum rather than a class? One might argue that their biradial rather than radial symmetry makes them distinct; yet the anthozoans also have a characteristic deviation from purely radial symmetry.

Passing now to the rest of the Metazoa (Bilateria) we may begin with the phylum Platyhelminthes, or flatworms. This group has undergone an enormous adaptive radiation of both free-living and symbiotic forms, including the trematodes and tapeworms. The rank of the various subunits would form an interesting topic for discussion, but we will here consider only what has been done with certain organisms when it has been uncertain whether or not they fit in. In particular, we have the Gnathostomulida, little worms which are still of uncertain position but which lack the synapomorphies that would have placed them within the Platyhelminthes. Before this was realized, however, there was no real problem treating them as flatworms. They sort of look like flatworms. They lack an anus. And in general they would make a nice little order within the class Turbellaria if they could reasonably be stuffed in there, though they cannot.

Now let the fun begin. The so-called pseudocoelomates are an assemblage the naturalness of which has always been seriously questioned, even though a good case can be made for some of them being close relatives, and even though the subunits are more easily defended. It would be nice if they all really had a characteristic body cavity called a pseudocoelom that resembles the true coelom of eucoelomates, and if the two kinds of cavity really and truly had separate evolutionary origins, the one not being derived from the other. In that case, we would have a good synapomorphy for the pseudocoelomates and one for the eucoelomates. All this is dubious at best, and even so it would leave us the question as to how to rank this taxonomic entity. A true coelom supposedly can be identified by an epithelial lining; a pseudocoelom supposedly lacks such things, whether primitively or secondarily being a major bone of contention. If organs were properly recognized as individuals that evolve, there would be far less muddled effort to force such defining properties on them.

Be this as it may, the pseudocoelomate assemblage as a whole, with a couple of exceptions, has been treated as a phylum Aschelminthes. But although some of the groups of aschelminths show evidence of close relationship, the tree of the group as a whole has long remained unclear. The "solution" has been to rank each of the groups not as classes but as distinct phyla. This even when, for example, Nematoda is obviously quite close to Nematomorpha. Here the reason for high rank of the taxa is clearly that nobody can decide how to classify the group in its entirety. In other words, the basic criterion of rank is, again, ignorance. Kristensen (1983) made the newly-discovered Loricifera a phylum, instead of calling it *incertae sedis*; but if, for the sake of argument, we accept the opinion of Neuhaus (1994) that Priapulida + Loricifera + Kinorhyncha form a holophyletic group, then the three together could make a nice little phylum with three classes.

The remaining animal phyla are often called the "eucoelomates" on the basis of their body cavity. A few of these phyla have some developmental similarities that probably reflect a common ancestor, and if so would form a natural supraphyletic taxon Deuterostomia: Chaetognatha, Echinodermata,

Hemichordata, and Chordata. The Chaetognatha, or arrow-worms, are very abundant creatures, but like many groups that are almost exclusively planktonic, they are not very diverse. Phyletic rank for them is partly due to their sharing very little anatomically with any other animals, so that if they are the sister group of any other phylum, they must have branched off very early, or perhaps undergone great secondary modification.

The phylum Echinodermata, again, consists of animals that have very little in common, as adults, with any others. The primitively bilateral symmetry has been effaced by a secondary but superficial radial symmetry, and there are peculiar features, such as a water-vascular system and tube feet, unique to the group. Furthermore, the classes both extant and fossil have undergone a vast radiation. So there are plenty of reasons, however imprecise they may be, for ranking this taxon as a phylum.

The Hemichordata are a small group of soft-bodied marine worms. Probably they are the sister group of our own phylum, Chordata, but perhaps they branched off earlier. It may even be that they are closer to Echinodermata, as a recent molecular study suggests (Halanych, 1995), or that some of them are; we don't know what the ancestral deuterostome was like. The phylum is divided into two classes, Enteropneusta and Pterobranchia, the former of which are solitary burrowers and the latter usually colonial animals that seem to be somewhat simplified. The problematic fossils called graptolites used to be treated as a class when treated as chordates, but they now appear to have been pelagic pterobranch colonies. Rank of class seems simply a matter of convenience. But why not, just for the sake of argument, grant that they are the sister group of the Chordata, and make them a subphylum within Chordata? The reason for not treating them as chordates is that a weak consensus has it that they do not have a notochord; though one can reasonably defend the homology of a structure called a stomochord to our own notochord. So the basic argument would have to be that a notochord, and one with a particular structure, is the "essence" of a chordate. In other words, Hemichordata is a phylum largely because of ignorance and bad metaphysics.

The lophophorate phyla (Phoronida, Brachiopoda, and Ectoprocta = Bryozoa in the strict sense) have sometimes been treated as a single phylum Tentaculata. They are easily recognized, and the major features of the body can be readily homologized. Diversity, such as it exists within the "phyla," is mainly at lower levels, suggesting that it would be easy to treat them as classes. The Brachiopoda somewhat resemble clams, so why not give them the same rank as the class Bivalvia within the Mollusca? Partly because of tradition: they were removed from Mollusca, and the simplest thing to do under such circumstances is to kick them upstairs. Partly because of ignorance: people aren't quite sure that Tentaculata is monophyletic. The phoronids are a very homogeneous little group of tubiculous worms: they would make a nice family of "sedentary" annelids were that their position. Ectoprocta are routinely treated as a class by those who want to put them together with Endoprocta in a single phylum Bryozoa.

The remainder of the Metazoa are the overwhelming majority, and all of them very likely had a pattern of segmentation rather like that which is found in modern annelids, though not everybody accepts this, because if so, it must have been reduced in some cases and in others even lost without a trace. Besides the annelids proper, we have Mollusca, Sipuncula, Pogonophora, Arthropoda, and a few minor groups to be mentioned. The Annelida have been variously circumscribed, and confusingly "defined" without it being clear whether it is on a morphological or a genealogical basis. Lately it has become increasingly apparent that Annelida is a paraphyletic grade from which various things have been derived. Be this as it may, everybody likes to put the Polychaeta and Clitellata (Oligochaeta + Hirudinea) in Annelida, so the group definitely contains the bristleworms, earthworms and leeches, and this gives us two or three classes. Echiura makes a nice additional class or subclass of aberrant polychaetes, but because the ontogeny does not recapitulate a segmented stage, they are often treated as a phylum. One could say the same for Sipuncula, except that there is even less evidence for previous segmentation, and their cleavage patterns suggest that they may be the sister group of the mollusks. Virtually everybody treats sipunculans as a phylum. Cutler and Gibbs (1985) divided Sipuncula into two classes, but evidently did so because making the group a phylum made that step easy.

The Pogonophora are aberrant worms from deeper waters that have only gradually been shown to have a lot of annelidan characters. Tradition has kept them a phylum, though because there are two distinct groups, they have been made into two distinct phyla by one of the leading authorities, who gives no explicit arguments for doing so (M. Jones, 1985; Gardiner and Jones, 1994). It seems doubtful that anybody else would object to making Pogonophora a class of Annelida with two subclasses or even orders. Molecular evidence has it that the two clades form a monophyletic assemblage (Winnepenninckx et al., 1995; see also Cullsen-Cenic and Flügel, 1995).

The usual interpretation of Arthropoda is that these animals are annelids (morphologically speaking) that have evolved an exoskeleton and otherwise become substantially modified. As was mentioned in an earlier chapter, the claim that they are polyphyletic has never been substantiated by showing that they are independently derived from two non-arthropod lineages. This eliminates one excuse for raising several arthropod clades to phylum level. What has been shown is that the features that are characteristic of modern arthropods display a lot of parallel evolution, something which is by no means unusual. Some people don't feel comfortable with an arthropod the legs of which are not exactly jointed. And this seems to be what justifies placing two rather homogeneous groups of uncertain affinity, which are generally treated as transitional between Annelida and Arthropoda, in phyla of their own: Tardigrada and Onychophora. Those who are less typological in their thinking have no real difficulty doing something such as dividing the phylum Arthropoda into Protarthropoda and Euarthropoda and putting tardigrades and onychophorans in the former.

It is traditional to treat the highest animals last, and as a malacologist I am of course pleased to end this survey with the mollusks. Ever since Cuvier divided the animal kingdom into four types, the mollusks have remained one of its primary subdivisions, and everybody seems happy with treating it as a phylum with a few more or less neatly delimited classes such as Gastropoda and Bivalvia. The only disagreement I have encountered comes from Mangold-Wirz and Fioroni (1970), who think that the Cephalopoda are remarkably modified and want to express this divergence in terms of higher rank. There seems to be no objective grounds one way or another, though it creates a paraphyletic taxon. We might add that some of the gastropods are very different, too. The gymnosomatous pteropods, for example, were once treated as cephalopods.

All this ranking seems very subjective, but I do think that we might use the Mollusca as a kind of yardstick to suggest what people are trying to do when they seek to equilibrate the rank of taxa. Although I did not at that time defend the notion, I suggested that the aim of ranking was a system such that things placed at the same level would be amenable to the same kinds of comparison (Ghiselin, 1981b). For example, if we compare an adult mollusk with an adult annelid or arthropod, we can generally homologize the basic organ systems, and we find that they are arranged in more or less the same way. The nervous system and circulatory system are good examples; however, the latter has expanded to become a haemocoelic cavity, and the coelom evidently has become reduced in extent, quite independently in both the mollusks and arthropods. So the homologies are rather general. A lot of parts that occur within a phylum have no obvious homologues outside of the phylum; the molluscan shell is an obvious example, as is the peculiar foot, a tongue-like feeding structure (radula), and the gill (ctenidium). So the homologies within phyla are more detailed. In comparing phyla that are more distantly related than annelids, mollusks, and arthropods, we may be able to homologize parts as tissues, but the arrangement of parts may have very little in common, the nervous systems of chordates and hemichordates, for example, are single and dorsal rather than paired and ventral.

Summing up, what I think people are trying to do when they rank taxa is to equilibrate on the basis of some criterion of degree of generality, but the procedure is anything but rigorous. The entities in question are "more or less general" in so many diverse, incommensurable, and often purely subjective ways that ranking them in that manner has little if any scientific value. This in addition to the point that the properties in question are largely a matter of tradition, artifacts of extinction, and how much information we happen to possess about one group or anther.

One might wish to say that the entities ranked at various levels have more or less general "body plans"—not necessarily translating this expression into German or attaching any idealistic connotation to it. If all one means is that as one proceeds upward in a hierarchical system, the higher taxa have more widespread diagnostic features than their components do, this is true. But it is a mere truism, and it tells us nothing about criteria of equivalence among taxa of the same rank. And the mere fact that features happen to be diagnostic of a taxon— at any level—does not provide us with the kind of information that is needed for understanding evolution. Since this point has been widely misunderstood, and a failure to appreciate it has given rise to many errors in the study of evolutionary mechanisms, we need to consider it in a bit more detail.

A diagnostic feature, of course, is one that is present in all the elements of a taxon and serves as a kind of differentia that distinguishes them from the elements of other groups. We have to qualify the statement, because it need not be present in all the organisms, for they might be absent in juveniles and present only in one of the two sexes. And in principle there is no reason beyond convenience why it could not be a genetically variable character. But why should parts or properties have this peculiarity? For any of a number of reasons, and that is precisely the problem. One possibility is that the feature was present in the common ancestor of the taxon in question. As a methodological assumption when working out genealogies, this is quite reasonable. However, where there is abundant material, all sorts of features turn out to be subject to evolutionary parallelism. Some of the features that are now widespread may be "general," because the ancestor, as a matter of contingent, historical fact, invaded a particular habitat in which that feature was highly advantageous, and it evolved separately in diverse lineages. If one assumes that a whole suite of such features was present in the common ancestor, then one gets the false impression that they have been there since the beginning of the existence of the group, perhaps arising suddenly by a kind of "macromutation" in the common ancestor itself.

Features are "diagnostic" in a broader sense if they were indeed present in the common ancestor but have been secondarily lost in certain lineages. This, we may recollect, was the basis of Platnick's notion that snakes really have limbs. But if a feature has been secondarily lost without a trace, then it becomes very easy to assume that it was not present in the common ancestor. Some phylogeneticists take it as a methodological assumption that parts never disappear without leaving a trace. They are obviously invoking negative evidence; the logical fallacy should be obvious. Furthermore, such facts as the disappearance of teeth from the jaws of turtles makes it clear that the reasoning is unreliable, but such is the life of the intellect. At any rate the methodological assumption that parts don't get lost without a trace often gets converted into a bogus "fact."

One might wonder if diagnostic features are something special, and that this gives reality to the higher categories in the system. Unfortunately the logic here is circular. It is as with the notion that taxonomically important characters do not vary, as Darwin (1859:46) pointed out. When systematists diagnose taxa, they search the list of known characters and select those that do not vary, are universally present within the group, and are absent in those organisms from which they are to be differentiated. A straightforward error that results from not appreciating this point is to get the impression that change occurs only at the time of speciation. Similarly, the fossil lineages have been broken up into "chronospecies" on the basis of differences having been detected. Small wonder then that Gould and Eldredge claimed that the diagnostic features of higher taxa are something special and that they need to be explained by some previously unrecognized evolutionary mechanism (for critique see Levinton, 1988; Levinton and Simon, 1980; Briggs, Fortey & Wills, 1992; Wills, Briggs and Fortey 1994).

As already pointed out, the empirical evidence, so far as it is available for organisms with a rich fossil record, shows that a lot of evolution does occur between speciation events. In the case of extant organisms, another kind of evidence leads to the same conclusion. Stanley (1975b; see also Stanley, 1979) introduced the term 'species selection' for a kind of supra-organismal selective process that he believed could account for much of macroevolution. He tried to explain the existence of sex as a consequence of such a process (Stanley, 1975a). I had already anticipated such a model in my own efforts to refute "group selectionist" explanations for those phenomena (Ghiselin, 1974a). The empirical evidence clearly shows that sex varies geographically within populations and is distributed in close correlation with ecological variables. Similar arguments have been marshalled against efforts to explain size trends as due to events that occur at the time of speciation (Maurer, Brown & Rusler, 1992). There are plenty of clines within species. But it is all too easy to dismiss each and every counterexample as not what one had in mind. On the other hand, the individuality thesis at least suggests the possibility that species might be selected, and this is one reason why persons interested in macroevolution have paid considerable attention to it (see Damuth, 1985; Damuth and Heisler, 1988; Lloyd and Gould, 1993).

The notion that the diagnostic features of higher taxa are something special was fundamental to the misguided thinking of such advocates of macromutation as Schindewolf (1950). As one passes upward in the taxonomic hierarchy along a single lineage, of course one finds that the groups one is dealing with are successively older. And it stands to reason that the time of origin of the higher taxa was earlier than that of the lower taxa. If one believes that the diagnostic features embodied in a pictorial diagnosis (*Bauplan*) correspond to some kind of "amount of difference," then it may appear that qualitatively and quantitatively different events took place at the time of common ancestry of phyla and classes, on the one hand, and genera and species, on the other. But there is no reason whatsoever to believe that, and the empirical evidence, where available, shows that there has been what Heberer (1959) called "additive typogenesis," the diagnostic features being added sequentially.

The point that higher taxa are "older" than lower ones has sometimes been adduced as an argument that there has been something peculiar about evolutionary lability in the past when compared to the present. Rather than provide an explicit critique of Gould's (1989) *Blunderful Life*, I will only point out that so far as the individuality thesis goes, Gould deserves high praise for emphasizing the importance of historical contingency in evolution, but he continued to treat the historical entities as if they were classes, with some unfortunate consequences. In dealing with the macroevolutionary phenomena in question, we need to consider why it is that some features have been more conservative than others, and especially why the more conservative ones make good diagnostic characters. Ideally, a good "marker" for a natural (holophyletic) group is a derived character, one that cannot possibly have arisen more than once and that, once gained, cannot possibly be lost. Another way to put it is to say that homoplasy (convergence, including parallelism, and reversion) is impossible. In Chapter Fifteen, certain kinds of chromosome rearrangements were treated as good examples of just that. Where such conditions have been met, working out genealogies is easy, given enough material. However, in practice we do not have such ideal materials, and the lability or conservativeness of a feature is a matter of degree and of historical contingency as well.

The amount of change that may be observed in any lineage is partly a matter of how much time has been available for change to occur. It also depends upon causal influences, some of which are intrinsic and some of which are extrinsic to the organisms; the former include such things as developmental constraints and the coadaptation of parts, the latter what is often referred to as the "environmental" circumstances which may exert strong or weak selection pressure. There does seem to be a kind of "inertia" that results from developmental pathways not readily being restructured and from organ systems having a limited range of functionally workable configurations. However, there would seem to be a full continuum in the degree of such intrinsic lability, and it may vary from group to group. But what kind of change, and how much of it takes place, will be contingent upon the extrinsic influences, especially the environmental circumstances. So the fact that an archaic feature is retained might mean that it is crucial to some kind of developmental mechanism, or it might mean that the organisms have never left the environment in which it has been advantageous. Among the latter we may include adaptations in groups that happen not to have diversified into environments where these were no longer useful, as well as adaptations that are useful in a range of environments other than those in which they originated.

From what we know about developmental mechanics, it seems very likely indeed that the notochord, which is diagnostic of Chordata, is retained because it has become bound up with important morphogenetic mechanisms. But in many other cases—notably the sort of segmentation with teloblastic growth that we find in annelids—this has by no means been established upon an experimental basis well enough to rule out, say, its earlier existence in sipunculans or echiurans. Spiral cleavage is widespread among animal phyla, but its loss without a trace is manifest in cephalopods and some other groups. Larval types such as the trochophore are very useful in connecting phyla, but they have repeatedly been lost, especially upon invading the land.

What makes for a good diagnostic feature is even more complicated, however. Sometimes a good diagnostic character arises because different lineages have diverged along alternative pathways of adaptation (Bock, 1959). They have responded to a functional or environmental challenge in different ways, either of which evidently works equally well, but once a lineage has adopted the one or the other, there is no turning back or switching over (at least under ordinary circumstances). A familiar example is the fourth aortic arch of higher tetrapods. In mammals the one on the left remains, and the one on the right has been lost; in birds the one on the right remains, and the one on the left has been lost. Another familiar example is the nervous system being single and dorsal in Chordata but paired and ventral in annelids and other Articulata. In such cases we have, again, something very much like the fact that in some countries everybody drives on the left side of the road, whereas in others everybody drives on the right. As a matter of contingent fact, everybody in Japan drives on the left. But are we to wave a metaphysical banner proclaiming that brute fact from the top of Fujiyama? Is driving-on-the-leftness part of the essence of Japaneseness? It may be that the Japanese are pretty much stuck with driving on the left side of the road. On the other hand, we have the example of Sweden, which did make the change, albeit at much expense and inconvenience. Whether one gets irrevocably committed here may be a matter of degree.

This implies that we had better be rather skeptical of suggestions that maybe evolution has lost most of its flexibility due to the genome having become congealed. It does seem that there have been bursts of adaptive radiation at certain periods in the history of life. And yet the filling up of the environment with large numbers of organisms would tend to restrict the opportunities for unlimited diversification. Diversification intensifies when conditions have ameliorated after a mass extinction, so extrinsic factors seem highly likely.

The notion that life has lost its creativity is partly, at least, a taxonomic artifact. The closer we get to the present, the more likely indications of relationship are to be preserved, the less likely they are to have been lost. The sample is biased, because modern forms that have deviated considerably very often retain diagnostic features that link them to their closest relatives, so that the differences may or may not be recognized by high categorical rank. A fine example is provided by the Concentricycloidea, a group of deep-sea echinoderms originally described as a new class (Baker et al., 1986). The rank of class was justified on the basis of aberrant morphology, for the gut in these creatures has been lost, and the tube feet are arranged in a circle, quite unlike anything known in any other echinoderms. Nonetheless, the describers pointed out resemblances to asteroids and ophiuroids. Subsequently, convincing evidence was adduced to show that the Concentricycloidea are highly aberrant sea-stars, cladistically falling well within the group (F. W. E. Rowe et al., 1988; Belyaev, 1990). So should another class, order, family, or just a genus have been added to the list of echinoderms?

If there are many cases of overspecialization leading to evolutionary dead ends, at least one mechanism has been proposed to circumvent that, and this involves getting rid of the later stages in ontogeny. It may well be that quite a number of higher taxa have been produced by some kind of juvenilization, especially progenesis in which sexual maturation occurs early and the later stages of ontogeny are eliminated (Harding, 1954; de Beer, 1958). As a mechanism for giving rise to new phyla, it seems at least plausible for Sipuncula and Mollusca. As a mechanism for giving rise to classes, it seems at least plausible for Holothuroidea within the Echinodermata. Within the Crustacea, the Maxillopoda are thought to have arisen by progenesis (Newman, 1992). Progenesis in general is by no means an unusual phenomenon. As Darwin was well aware, it provides a straightforward mechanism for the evolution of dwarf males in cirripedes (Newman, 1980) and certain other animals where the females remain unaltered (Rieger, Haszprunar and Schuchert, 1991). Hence attributing the origin of higher taxa to it seems eminently reasonable.

This is not the place to contest the empirical issues surrounding macroevolution and its supposed mechanisms. The only point to be made here is that if we are to achieve progress in this area, our classifications need to be based upon something other than tradition, ignorance, and bad metaphysics. Whatever phyla are, they are not, as Gould (1989) puts it, "designs for living," an expression that is all too cute and doesn't really mean anything. Phyla at best are lineages, i.e., the descendants of a common ancestor, that some people for any of a number of reasons happen to have ranked between the class and the kingdom in a Linnaean hierarchy. Whatever may be the supraspecific level at which a taxon is ranked, the criteria used to differentiate it from other groups have no particular biological significance. As a matter of contingent fact, they do happen to be good indicators of relationship, but this too may be for any number of reasons.

So if we want to understand the history of life on earth and explain it properly, we need something other than generalizations about classes of genera, families, orders, classes, phyla, and the like. We need an explicit history of the lineages themselves, a kind of scenario that tells us what really happened to real organisms, and real populations thereof, in real environments. To the extent that development has prevented, constrained, or facilitated changes, this will have to be documented, and not just presupposed. Much the same can be said for the anatomical, physiological, and environmental aspects of the picture. If this seems a big order, nonetheless it means that paleobiologists are not about to run out of problems.

Furthermore, we can solve all sorts of problems without recourse to equivalent categorical rank. The causes of extinction and of adaptive radiation are a case in point. With respect to extinction, it has often been reported that the largest, or the most morphologically "bizarre" organisms, are the ones least apt to survive. Taxa that have recovered from extinctions often re-evolve superficially (n.b.) similar morphologies, as has been nicely documented by Urbanek (1996) for the graptolites. Here is not the appropriate place to evaluate such empirical claims or to attempt to interpret them, perhaps in ecological (or economic) terms. My point is only that the apparent patterns can be analyzed irrespective of whether they characterize, say, an order or a phylum, and raising or lowering the rank of a taxon should make no difference whatsoever. The nomothetic generalizations that we seek are about classes of radiations within clades, not extinctions or radiations of clades that happen to be ranked at any particular level. And the goal of history should be to explain, in terms of laws of nature, why it was that the individual species were able to speciate and radiate, on the one hand, or became extinct, on the other, under particular circumstances.

One might speculate just a bit about what we might expect from such an enterprise. For one thing, the kind of arms race that Vermeij (1978, 1987) has documented, with mollusk shells becoming ever more resistant to ever stronger predators such as crabs, fits in quite nicely with the Darwinian view of competition-driven directional change. And it may well be that, as Vermeij (1995) suggests, global "macroeconomics" determines what kind of competition prevails.

Furthermore, Darwin's view that progress is a real phenomenon makes a great deal of sense if, as I have lately suggested, it is viewed as the analogue of technological progress (Ghiselin, 1995a). If this be so, then we ought to be able to write a history of the living world in which the rise and prosperity of lineages is accounted for by means of important functional innovations—an economic history much like the history of industrial and technological innovation (Mokyr, 1990).

Such narrative is already available. For example, Kluge, Gauthier, and Rowe (1988) have accounted for important trends in the Tetrapoda in terms of the animals' acquiring the ability to run and breathe at the same time. Such innovation might reasonably be compared to the invention of a kind of horse collar, which allowed the animal to draw a heavy load without cutting off the air supply, thereby greatly increasing the productivity of mediaeval farmers.

I stress the economic aspect because it offers an alternative to older scenarios that tried to explain things in terms of survival of the fittest, rather than reproductive success in a competitive natural economy. For example, Romer (e.g., 1945), in his very successful textbooks, cast his explanations of the evolution of vertebrate terrestriality in terms of the animals being able to survive when oxygen in the water gets scarce and to slither from pond to pond in times of drought. As someone familiar with marine pulmonate gastropods, it strikes me that the ability to breathe air allows an animal to feed and reproduce at times and in places where non-amphibious creatures must remain inactive. So I would suggest that we might rewrite the scenario with the emphasis upon taking advantage of "business opportunities" rather than just keeping the enterprise intact. The Synthetic Theory quite generally conceptualized natural selection in such purely negative terms, as if it were merely culling out the unfit. And it treated competition as if it were something to be avoided in order to keep from going extinct. Such misconceptions are still remarkably widespread.

If Darwin's ideas about the natural economy were better understood, therefore, we would have much better prospects for an understanding of macroevolution, including its static, progressive, and regressive aspects. On the other hand, such an understanding has long been hampered by lack of appreciation for the fundamental principles of systematics. As Darwin was profoundly aware, a major difficulty in dealing with progressive evolution is the problem of incommensurability. It is the same problem that we have encountered with similarity and with the equivalency of categorical rank. We can say that microscopes were improved by means of the iris diaphragm, the achromatic lens, and other features, but there is no extensive unit of improvement here. And by the same token, we can say that both electron microscopes and computers have been improved over the last few decades, yet there is no common scale upon which we can compare the two. The same may be said of the progressive changes that have occurred in eyes, brains, and other organs. But the fact of the matter is that such innovations do tend to accumulate, and furthermore, the history of the world makes a great deal more sense when we can document what these are and what role they have played. So rather than try to measure the unmeasurable, our goal should be to document that which can be documented.

18

Toward a Real History of Life

"Light will be thrown on the origin of man and his history."

—Darwin, *Origin of Species*, first edition, page 488

Science is a body of knowledge: a cohesive individual and a whole with interdependent parts. It is no mere list, or dung-hill, of propositions. I affirm this at a time when the ideal of the unity of science has fallen upon hard times, and perhaps with good reason. An older philosophy of science tried to formulate its conceptions of what science is on the assumption that history is a mere curiosity. Science consisted of timeless laws and nothing else, even though the attainment of scientific knowledge was utterly unthinkable without reference to particulars. Such views were of course incoherent; one looks downright silly affirming that astronomy is a science out of one side of one's mouth while denying it out of the other. Failing to find a common theory or a common language, many philosophers of science have been tempted to just give up. Anything to avoid thinking about metaphysics.

If we admit to the interconnectedness of the various kinds of science and to their historical character, and think of them in processual terms, they can readily be made out as evolving together. One might argue this proposition upon an historical basis, and some have, but here we will approach the issues in a more strictly philosophical spirit. We will attempt to develop a "dual aspect" theory for biological concepts, in which phenomena are envisioned both from the point of view of history and from the point of view of law. Although keeping these viewpoints conceptually distinct may be desirable for the sake of analysis, to do so in the context of research may have substantial disadvantages.

The notion of a "dual aspect theory" has been used with respect to the philosophy of psychology, where it is proposed that the mind is neither distinct from nor identical with the body, but the two instead are different "aspects" of the same thing. Architects often draw the "plan" and "elevation" of a single building, depicting it from two viewpoints, or aspects. A point of view is neither true nor false, although some points of view are more informative or less misleading than others. They may indeed be complementary, especially if they provide us with different information.

One point that might be clarified a bit by such an analysis is the notion that the biological species concept is "non-dimensional," a claim that appeals to some people but never made much sense to me. I suppose that in the manner in which it is usually defined, the biological species tends to be conceptualized from a localized and atemporal point of view. Yet there seems to be no good reason why we must treat a biological species, an organism, or any other individual from a "synchronic" aspect, as if it were definitionally a "slice in time." People are perhaps not of one mind, however, whether they want to deal with species as if they had a history or as if they did not, and very likely they shift from the one to the other without giving it much thought. Such ambiguity would seem to be rather common in biological discourse, and of course it is apt to give rise to confusion, though not inevitably. At least some terms may be easier to understand if we bear this in mind, and we now proceed to consider several of them, namely 'function', 'adaptation', and 'fitness'.

My point with respect to 'function' should not be confused with the less subtle notion that the term is used equivocally. Obviously the term has a mathematical sense as well as a biological one, but this would seem rarely if ever to have confused anybody. Likewise when Bock and von Wahlert (1965:274) chose to define the function of a feature as "predicates which indicate all physical and chemical properties arising from its form," they stipulated in a way that has found few if any followers and is not apt to be confounded with what most of us consider the real thing.

However, even with respect to ordinary biological usage, it may not be obvious to everybody that there are two quite different senses of 'function'. For example, "functional anatomy" addresses the question of how things function—how they work. This is by no means the same question as what might be the function of a part—what it contributes to such functionings. However, the two sorts of question will generally be addressed in the same discourse, for an understanding of the former may be necessary to deal effectively with the latter. How a whole functions is important context with respect to what a part does. We might say that the one deals with the mechanical, the other with the economic, aspects of the same basic phenomena.

In attempting to come up with a clear and lexical definition of 'function' in the latter sense, it has long been recognized that just any attribute of a feature will not do (Hempel, 1959, re-issued in Hempel, 1965; Nagel, 1977). Haemoglobin makes the blood red, but that is not its function. It is what George Williams (1966) termed an "effect." So the philosophical task has been to come up with criteria that will specify only those consequences of a part or other feature being present that are really implied by the term being used. Efforts to explain function in terms of necessary and sufficient conditions for the existence or proper working of the organism have been attempted, but they are not altogether satisfactory, as is generally acknowledged. For one thing, the presence of a feature that subserves a function provides only a weak sort of necessity: an organism can get along without some pretty important organs, an anus, for example, having been lost in various brachiopods, echinoderms, and other animals. And various alternative arrangements might have the same basic effect, so one has to invoke a list of alternatives if necessary and sufficient conditions are going to work (Ghiselin, 1974a). It is largely because the functional requirements of an organism can be met in so many different ways that it is so difficult to "reduce" biology to lower integrational levels.

Efforts to make function an evolutionary concept have been attempted by various authors (Wright, 1973; Bigelow & Pargetto, 1987) but these have encountered considerable resistance (Boorse, 1976; Prior, 1985; Millikan, 1989; Amundson and Lauder, 1994). It would be all well and good if we could define the function of a part as whatever consequences of its presence caused it to evolve through natural selection. But this runs into difficulty when we consider the extent to which evolution has occurred by changes in function. What something does now has nothing in particular to do with what it did in the past. If I use my old computer as a paper weight, the only thing that matters is that I use it for that, not what I acquired it for in the first place. A well-argued scenario showing that a feature has evolved a given function through natural selection provides good evidence for its function being just that. But one could equally well argue for that function being the correct one on the basis that it is currently being maintained by selection, and in spite of its having originally evolved with some quite different function.

It is no secret that pre-evolutionary biologists were perfectly capable of thinking in functional terms, both explaining the workings of organs and organ systems and ascribing functions to parts. William Harvey's *De Motu Cordis*, which was first published in 1628, is a fine example (translation in Harvey, 1847). Darwinian biology helped us to think historically rather than teleologically about such matters, but the heart is still the pump that Harvey said it was, and the valves in the veins that tipped him off to the circulation of the blood do indeed have the function that he thought they did. And the kind of reasoning by analogy with machines that guided his investigations is still with us. Both senses of 'function' can be applied to machines as well as organisms, and a "mechanist" might well wish to argue that the organisms are just a special case.

It seems to me that in our efforts to be philosophical, we are reading more into the concept of function than is really there. The physiological specialization of parts is a brute fact that can scarcely have escaped the attention of thoughtful anatomists from the earliest times. And the effects of injury or experimental manipulation on different parts of the body can only reinforce such an impression. The principle of division of labor was first elaborated in economics by Adam Smith (1776) and later applied to biology by Henri Milne Edwards (especially 1851). Their understanding of it was rather poor, however, and economists and biologists alike tended to assume that dividing labor always resulted in more effective physiology. Later it came to be understood that there are advantages and disadvantages to both the division and the combination of labor (Ghiselin, 1974a, 1978). In air-breathing vertebrates there has been a gradual separation of the respiratory and digestive systems, the advantage of which is apparent whenever some food or water makes its way into our lungs. On the other hand, although the palate allows us to breathe and chew at the same time, we have to expend a certain number of calories daily in support of it. Using the mouth as an organ of communication as well as of mastication provides a distinct economy, and being able to talk with our mouths full would probably not be worth the anatomical modifications that would be necessary.

One really doesn't have to understand the why of it to recognize the brute fact that physiological labor is divided, and to want to know the particulars of specialization. So when somebody asks what the function is of a given organ, the question is really about how labor is divided within the organism. It is the same sort of question that somebody might want answered when interviewing for a job at a university: who are the administrators, who are the professors, what do they teach, and so forth. If anything more is demanded, it is not some historical scenario about the social, economic, or other forces that have brought about this state of affairs, however interesting this may be and however important for answering other kinds of questions. A function statement pure and simple does not address that issue.

On the other hand, the function statement does presuppose something, namely a kind of causal relationship. The organism is assumed to have an economic character, and its parts to be mechanically configured in a manner that is appropriate to the kinds of work that are done. This basic assumption might be justified upon the basis of evolutionary theory, though the fact that we human beings design machines underscores the point that teleology is at least a logically possible alternative. The definitional point that counts is only that things are organized that way, and what brought them into being is irrelevant with respect to what the term means. The causal nexus here is particularly interesting, because a part has a function only in the context of some larger whole of which it is a component. Amputate one of a pair of kidneys, and it no longer has a function. Amputate both, and the entire organism will cease to function. As these examples illustrate, the two senses of function are applied at different integrational levels in a whole-part hierarchy. And they are complementary. A functional explanation places the functions of the parts in the context of the functioning of the whole. It tells us, for example, that an ultrafiltrate solution is produced in one part, and then water is resorbed in another. Or it tells how the entire kidney maintains the organism's water balance.

In discussing 'function' I have proceeded as if the term were purely an anatomical one, and in that science the term is generally restricted to such parts as organs. We anatomists rarely use it for other ontological categories, though this mainly reflects what interests us. And some of these other usages make the idea that a function refers to a part of a whole seem more than just a bit forced at times. For example, a territory is said to have a function, such as allowing a bird to acquire a mate; but it is not obvious what is the larger whole of which the territory is a part. A territory, however, is not a part of the ontological hierarchy of which the organism and the organ are members. Rather it is a place that an organism defends from other organisms because it is a valuable resource—a piece of property. We would hardly call such a place a part of an organism. So if I am going to apply the analysis of function suggested here, I will need to invoke some supra-organismal economic whole, consisting of the organism together with the resources that it controls. Common sense tends to speak against this, but consider a couple of examples that make it seem a bit more palatable. A snail secretes a shell: is the shell a part of the organism? Most malacologists would probably say that it is. What happens when the snail dies, and a hermit crab appropriates the shell? Is the shell now a part of the hermit crab? What of a spider together with her web? They form a very effective economic unit, however physiologically distinct.

If these maneuvers seem a bit odd, they nonetheless suggest where the problem lies and also one possible way to solve it. What applies in a pretty straightforward manner within a given ontological category may pose difficulties when we switch to another category or try to combine categories. Perhaps we should opt for alternative solutions. We always have the possibility that the word is used in various senses, and an analysis that makes sense for anatomy may not work for behavior.

The mere fact that there is no good reason to define 'function' in terms of evolutionary origin does not mean that function should be studied apart from the context of evolutionary history. On the contrary. What we need is a form of narrative history in which function occupies its appropriate place. Such narrative history began to emerge in the nineteenth century. The little book by Anton Dohrn (1875) on the origin of vertebrates and the principle of succession of functions is a sufficiently interesting example that I went to the trouble of translating it into English (Ghiselin, 1994b). Although Dohrn's hypothesis of an annelidan ancestry for the origin of vertebrates has been reduced to little more than an historical curiosity, the principles by means of which he attempted to reconstruct the history of our lineage, and also to explain how new structures originate and are transformed, have by no means lost their interest. He argued that parts evolve according to an orderly and law-like process in which one function gradually replaces another (*Funktionswechsel*). So gill arches become jaws, and so forth. In such scenarios the functions of parts are by no means defined in terms of what selection pressures brought them into being. In his book on orchids, Darwin (1862) emphasized the fact that structures had repeatedly changed their functions. He argued that selection, as it were, seizes upon whatever, as a matter of contingent fact, happens to be available.

Nothing but confusion, therefore, would be added by treating function as if it were an historical concept. It is a purely diachronic attribute, not a process, and the goal of evolutionary anatomy is to place it in historical context. On the other hand, the term 'adaptation' does imply history, for in one of its major senses it does indeed mean the product of a process. To be sure, it also has two other senses (Ghiselin, 1966c). One of these senses is the condition of adaptation. Another is the part or attribute that confers adaptation upon the adapted organism. Generally when there is adaptation, this is due to the presence of adaptations as a consequence of adapting, giving us a third sense. So I would argue that adaptation is the historical concept, not function, and the philosophers who favor an "etiological" definition of 'function' have been confusing two kinds of discourse. Evidently this is because they have failed to understand how evolutionary biologists deal with history. On the other hand, the fact that adaptations can arise fortuitously through what is often called "pre-adaptation," and then get maintained by selection, would seem to preclude simplistic definition in terms of the product of the process. (Substituting "adaptational predisposition" for "pre-adaptation" might help to avoid teleological connotations and to avoid some even less satisfactory neologisms that might as well be passed over in silence.)

Once we have dismissed such philosophical confusion about 'function' and 'adaptation' we can deal in a relatively straightforward manner with 'fitness'. It is interesting that Darwin had serious doubts about Spencer's expression "the survival of the fittest," even though he acknowledged that there were some advantages to it. Unfortunately, Darwin didn't really explain what is wrong with the term. The difficulties that have vexed others, such as its allegedly tautological character, may or may not have bothered him. At any rate the term is hard to define.

The usual approach during the last few years has been to treat 'fitness' as a "dispositional term" like poisonousness. And this does away with the charge that all it means is having survived. For a fit organism is not one that survives and does all those other things, but rather one that is endowed with the capacity, proclivity, tendency, or disposition to do so. Therefore, as a matter of contingent fact, it often happens that the less fit organism is the one that survives. The race is not always to the swift. So we get a "propensity" view which has had some able supporters (Brandon, 1978, 1990; Mills and Beatty, 1979; Beatty, 1980, 1992; Beatty & Finsen, 1989) as well as opponents (Rosenberg & Williams, 1986; Settle, 1993; Van der Steen, 1994).

An additional problem vexes those who use the concept of fitness, insofar as there is a question of how it works at different levels and under changing circumstances. The problem is most apparent when one realizes that in sexual selection, the "fitness" of a male organism relative to his conspecifics may actually decrease the "fitness" of his species relative to its competitors. And we get a kind of "Cambridge change" when supposedly fitness is lost as the environment fills up with increasingly well-adapted competitors.

Fitness is rather like wealth in this respect. If we have what are sometimes called "wasting assets" that are always undergoing depreciation, it makes a certain amount of sense to say that average wealth is always decreasing, and we would hardly object to the view of R. A. Fisher that because of mutation pressure, the average fitness of a population is always decreasing. Granting the laws of thermodynamics, it must; yet this very general proposition tells us little of real interest. But what of his claim that the environment is always deteriorating, because other species are becoming more fit, and therefore that fitness is decreasing? If wealth is defined absolutely, as a capacity to utilize resources, we get something very different from wealth defined relatively, by comparing one person with his neighbor. If the richest man in town goes bankrupt, there is a Cambridge change, such that the second-richest man becomes the richest, but in absolute rather than relative terms, everybody in town might well be worse off, especially if they had lost a source of capital or philanthropy. If everybody tried to enrich himself, strictly in relative terms, solely by reducing the absolute wealth of his neighbors, everybody would beggar himself as he beggared his neighbor. And when one person steals from another, his wealth is increased and perhaps his fitness as well, but wealth is thereby redistributed and not created, as it is in the manufacture and distribution of goods and services. In the theory of natural selection, including sexual selection, the trouble is that what really makes for fitness is maximizing the analogue of relative wealth, something which may or may not be accomplished by maximizing the analogue of absolute wealth.

Perhaps there are no insuperable difficulties here for those who address certain kinds of problems in theoretical population genetics (see Dunbar, 1982; Burt, 1995). They are not necessarily concerned with situations in which historical contingencies are important. They may be looking at the nomothetic aspect of evolutionary biology. Or they may be making some simplifying assumptions. Some of us, however, have to deal with the real world, and to succeed we must avoid the pitfalls one way or another. There are various ways in which this may be accomplished. Just because there is no such thing as "overall similarity," we needn't despair of creating a classification system in which there is a strong correlation among the kinds of properties that interest us. The mere fact that there is no single thing called wealth does not invalidate all efforts to estimate the prosperity of ourselves and our neighbors. And even if fitness is rife with incommensurabilities, biologists have to get on with their business.

In evolutionary biology, survival is one thing, reproduction another, and sometimes they are said to be "components" of fitness. This notion completely misses the mark, for in a more fundamental sense, reproduction is everything, and survival is nothing, except insofar as it contributes to reproductive success. In terms of bioeconomic theory, organisms are capital goods whose sole business it is to produce more capital goods that are much like themselves. All that matters is the return upon investment, with reproduction the sole currency. Whether to allocate resources toward maintenance or production is determined by what makes for reproductive success, and absolutely nothing whatsoever else. Darwin's little male barnacles are a case in point. For them, the struggle for existence gets reduced to nothing but development, dispersal, and copulation. They epitomize both the essence of masculinity and the purpose of life.

In thinking about such matters, I naturally considered what role the concept of fitness might have played in my own work, and was driven to the conclusion that it had played remarkably little, if any. In fact, my theory of the evolution of sex-switches was cast not in terms of fitness, but in terms of reproductive success. And here is the point. The theory asserts that when certain conditions are met, certain things will happen: this is the nomothetic aspect. But the theory also asserts that as a matter of contingent fact, those conditions have actually been met. This is provided by the historical aspect of the theory. In other words, the theory incorporates both laws of nature and an historical narrative. It is a theory about the real world and is justified in terms of what actually did happen, not by what might have happened had conditions been otherwise. There is no problem with particular organisms failing to reproduce in spite of their propensity to do so. This way of getting around the problems with respect to fitness agrees quite well with some recent ruminations by Wim van der Steen (1991, 1994), who suggests that evolutionary biology is natural history, and the philosophy of physics is largely irrelevant to the issues at hand. One might want to say that the dogs have been barking up the wrong tree. Or perhaps it would be better to say that they haven't been barking up a tree at all, when they should have been barking up a phylogenetic one.

It seems to me that all legitimate work in evolutionary biology has such a dual character. It is compounded of both history and law, and it aims to test hypotheses having to do with both its idiographic and its nomothetic aspect. We often lose sight of this when making excursions into the world of pure possibility. But one can also forget that what actually happened is not the only thing that interests us. To understand the world, we need to comprehend both aspects of reality, and that is why history is so important.

Much more is gained from adopting such a perspective than simply being able to dismiss the concept of fitness as irrelevant to the scientific research of real evolutionary biologists. Among other things, we might extend such considerations far beyond the limits of evolutionary biology. In consequence Darwinism sets the universal standard of excellence among the natural sciences.

At least a great deal that we evolutionists do now makes a great deal more sense. We need not be altogether puzzled when we find that Darwin, his basic evolutionary theory well in hand, spent eight years producing a monograph on the Cirripedia that ran well over a thousand pages long. Darwin was not just the Newton of the grass-blades; he was their Copernicus. The testing of evolutionary laws is carried out by relating them to particulars, such as species, lineages, and locations in space and time. The relevant data have to be made available and arranged in such a fashion that they bear upon the issues at hand. The organisms and populations have to be placed in the proper historical context if they are to tell us anything. And the more detailed the genealogy, the better the data can be made to confront the hypotheses. So it stands to reason that a serious student of evolutionary mechanisms either does the historical work himself, relies upon his colleagues, or leaves the questions unanswered. Furthermore, because the hypothesized laws under consideration are broadly applicable, it makes sense that those who seek them will attempt to evaluate them against as wide a range of materials as possible. Hence the need for comparative biology, rather than an effort to study a few "model systems" that might well be called domesticated essences.

Given that biology has such an historical character, and biology is by no means unique in this respect, it makes a great deal of sense that its hypotheses should be evaluated on the basis of their explanatory value. The claims that evolutionary biologists make are not just that the hypotheses in question are not false, but that they provide a legitimate causal explanation for known phenomena. They assert that certain events have in fact obtained, and furthermore that the events in question have been governed by stipulated laws. This point has been very poorly understood by persons with the wrong philosophical cast of mind. Darwin was pleasantly surprised when Herbert Spencer noticed, as no other commentators had, that the long argument in the *Origin* was based upon the epistemological principle that a theory that explains such a wealth and diversity of facts is in all probability true. Traditional philosophy of science, then as now, has tended to spurn such arguments. So much the worse for traditional philosophy of science.

According to the traditional view, as favored by Hempel, Popper, and others, prediction and explanation are essentially the same thing. And yet a great deal of evolutionary biology is directed towards explaining states of affairs that nobody would hope to have predicted. Of course evolutionary biology is full of laws that predict as well as explain. But its claims are also about the particulars of history, and the laws only provide some limitations, such that various possibilities might be ruled out. One has to know the particulars, but the laws are helpful in inferring the course of history itself. So there is a kind of dialogue or confrontation between the two that goes on in the thought processes of the scientist.

In such an endeavor it is necessary to dissect out the separate causal influences of history and of law. It is here that the individuality thesis proves so useful, for the two have so often, and so perniciously, been confounded. To this end, it is desirable for those who study the history and the laws to work in close collaboration, or better still for individual scientists to develop the capacity to combine the two. The division of labor can be carried too far.

Furthermore, by thinking historically, biologists are able to purge their science of teleology (Ghiselin, 1974a). And this means explaining what has happened within the context of an historical narrative (Mayr, 1983b). If we provide an account of how a feature of an organism has come into being, complete with the ancestral conditions, the initial and subsequent conditions of existence, the manner in which the configuration of the organism affected reproductive competition, and whatever other contingencies, laws, and principles might be germane, then we can cast off a vast burden of worthless metaphysical baggage. However difficult it may be to avoid talking about what something is "for," what "good" it is, for whose "sake" it exists, or in whose "interest" it is done, we can avoid fallacies of question framing that are implicit in such locutions.

But the basic implication of all this—that we don't need to think teleologically at all—has not gone over very well with some philosophers (Lennox, 1992, 1993, 1994; cf. Ghiselin, 1994a). They occasionally cite some biologists, such as Ayala (1970), who like purpose-talk, while ignoring real teleologists, such as Driesch (1893), on the one hand, and the serious criticisms of teleology of evolutionary biologists, such as Mayr (1983a, 1992c) and Ghiselin (1974a), on the other. And, insofar as they perhaps define 'teleology' in such broad terms that it means whatever they want it to mean, they perhaps have a point. On the other hand, there was a very good reason why Pittendrigh (1958) tried to clean up the language by introducing the term 'teleonomy' for what some philosophers might call "good teleology." Even Brandon (1990), who shows how we can recast everything in mechanistic terms, is indulgent with respect to "what for" questions and willing to call the legitimate answers thereto "teleological." I suspect that Brandon would change his mind if he took a somewhat harder look at the historical branches of biology.

At issue, however, is not just how one defines a term. Dennett (1983, 1995) was most offended when I insisted that we need not ask what is "good" about the various features of organisms (Ghiselin, 1983; see also Ghiselin, 1984b). He insists that we are compelled to take what he calls the "intentional stance," according to which one tries "to figure out *what the designers had in mind*" (Dennett, 1995: 229–30). What about my backaches? What did the designers have in mind? Why is it "good" to have them? Rather than give stupid answers to stupid questions, we need only observe that our ancestors shifted from quadrupedal to bipedal locomotion, and our spines were subjected to increased burden. Nothing was in mind, nothing was designed. The pain is not good. It is not even a necessary evil, but a contingent one, and even calling it an evil is a value judgment.

My rejection of the "intentional stance" did not occur in an intellectual vacuum, but rather as the consequence of having to deal with the realities of life in the field and the laboratory, and especially with respect to reproductive evolution (Ghiselin, 1969a, 1974a). If the history of post-Synthesis evolutionary biology has taught us anything, it is the folly of explaining this, that, and the other as having evolved "for the good of the species." This was, as mentioned earlier, a rediscovery of something that Darwin understood very well indeed. In his book on orchids, he ridiculed the natural theology of Paley (1802), while at the same time showing that supposedly "trifling" characters often had unexpected physiological significance (Darwin, 1862). Dawkins (1986) wrote a popular book that repeated what Darwin and others had said, but happily ignored not only Darwin's priority but the highly relevant contributions of Asa Gray and John Dewey (1910) (see Ghiselin, 1986b).

Unfortunately, it has been a bit harder to rediscover and appreciate the value of Darwinian historical narrative for the study of adaptive significance. The notion, so widely accepted in the early days of the Synthesis, that taxonomic characters are largely nonadaptive, rested upon nothing more than ignorance and lack of imagination (Robson, 1928). Latter-day critics of "adaptationism" all too often repeated the same old mistakes (Gould and Lewontin, 1979). Adaptation, maladaptation, degeneration, and progress are economic and historical concepts. The only way to identify these things is within the context of empirical economic history, and that means phylogenetics and explanatory narratives.

Our appreciation for the crucial role of historical narrative in scientific thinking helps us to understand why scientists often take an interest in the history of their subject, and provides good reason for their doing so. This in addition to the bad reasons, such as smearing one's academic adversaries. Science is itself an historical entity. It evolves as a result of something like a selective process—whether natural, sexual, or artificial we needn't specify. What is considered true or acceptable at a particular time is largely conditioned by historical causes. It may be nothing more than a matter of who wrote a textbook. In phylogenetics, traditions are among the most important characters, and all too often they are traditions masquerading as facts.

A good example is the phylum Brachiopoda, which, as was mentioned in Chapter Seventeen, has been problematic in the extreme. In recent textbooks the usual way of dealing with them and the other Tentaculata (lophophorates) is to treat them as somehow "intermediate" between the protostome and the deuterostome coelomates, often treating them as an early offshoot of the latter— the sister group of Echinodermata + Hemichordata + Chordata. Our early results with ribosomal rRNA sequence data indicated that the brachiopods are closely related to the annelids (Field et al., 1988; Raff et al., 1989). During the discussion at a Nobel symposium in which I defended this relationship (Ghiselin, 1989), Peter Ax said that the embryology of the coelom did not support the relationship. I answered that the published data showed various patterns of coelom development; the advocates of deuterostome affinities for brachiopods had in fact selected as typical the pattern that fitted their "archicoelomate" hypothesis. According to the archicoelomate hypothesis, sometimes called the enterocoel theory, of metazoan phylogeny, the ancestral metazoan has three pairs of coelomic cavities, and brachiopods and other tentaculates, as well as the deuterostomes, must have branched off before anything like annelid segmentation and multiple pairs of coelomic cavities had evolved (Remane, 1959; Emig, 1975; Siewing, 1972, 1979, 1980). Likewise suppressed have been data about the setae of brachiopods, which ultrastructurally are very much like those of annelids (Gustus & Clony, 1972; Storch & Welsch, 1972). In fact, the conspicuous setae in the larvae would appear to be one reason why many years earlier A. O. Kowalevsky (1874) had compared brachiopods to annelids. Neilsen (1985) has recently found in the brachiopod *Crania* not three but four pairs of coelomic cavities, each associated with groups of setae; it looks like an annelid larva rather than a tripartite hypothetical ancestral coelomate. Further work on rRNA has tended to support annelidan affinity of the brachiopods and other Tentaculata (Halanych et al., 1995). The issues here remain debatable, and the arguments are complicated because of over-interpretation of the molecules by some persons and suppression of the legitimate results by others.

On the other hand, one thing does seem clear: deja vu. Over a century earlier Edward Sylvester Morse (1873) explained how he had approached the embryology of brachiopods fully expecting to find that they were mollusks. And this because of the dead weight of tradition. Baron Cuvier had followed the usual practice of keeping the brachiopods in Mollusca, even though the resemblances were rather vague and superficial. So Morse had a difficult time overcoming the intellectual inertia necessary to gain a fair hearing for the annelid hypothesis. Nor was this unusual. When Kowalevsky (1866, 1871) found good evidence in the embryology of tunicates for their having a close relationship to vertebrates, von Baer (1873) made a concerted effort to show how difficult it was to derive them from such mollusks as clams. And we are still doing that sort of thing.

In logical reasoning it is highly desirable that one know that the premises are true and that the conclusions follow. In order to know whether or not the premises are true, it helps to know what the premises are. There is a real possibility of circular reasoning here. If the only reason for believing that the brachiopod setae are convergent with those of annelids is that the brachiopods are unrelated to annelids, one cannot legitimately argue that, because the setae are convergent, they are not evidence for a relationship between brachiopods and annelids (a synapomorphy). If one is going to avoid such circularities and other pitfalls, one actually has to find out what the premises are and what is their justification. As much as possible, things need to be rendered explicit. This means going back to the sources of our knowledge, including of course reinvestigating the organisms that interest us with new techniques both physical and intellectual, but also taking a hard and critical look at the sources through which we have derived our tradition. It would be all well and good, were we simply able to check up on the observations, but the labels that we put on our drawings are theory-laden in the extreme. To interpret the texts we need to know the assumptions, especially the tacit assumptions, and this means knowing a lot of context and background.

Since the parts of our knowledge are integral components of our knowledge as a whole, much gets lost when we attempt to treat them in isolation. In this sense, at least, we must reject pluralism and aim at synthesis and unified science. It would be foolish, however, to attempt to legislate how we shall seek the truth. In our efforts to attain rigor, let us not impose something akin to rigor mortis. Research should be conducted in a spirit of continued improvement in our methodology, aware of the fact that methodology is itself historically conditioned (Zanzi, 1991). In such endeavor much is to be said on behalf of the division of intellectual labor, but it bears repeating that the division of labor can be carried too far. The kind of historical knowledge that scientists need can only be provided through the active participation of scientists themselves in the study of their own subject. History is too important to be left to the historians.

We are at last in a position to conclude this work with a new approach to classifying the sciences. Earlier efforts along such lines have been so far from satisfactory as to make one hesitate. For example, the work of such nineteenth-century pioneers of sociology as Auguste Compte and Herbert Spencer might be mentioned as perhaps little more than historical curiosities. All too often the schemes reflect the priorities of the classifier more than anything else. Witness Saint Thomas Aquinas, who ranked Theology above all the other sciences both in dignity and in certitude (*Summa Theologica*, Book 1, Question 1, Article 5). Or Dilthey's division into *Geisteswissenschaften* and *Naturwissenschaften*. It makes history look more like one of the humanities than some of us would think desirable.

The approach to be considered will not be presented in the form of a detailed classification. Rather the underlying rationale will be sketched out and developed by means of a few pertinent examples. We may begin by incorporating the distinction that goes back to Wilhelm Windelband, between the nomothetic and the idiographic sciences. The Greek roots allow these terms to be roughly translated as law-propounding and person-writing, respectively. The choice of roots is not altogether satisfactory, but the terms are widely used and good enough for our purposes. Furthermore the distinction between the two fits in quite well with what has been said here about laws of nature, on the one hand, and history, on the other. But one division hardly makes for a classification. Therefore, let us divide the idiographic sciences, likewise along familiar lines, into the synchronic and the diachronic. Were this all that we did, however, we would leave out something very important, namely the sciences that interrelate the idiographic and the nomothetic, or the individuals and the laws. Without that, the nomothetic sciences are devoid of content, and the idiographic ones purely descriptive. Let us use the term 'synthetic' for those sciences that put the idiographic and nomothetic aspects of reality together. And let us proceed to a series of examples that flesh out the scheme.

We may begin with physics and the kind of astronomy that has been used by physicists and astronomers such as Copernicus, Galileo, Kepler, Newton, and Laplace. The laws propounded by such persons are too familiar to require comment, and astrophysics, insofar as it is purely concerned with laws, would be a nomothetic science. A purely synchronic and strictly idiographic astronomy would aim at the description of the universe at a particular moment in time; one might prefer to call it an "astrography." Purely synchronic sciences can be synthetic, as here defined, by relating the observational data to laws of nature, for example by explaining the shape of orbits in terms of gravitational attraction. Much of astronomy aims at just that. A diachronic but purely idiographic astronomy would attempt to describe the sequence of configurations of celestial bodies through time, and for that one might use the term 'cosmology', although it has other senses. Now, for a synthetic, diachronic astronomy, we would need an historical narrative that explains what has happened in terms of laws of nature as well as particular events. Where the entire universe is the object of interest, the term 'cosmogony' is traditional.

In geology, it is rather more difficult to enumerate the sciences that are purely nomothetic, though geophysics is an obvious example of one that tries to be. However, the idiographic ones are all the more instructive. Although a geological map tells us something about history, it really shows us where the rocks and formations are located at a particular time, and may therefore be said to be synchronic. Stratigraphy, insofar as it is purely descriptive, provides a fine example of a diachronic idiographic science. And plate tectonics, which provides an explanatory historical narrative, is synthetic as well as diachronic. It relates the configuration of the earth to what moves the parts of its crust around in conformity with laws of nature.

The study of organic beings has not infrequently been divided into neontology and paleontology, reflecting the distinction between the synchronic and diachronic aspects, respectively, of life through time. A synchronic and purely idiographic approach, which might be called "neontography" to emphasize its purely descriptive character, is particularly well exemplified by genomic sequencing, but the kind of cytography that records chromosome numbers and some kinds of gross anatomy would do. Once one starts tracing lineages by means of sequence data or chromosome numbers, the science becomes diachronic or historical, a kind of descriptive phylogenetics. A synchronic synthesis is of course possible; molecular biologists, cytologists, and some kinds of functional anatomists have attempted to explain things physiologically, or in terms of "proximate factors" as Herbert Spencer called them. It is proximate synthesis.

When we aim at diachronic synthesis with respect to lineages, populations, organisms, and their parts, our goal is an historical narrative that explains the successive configurations of living matter in terms of both individual events and laws of nature. It invokes "ultimate factors" and therefore may be termed "ultimate synthesis." The appropriate term for this science would seem to be 'phylogeny' by virtue of its traditional sense (Haeckel, 1866), and its concordance with terms such as 'cosmogony', which, etymologically at least, suggest coming into being, development, and evolution.

The same approach can be applied to the classification of the various branches of philosophy and intellectual history. Logic and the study of methodology are sciences that are generally thought of as nomothetic, attempting to deal with inference and investigation in general, apart from any particulars. However, there is supposed to be some connection between the laws and principles in the abstract and something idiographic that forms the subject matter of intellectual life. That subject matter may be conceived of, again, either synchronically or diachronically, the latter being the subject of intellectual history or, if synthetic, what might be called "epistemogeny," again aiming at an ultimate synthesis rather than just a proximate one.

The approach just suggested is "ultimate" only in the sense that it takes history into account. It should not be taken to mean treating the diachronic, the synchronic, the idiographic, the nomothetic, and the synthetic as if they represented historical stages. We are classifying the sciences, and we are classifying them functionally, as kinds of processes. A science is synthetic if it engages in synthesizing, not because the synthesis has been completed.

Once again we have to resist the old inductionist tradition, according to which we proceed from the simple to the complex, from the concrete to the abstract, from the data to the laws, and from the laws to the total picture. Such a travesty of science, again, is quite common among its practitioners. However, philosophy has often been cursed with the opposite sort of mistake, and some persons have supposed that a logician or pure methodologist can tell scientists how they ought to do their work. But knowledge evolves. The content of our scientific knowledge is not fixed, and neither are the rules that we must follow in order to obtain it. As knowledge evolves, everything has to be revised, and revised again. Synthesis does not put an end to investigation, but provides the basis for a new beginning.

When, in 1859, Darwin presented an abstract of his theory, he showed how a vast range of phenomena could be envisioned from a unitary point of view. There and in his other works, we find all sorts of examples and suggestions as to how such unification might be obtained, but above all else we find prospects for new investigations. The problems of heredity were evaded with magisterial skill, but nonetheless they cried out for solution. When, a century later, the architects of the Synthetic Theory assessed both their own accomplishments and those of their great predecessor, they were most impressed by the solidity of the foundations. A few decades later, we are all the more impressed. That is partly because we have gone back to the original sources and can better understand what Darwin was up to. Yet it is also because the theory has continued to evolve, through synthesis as much as anything else. We now have a much better conception of what the basic units are, of the roles that they play in the fundamental processes, and of why things like that matter. This puts us in a position to engage in a somewhat different kind of science, and a very different kind of philosophy.

APPENDIX:
APHORISMS, SUMMARY
AND GLOSSOGRAPHIC

Introductory note. This appendix provides an overview of the work and explanations of some important terms. The topics are arranged in conceptual rather than alphabetical order, but since the terms are in **boldface type** where they are defined, it can also be used as a glossary either by scanning the text or referring to the index.

Synthesis means putting things together so that they constitute a unitary whole.

Metaphysics is the natural science that deals with reality at its most fundamental level.

Ontology, or the theory of being, has to do with what and how things may exist, whereas **epistemology,** or the theory of knowledge, is concerned with what is known or knowable.

Absolute realism, herein embraced, holds that what is true or false has absolutely nothing whatsoever to do with what anybody knows or opines.

Classification in its most general and fundamental sense is the process of organizing knowledge.

Classification is a creative act that produces a classification and should not be confused with **identification,** which is finding a place for something in a pre-existing classification.

A **classification** arranges materials in a way that tells us something about them: a mere **list** has no such character.

Possibility is more fundamental than existence.

A **logical** possibility is consistent with the rules of valid inference.

A **physical** possibility is consistent with the laws of nature.

A **metaphysical** or **ontological** possibility is consistent with the categorical order of things.

A **taxonomic category** is a level in a hierarchical system of classification at which taxa are ranked.

An **ontological category** is one of the most general "headings" under which an entity can fall.

The distinction between **class** and **individual** is altogether fundamental: everything is one or the other.

Individuals are absolutely and without qualification **concrete,** whereas classes admit of abstractness, which admits of degree.

Individuals, unlike classes, can have no **defining properties,** and therefore their (**proper**) names can only be defined ostensively.

Individuals differ from classes in having no **instances,** and this is perhaps the best logical criterion that distinguishes them.

Individuals are **spatio-temporally restricted,** with a beginning and an end, whereas classes are spatio-temporally unrestricted.

Individuals are **incorporated** in other individuals, whereas classes are **included** in other classes.

Individuals may have the capacity to participate in **processes,** but for classes to do anything of the sort is metaphysically impossible.

Individuals are not referred to as such by the **laws of nature,** which are regularities of classes of individuals that when properly formulated are true of everything to which they apply.

Individuals can be, but do not have to be:
Persons, or other organisms
Substances
Cohesive
Distinctly bounded
Strikingly different from one another
Autonomous
Indivisible

Individuality is not **particularity,** for particularity and its contrast term **generality** apply both to individuals that are incorporated in other individuals and to classes that are included in classes: e.g., 'Germany' is more particular than 'Europe', and 'chair' is more particular than 'furniture'.

An individual does not unequivocally cease to exist when it ceases to be a member of some class and becomes a member of some other class, as for example, when it ceases to be a single organism and becomes a clone.

Herein we provisionally revise Aristotle's list of **ontological categories** to make it somewhat more of a classification rather than a mere list:
Substance
Process
Action
Affection
Place

Property
    Quantity
    Quality
    Relation
    Posture

As a rough approximation:

Substance is the physical substratum of individuals.

Processes are what individuals do.

Places are when and where individuals are located.

Properties are that in terms of which individuals change and manifest diversity.

Individuals may be classified according to their ontological category, such that there are individual substantial beings, individual processes (such as events), individual places, and individual properties.

Conventional, **substance metaphysics** treats substance as ontologically prior to the other categories and denigrates the others as parasitic.

Process metaphysics, which places what substances do upon at least an equal footing with the substances themselves, is perhaps more appropriate to the sciences that deal with living, thinking, and evolving.

Place metaphysics and property metaphysics are intriguing possibilities.

Numbers are purely abstract: two is not an individual, although there are individual dualities.

"**Character**" is ambiguous (equivocal), leading many to confound the parts of a whole with the properties of such parts.

A **definition** is the criterion for the application of a term to that to which it may refer.

A definition in the sense of a terse formula tends to leave out the fine print.

The definition may be expressed in quite different ways, paraphrased, or translated into another language, yet remain equivalent with respect to the meaning of the term.

Definitions may be lexical or stipulative. If **lexical,** they describe how a term is used. If **stipulative,** they propose how a term shall be used.

Ostensive definition connects a term to that to which it refers by "pointing" at it or one of its elements.

Intensional definition enumerates the defining properties that are necessary and sufficient for a term to apply.

Extensional definition enumerates the members of the class that a term designates.

Description enumerates the properties of things, irrespective of whether or not the properties in question are defining.

Diagnosis enumerates properties that are useful in identification.

Essences, minimally, are the defining properties of a class, but all too often they are burdened with some excess metaphysical baggage, such as being viewed as a kind of norm or the fundamental cause of a thing being what it is.

Essentialism, or **typology,** is conceiving of groups in terms of stereotypes, norms, and the like; it is especially pernicious when it attempts to treat those groups which are individuals as if they were classes.

A **hierarchy** is anything having **levels,** with the levels indicating a relation of **subordination** between the occupants of different levels.

A purely **subordinative** hierarchy is one, such that the individuals stand to each other in the relation of subordination to one another, but are neither parts of nor included in that to which they are subordinate.

An **incorporative** hierarchy is one, such that an individual at a given level is a part, or component, incorporated in an individual at a higher level.

An **inclusive** hierarchy is one, such that classes are included in classes of higher rank.

The individuals ranked at different levels in a hierarchy often do quite different things: organisms but not species procreate, and species but not organisms speciate, whereas genera do not do anything.

When species speciate, they produce a genealogical hierarchy made up of more and less incorporative lineages. Such lineages are individuals, even if they lack cohesion.

In the **Linnaean hierarchy** of formal taxonomy, the levels (such as phylum) are called **taxonomic categories,** and the entities ranked at such levels (such as Mollusca) are called **taxa.** The categories must be classes, whereas the taxa can be individuals.

Biologists recognize an incorporative hierarchy, such that populations are composed of lesser populations as well as organisms, and organisms are composed of such components as cells, and so forth.

The **biological species** is, roughly speaking, the most incorporative populational taxonomic category.

The **organism** is generally understood to be the unit of physiological autonomy.

A **species definition** is a definition of the taxonomic category "species" and aims to provide necessary and sufficient conditions for differentiating it from all other categories.

A species definition *is not* a definition of the name of an individual species.

The properties of organisms can be used to describe, or to diagnose, their species, but the name of the species itself has to be defined ostensively, and in formal taxonomy this is accomplished by type designation.

The following formula of the **biological species definition** is herein put forth as in some ways less bad than the available alternatives:

Biological species are populations within which there is, but between which there is not, sufficient cohesive capacity to preclude indefinite divergence.

Cohesive capacity is a propensity to stick together. Like other **dispositional terms,** such as 'potential interbreeding' and 'toxicity,' it is something that biologists find it hard to do without.

Putative alternatives to the biological species definition are largely:

1. paraphrases of the biological species definition;
2. proposals to have the term 'species' designate either more incorporative or less incorporative individuals;
3. efforts to treat species as if they were classes of similar organisms.

"Pluralistic" species concepts make the term 'species' equivocal: it can mean all sorts of things and therefore lacks the precision and clarity that is desirable in scientific discourse.

A species is *not* the place that it occupies in the economy of nature.

An asexual biological species is a contradiction in terms.

There is no compelling reason to insist that every organism should be a part of a species.

One can talk about the parts of species without calling any such parts species.

Languages, with their dialects and idiolects, are like species, with their subspecies and organisms: all such things are individuals.

Species may participate in evolutionary processes, as in the theory of species selection; this is possible because species are individuals, but whether they actually do so is an empirical issue.

Species have properties, e.g., sex-ratios, over and above those of their component organisms.

Species make a store of genetic resources available to their component organisms.

Because of sex, **populations** that can interbreed exist; their interbreeding is not to be confused with that of their component organisms, nor is their isolation one from another.

Reproductive isolation between species results from a lack of **compatibility** between the parts.

Compatibility is the capacity of individuals to function together.

Integration is the linking up of parts into a whole that functions as a system of interacting and interdependent subunits.

Coadaptation is the possession of features that allow different individuals to interact effectively.

Objective means with respect to the object known, rather than to the knowing subject. With respect to systems of classification, it means those that correspond to a reality that exists without reference to our apprehension of it.

Natural systems arrange material in a way that displays the objective order among things; artificial systems do otherwise, often for reasons of convenience to the user.

Homology and **analogy** are relations of **correspondence,** not similarity.

A **homologue** is a part of a part of an individual, whereas an analogue is a part of a member of a class.

Homologues can be traced back to a common ancestral precursor, or in the case of iterative homology, to a common causal influence that is spatio-temporally restricted.

Analogues are the products of convergent evolution, and classes thereof are not spatio-temporally restricted, so that a pair of them might exist anywhere in the universe. The correspondence generally has to do with the same laws of nature being operative; that cause may or may not have something to do with function, but at any rate common function is not defining of the relation.

Laws of nature are its regularities that correspond to physical necessity.

Laws of nature make no reference to any particular individual, but rather generalize about classes thereof.

Because taxonomic statements are about individuals, they are descriptions of contingent historical facts and are not laws of nature, even if highly reliable in forecasting.

Laws of nature about species are about kinds of species, for example, about in-bred species in general, and not about any particular species.

Laws of nature about embryology are about kinds of developmental processes, not about the historical aspect.

Laws of nature can be used to test hypotheses about individual historical events, at the very least by restricting the number of scientifically acceptable possibilities.

Darwin's biology is every bit as well established as Einstein's physics, and if either of them lacks laws, we had better say the same of both.

Misunderstandings about ontology often lead to misinterpretations of the fossil record and much else.

Clades are the individual branches in a **genealogical nexus.**

The classification systems that are now in use may or may not correspond to the genealogical order of things, and if they do not so correspond it may be better to work with tree-like diagrams rather than named taxa.

A **holophyletic,** or **strictly monophyletic,** taxon crudely defined is one that incorporates the common ancestor of the group and all of its descendants; in other words, it is an entire clade.

A **polyphyletic** taxon is one that leaves out the common ancestor; in other words, it is more than one entire clade.

A **paraphyletic** taxon is one that leaves out some of the descendants of the common ancestor; in other words, it is a clade minus some of its components.

Systems with paraphyletic taxa may be simpler, better in accord with vernacular language, and more conveniently expressive of features deemed important (such as major changes in organization).

When all taxa are strictly monophyletic, the amount of genealogical information is maximized and the level of ambiguity is minimized.

Pseudoextinction is an apparent but not real cessation of existence of a taxon that results from use of paraphyletic taxa; a lineage that is classified apart from the whole is misinterpreted as if it had ceased to exist, whereas all that it has really done is evolve.

Features that are useful in diagnosing a group are selected because of their invariability, and treating these as "defining" gives a false impression that change has been abrupt or discontinuous.

Above the species level, the categories are by no means equivalent, owing to the fact that the criteria for ranking are subjective and incommensurable.

In theory, the criterion for assignment of categorical rank is amount of difference, or degree of similarity.

In practice, the criteria for assignment of categorical rank are all too often tradition, ignorance, and bad metaphysics.

An individual may be viewed from a **synchronic aspect** (a slice in time) or a **diachronic aspect** (through time), but its ontological status is thereby unaffected.

The individuality of species, and of much else as well, implies that history should take a much more dominant role in biological thought than it has.

The long-term goal of evolutionary biology is an **historical narrative** that accounts for the way things are in terms of individual events and laws of nature; systematic biology gives the individuals by virtue of which a science is about something.

Teleology is a bad substitute for an historical explanation.

Function addresses the question of how things work, including how labor is divided within the body; it is not an historical concept.

Adaptation implies, at least covertly, an historical narrative that accounts for its existence; it is an historical concept.

Fitness, as the term is commonly understood, is a dispositional term that applies to what might happen at a given moment, but it has no narrative aspect.

Explaining the phenomena in terms of **reproductive success,** and relating them to a real historical narrative renders the problems of "fitness" moot.

The sciences may be classified thus:

Idiographic sciences are descriptive of contingent facts.

Nomothetic sciences propound the laws of nature.

Synthetic sciences put the contingencies and the laws together.

Synchronic sciences deal purely with things at a particular moment in time.

Diachronic sciences deal with things through time.

A synchronic, purely idiographic science describes a state of affairs at a given time.

A diachronic, purely idiographic science describes states of affairs through time.

A synchronic, synthetic science relates a given state of affairs to laws of nature. Because it deals only with proximate factors, it is capable of what we may call "proximate synthesis."

A diachronic, synthetic science provides an historical narrative that relates sequences of events to laws of nature. Because it deals with both **proximate** and **ultimate factors,** it is capable of what we may call "ultimate synthesis."

Ultimate synthesis is the real goal of evolutionary biology.

REFERENCES

Agassiz, L. 1962. Essay on Classification [2nd (1859) ed.]: Cambridge, Harvard University Press, Edited and with an introduction by Edward Lurie, xxxiv + 268 pp.

Ager, D. V. 1988. Extinctions and survivals in the Brachiopoda and the dangers of data bases, *in* ch. 4 *of* Larwood, G. P., ed., Extinction and Survival in the Fossil Record: Oxford, Clarendon Press, pp. 89–97; Systematics Association Special Volume No. 34.

Akam, M. 1995. *Hox* genes and the evolution of diverse body plans: Philosophical Transactions of the Royal Society of London, B, v. 349, pp. 313–19.

Alberch, P., Gould, S. J., Oster, G. F., and Wake, D. B. 1979. Size and shape in ontogeny and phylogeny: Paleobiology, v. 5, pp. 296–317.

Alchian, A. A. 1950. Uncertainty, evolution and economic theory: Journal of Political Economy, v. 58, pp. 211–21.

Amundson, R., and Lauder, G. V. 1994. Function without purpose: the uses of causal role function in evolutionary biology: Biology and Philosophy, v. 9, pp. 443–69.

Anderson, D. T. 1973. Embryology and Phylogeny in Annelids and Arthropods: Oxford, Pergamon, xiv + 495 pp.

Andersson, M. 1994. Sexual Selection: Princeton, Princeton University Press, xx + 599 pp.f

Appel, T. A. 1987. The Cuvier-Geoffroy Debate: French Biology in the Decades before Darwin: New York, Oxford University Press, xii + 305 pp.

Arber, A. 1937. The interpretation of the flower: a study of some aspects of morphological thought: Biological Reviews of the Cambridge Philosophical Society, v. 12, pp. 157–84.

———. 1950. The Natural Philosophy of Plant Form: Cambridge, Cambridge University Press, xiv + 247 pp.

Armstrong, D. M. 1978. A Theory of Universals, Volume II *of* Universals and Scientific Realism: Cambridge, Cambridge University Press, vii + 190 pp.

———. 1983. What is a Law of Nature?: Cambridge, Cambridge University Press, x + 180 pp.

Ashlock, P. D. 1971. Monophyly and associated terms: Systematic Zoology, v. 20, pp. 63–69.

Atran, S. 1990. Cognitive Foundations of Natural History. Towards an Anthropology of Science: Cambridge, Cambridge University Press, xii + 360 pp.

Avise, J. C., Quattro, J. M., and Vrijenhoek, R. C. 1992. Molecular clones within organismal clones: mitochondrial DNA phylognies and the evolutionary histories of unisexual vertebrates: Evolutionary Biology, v. 26, pp. 225–46.

Ax, P. 1987. The Phylogenetic System: the Systematization of Organisms on the Basis of their Phylogenesis [2nd ed.]: New York, Wiley, xiii + 340 pp.; Translated by R. P. S. Jeffries.

Ayala, F. J. 1970. Teleological explanations in evolutionary biology: Philosophy of Science, v. 37, pp. 1–5.

Ayers, M. R. 1974. Individuals without sortals: Canadian Journal of Philosophy, v. 4, pp. 113–48.

———. 1981. Locke versus Aristotle on natural kinds: The Journal of Philosophy, v. 78, pp. 247–72.

Baer, K. E. von. 1828. Über Entwickelungsgeschichte der Thiere. Beobachtung und Reflexion. Erster Theil: Königsberg, Gebrüder Bornträger, xxii + 271 pp.

———. 1873. Entwickelt sich die Larve der einfachen Ascidien in der ersten Zeit nach dem Typus der Wirbelthiere:: Mémoires de l'Académie Impériale des Sciences de Saint-Pétersbourg, v. (7)19, pp. 1–35; Taf. XII.

Baker, A. N., Rowe, F. W. E., and Clark, H. E. S. 1986. A new class of Echinodermata from New Zealand: Nature, v. 321, pp. 862–64.

Baker, J. R. 1988. The Cell Theory: a Restatement, History, and Critique: New York, Garland Publishing, Inc., 155 pp.

Baker, R. 1967. Particulars: bare, naked, and nude: Nous, v. 1, pp. 211–12.

Balme, D. M. 1987. Aristotle's biology was not essentialist, *in* ch. 11 *of* Gotthelf, A., and Lennox, J. G., eds., Philosophical Issues in Aristotle's Biology: Cambridge, Cambridge University Press, pp. 291–312.

Balsamo, M. 1992. Hermaphroditism and parthenogenesis in lower Bilateria: Gnathostomulida and Gastrotricha, *in* Dallai, R., ed., Sex Origin and Evolution: Modena, Mucchi, pp. 309–27.

Bambach, R. K. 1985. Classes and adaptive variety: the ecology of diversification in marine faunas through the Phanerozoic, *in* ch. 6 *of* Valentine, J. W., ed., Phanerozoic Diversity Trends: Profiles in Macroevolution: Princeton, Princeton University Press, pp. 191–253.

Bancroft, F. W. 1903. Variation and fusion of colonies in compound ascidians: Proceedings of the California Academy of Sciences, v. 3, pp. 137–86.

Barlow, N. 1959. The Autobiography of Charles Darwin 1909–1882. With Original Omissions Restored: New York, Harcourt, Brace, 253 pp.

Baroni-Urbani, C. 1979. The causes of evolution: converging orthodoxy and heresy: Systematic Zoology, v. 28, pp. 622–24.

Bateson, P. 1980. Optimal outbreeding and the development of sexual preferences in the Japanese quail: Zeitschrift für Tierpsychologie, v. 53, pp. 231–44.

———. 1982. Preferences for cousins in Japanese quail: Nature, v. 295, pp. 236–37.

———. 1983. Optimal outbreeding, in Bateson, P., ed., Mate Choice: Cambridge, Cambridge University Press, pp. 257–77.

Beatty, J. 1980. Optimal-design models and the strategy of model building in evolutionary biology: Philosophy of Science, v. 47, pp. 532–61.

———. 1992. Fitness: theoretical contexts, in Keller, E. F., and Lloyd, E. A., eds., Keywords in Evolutionary Biology: Cambridge, Harvard University Press, pp. 115–19.

Beatty, J., and Finsen, S. 1989. Rethinking the propensity interpretation: a peek inside Pandora's Box, in Ruse, M., ed., What the Philosophy of Biology Is: Essays Dedicated to David Hull: Dordrecht, Kluwer Academic Publishers, pp. 17–30.

Beckner, M. 1974. Reduction, hierarchies and organicism, in ch. 10 of Ayala, F. J., and Dobzhansky, T., eds., Studies in the Philosophy of Biology: Reductionism and Related Problems: Berkeley, University of California Press, pp. 163–77.

Becquemont, D. 1992. Darwin, Darwinisme, Évolutionisme: Paris, Kimé, 354 pp.

Bell, G. 1982. The Masterpiece of Nature: the Evolution and Genetics of Sexuality: Berkeley, University of California Press, 635 pp.

———. 1988. Sex and Death in Protozoa: the History of an Obsession: Cambridge, Cambridge University Press, xiv + 199 pp.

Belyaev, G. M. 1990. Is it valid to isolate the genus Xyloplax as an independent class of echinoderms?: Zoologicheskii Zhurnal, v. 69, pp. 83–96; In Russian.

Berg, L. S. 1926. Nomogenesis or Evolution Determined by Law: London, Constable, Translated from the Russian by J.N. Rostovtsov, xxiv + 477 pp.

Bergquist, P. R. 1985. Poriferan relationships, in Conway Morris, S., George, J. D., Gibson, R., and Platt, H. M., eds., The Origins and Relationships of Lower Invertebrates, 28 of Systematics Association Special Volume: Oxford, Oxford University Press, pp. 14–27.

Berlin, B. 1973. Folk systematics in relation to biological classification and nomenclature: Annual Review of Ecology and Systematics, v. 4, pp. 259–71.

———. 1992. Ethnobiological Classification: Principles of Categorization of Plants and Animals in Traditional Societies: Princeton, Princeton University Press, xvii + 335 pp.

Berlin, B., Breedlove, D. E., and Raven, P. H. 1966. Folk taxonomies and biological classification: Science, v. 154, pp. 273–75.

Berlin, B., and Kay, P. 1969. Basic Color Terms: Their Universality and Evolution: Berkeley, University of California Press, xi + 178 pp.

Bernier, R. 1984. The species as an individual: facing essentialism: Systematic Zoology, v. 33, pp. 460–69.

Bernstein, H. 1977. Germ line recombination may be primarily a manifestation of DNA repair processes: Journal of Theoretical Biology, v. 69, pp. 371–80.

Bernstein, H., Hopf, F. A., and Michod, R. E. 1988. Is meiotic recombination an adaptation for repairing DNA, producing genetic variation, or both?, *in* ch. 9 *of* Michod, R. E., and Levin, B. R., eds., The Evolution of Sex: an Examination of Current Ideas: Sunderland, Sinauer Associates, pp. 139–60.

Bierzychudek, P. 1985. Patterns in plant parthenogenesis: Experientia, v. 41, pp. 1255–64.

Bigelow, J., and Pargetter, R. 1987. Functions: The Journal of Philosophy, v. 84, pp. 181–96.

Black, M. 1971. The elusiveness of sets: Review of Metaphysics, v. 24, pp. 614–36.

Bock, W. J. 1959. Preadaptation and multiple evolutionary pathways: Evolution, v. 13, pp. 194–211.

———. 1964. Bill shape as a generic character in the cardinals: Wilson Bulletin, v. 76, pp. 50–61.

———. 1973. Philosophical foundations of classical evolutionary classification: Systematic Zoology, v. 22, pp. 375–92.

———. 1979. The synthetic explanation of macroevolutionary change—a reductionist approach: Bulletin of The Carnegie Museum of Natural History, v. 13, pp. 20–69.

———. 1986. Species concepts, speciation, and macroevolution, *in* Iwatsuki, K., Raven, P. H., and Bock, W. J., eds., Modern Aspects of Species: Tokyo, Tokyo University Press, pp. 31–57.

———. 1991. Levels of complexity and organismic organization, *in* Lanzavecchia, G., and Valvassori, R., eds., Form and Function in Zoology, 5 *of* Selected Symposia and Monographs U.Z.I: Modena, Mucci, pp. 181–213.

———. 1994. Ernst Mayr, naturalist: his contributions to systematics and evolution: Biology and Philosophy, v. 9, pp. 267–327.

Bock, W. J., and Wahlert, G. von. 1965. Adaptation and the form-function complex: Evolution, v. 19, pp. 269–99.

Bonde, N. 1984. Primitive features and ontogeny in phylogenetic reconstructions: Videnskabelige Meddelelser fra Dansk naturhistorisk Forening, v. 145, pp. 219–36.

Boorse, C. 1976. Wright on functions: Philosophical Review, v. 85, pp. 70–86.

Boss, K. J. 1978. Taxonomic concepts and superfluity in bivalve nomenclature: Philosophical Transactions of the Royal Society of London, v. B284, pp. 417–24.

Bowler, P. J. 1979. Theodor Eimer and orthogenesis: evolution by definitely directed variation: Journal of the History of Medicine and Allied Sciences, v. 34, pp. 40–73.

———. 1983. The Eclipse of Darwinism: Anti-Darwinian Evolution Theories in the Decades around 1900: Baltimore, The Johns Hopkins University Press, xi + 291 pp.

———. 1984. Evolution: the History of an Idea: Berkeley, University of California Press, xiv + 412 pp.

Boyden, A. 1943. Homology and analogy: a century after the definitions of "homologue" and "analogue" of Richard Owen: Quarterly Review of Biology, v. 18, pp. 228–41.

———. 1947. Homology and analogy. A critical review of the meanings and implications of these concepts in biology: American Midland Naturalist, v. 37, pp. 648–69.

Brady, R. H. 1994. Pattern description, process explanation, and the history of morphological sciences, in ch. 2 of Grande, L., and Rieppel, O., eds., Interpreting the Hierarchy of Nature: from Systematic Patterns to Evolutionary Process Theories: San Diego, Academic Press, pp. 7–31.

Brandon, R. N. 1978. Adaptation and evolutionary theory: Studies in the History and Philosophy of Science, v. 9, pp. 181–200.

———. 1990. Adaptation and Environment: Princeton, Princeton University Press, xi + 214 pp.

Briggs, D. E. G., Fortey, R. A., and Clarkson, E. N. K. 1988. Extinction and the fossil record of the arthropods, in ch. 9 of Larwood, G. P., ed., Extinction and Survival in the Fossil Record: Oxford, Clarendon Press, pp. 171–209; Systematics Association Special Volume No. 34.

Briggs, D. E. G., Fortey, R. A., and Wills, M. A. 1992. Morphological disparity in the Cambrian: Science, v. 256, pp. 70–73.

Brooks, D. R. 1988. Macroevolutionary comparisons of host and parasite phylogenies: Annual Review of Ecology and Systematics, v. 19, pp. 235–59.

Brooks, D. R., and Brandon, S. M. 1988. Coevolution and relicts: Systematic Zoology, v. 37, pp. 19–33.

Brooks, D. R., and McLennan, D. A. 1991. Phylogeny, Ecology, and Behavior: a Research Program in Comparative Biology: Chicago, University of Chicago Press, xii + 434 pp.

Brown, C. H. 1984. Language and Living Things: Uniformities in Folk Classification and Naming: New Brunswick, Rutgers University Press, xvi + 306 pp.

Brown, W. L., Jr. 1987. Punctuated equilibrium excused: the original examples fail to support it: Biological Journal of the Linnaean Society, v. 31, pp. 383–404.

Browne, R. A. 1992. Population genetics and ecology of *Artemia*: insights into parthenogenetic reproduction: Trends in Ecology and Evolution, v. 7, pp. 232–37.

Bruton, D. L., Jensen, A., and Jacquet, R. 1985. The use of models in the understanding of Cambrian arthropod morphology: Transactions of the Royal Society of Edinburgh (Earth Sciences), v. 76, pp. 365–69.

Buck, R. C., and Hull, D. L. 1966. The logical structure of the Linnaean hierarchy: Systematic Zoology, v. 15, pp. 97–116.

Buerton, P. J. 1995. How is a species kept together?: Biology and Philosophy, v. 10, pp. 181–96.

Bull, J. J., and Charnov, E. L. 1985. On irreversible evolution: Evolution, v. 395, pp. 1149–55.

Bulmer, R. 1967. Why is the cassowary not a bird? A problem of zoological taxonomy among the Karam of the New Guinea highlands: Man, v. 2, pp. 5–25.

Burchfield, J. D. 1975. Lord Kelvin and the Age of the Earth: New York, Science History Publications, xii + 260 pp.

Burke, M. B. 1980. Cohabitation, stuff and intermittent existence: Mind, v. 89, pp. 391–405.

Burt, A. 1995. The Evolution of Fitness: Evolution, v. 49, pp. 1–8.

Bush, G. L. 1975. Modes of animal speciation: Annual Review of Ecology and Systematics, v. 6, pp. 339–64.

———. 1981. Stasipatric speciation and rapid evolution in animals, *in* ch. 10 *of* Atchley, W. S., and Woodruff, D. S., eds., Evolution and Speciation: Essays in Honour of M.J.D. White: Cambridge, Cambridge University Press, pp. 201–15.

———. 1982. What do we really know about speciation?, *in* ch. 7 *of* Milkman, R., ed., Perspectives on Evolution: Sunderland, Sinauer Associates, pp. 119–28.

————. 1994. Sympatric speciation in animals: new wine in old bottles: Trends in Ecology and Evolution, v. 9, pp. 285–88.

Buss, L. W. 1987. The Evolution of Individuality: Princeton, Princeton University Press., xv + 203 pp.

Butler, P. M. 1982. Directions of evolution in the mammalian dentition, *in* ch. 7 *of* Joysey, K. A., and Friday, A. K., eds., Problems of Phylogenetic Reconstruction: London, Academic Press, pp. 235–44.

Butlin, R. 1987a. Speciation by reinforcement: Trends in Ecology and Evolution, v. 2, pp. 8–13.

Butlin, R. K. 1987b. Species, speciation, and reinforcement: American Naturalist, v. 130, pp. 461–64.

Butlin, R. 1989. Reinforcement of premating isolation, *in* ch. 7 *of* Otte, D., and Endler, J. A., eds., Speciation and its Consequences: Sunderland, Sinauer Associates, pp. 158–79.

Cahn, T. 1962. La Vie et l'Oeuvre d'Étienne Geoffroy Saint-Hilaire: Paris, Presses Universitaires de France, 318 pp.

Calman, W. T. 1940. A museum zoologist's view of taxonomy, *in* Huxley, J., ed., The New Systematics: Oxford, Oxford University Press, pp. 455–59.

Campbell, D. T. 1974. Evolutionary epistemology, *in* Schilpp, P. A., ed., The Philosophy of Karl Popper: La Salle, Open Court, v. 1, pp. 413–63.

Caplan, A. L. 1981. Back to class: a note on the ontology of species: Philosophy of Science, v. 48, pp. 130–40.

————. 1981. Pick your poison: historicism, essentialism, and emergentism in the definition of species: The Behavioral and Brain Sciences, v. 4, pp. 285–86.

Caplan, A. L., and Bock, W. J. 1988. Haunt me no longer: Biology and Philosophy, v. 3, pp. 443–54.

Carnap, R. 1956. Meaning and Necessity: a Study in Semantics and Modal Logic [2nd ed.]: Chicago, University of Chicago Press, viii + 258 pp.; First Edition, 1947.

————. 1967. The Logical Structure of the World, and Pseudoproblems in Philosophy: Berkeley, University of California Press, Translated by Rolf A. George, xxvi + 364 pp.

Carson, H. L. 1957. The species as a field for gene recombination, *in* Mayr, E., ed., The Species Problem: Washington, American Association for the Advancement of Science, pp. 23–38.

————. 1982. Speciation as a major reorganization of polygenic balances, *in* Barigozzi, C., ed., Mechanisms of Speciation: New York, Alan R. Liss, pp. 411–33.

————. 1987. The genetic system, the deme, and the origin of species: Annual Review of Genetics, v. 21, pp. 405–23.

————. 1989. Genetic imbalance, realigned selection, and the origin of species, *in* ch. 17 *of* Giddings, L. V., Kaneshiro, K. Y., and Anderson, W. W., eds., Genetics, Speciation, and the Founder Principle: New York, Oxford University Press, pp. 345–62.

————. 1990. Evolutionary process as studied in population genetics: clues from phylogeny: Oxford Surveys in Evolutionary Biology., v. 7, pp. 129–56.

Carson, H. L., and Templeton, A. R. 1984. Genetic revolutions in relation to speciation phenomena: the founding of new populations: Annual Review of Ecology and Systematics, v. 15, pp. 97–131.

Cartwright, N. 1983. How the Laws of Physics Lie: New York, Oxford University Press, x + 221 pp.

————. 1989. Nature's Capacities and their Measurement: Oxford, Clarendon Press, x + 268 pp.

Casson, R. W. 1983. Schemata in cognitive anthropology: Annual Review of Anthropology, v. 12, pp. 429–62.

Chaline, J. 1987. Arvicolid data (Arvicolidae, Rodentia) and evolutionary concepts: Evolutionary Biology, v. 21, pp. 237–310.

Chaline, J., and Laurin, B. 1986. Phyletic gradualism in a European Plio-Pleistocene *Miomys* lineage (Arvicolidae, Rodentia): Paleobiology, v. 12, pp. 203–16.

Chao, L. 1992. Evolution of sex in RNA viruses: Trends in Ecology and Evolution, v. 7, pp. 147–51.

Charnov, E. L. 1982. The theory of Sex-Allocation: Princeton, Princeton University Press, x + 355 pp.

————. 1993. Life History Invariants: Some Explorations of Symmetry in Evolutionary Ecology: Oxford, Oxford University Press, xv + 167 pp.

Chatterton, B. D. E., and Speyer, S. E. 1989. Larval ecology, life history strategies, and patterns of extinction and survivorship among Ordovician trilobites: Paleobiology, v. 15, pp. 118–32.

Cherry, L. M., Case, S. M., and Wilson, A. C. 1978. Frog perspective on the morphological difference between humans and chimpanzees: Science, v. 200, pp. 209–11.

Chisholm, R. M. 1973. Parts as essential to their wholes: Review of Metaphysics, v. 26, pp. 581–603.

Chomsky, N. 1980. Rules and Representations: New York, Columbia University Press, viii + 299 pp.

Christen, R., Ratto, A., Baroin, A., Perasso, R., Grell, K. G., and Adoutte, A. 1991. Origin of metazoans: a phylogeny deduced from sequences of the 28S ribosomal RNA, *in* Simonetta, A. M., and Conway Morris, S., eds., The Early Evolution of Metazoa and the Significance of Problematic Taxa: Cambridge, Cambridge University Press, pp. 1–9.

Claridge, M. F. 1995. Species concepts and speciation in insect herbivores: planthopper case studies: Bollettino di Zoologia, v. 62, pp. 53–8.

Clarke, B. L. 1981. A calculus of individuals based on "connection": Notre Dame Journal of Formal Logic, v. 22, pp. 204–18.

Cole, F. J. 1944. A History of Comparative Anatomy from Aristotle to the Eighteenth Century: London, Macmillan, viii + 524 pp.

Coleman, W. 1962. Lyell and the "reality" of species: Isis, v. 53, pp. 325–38.

Colless, D. H. 1967. The phylogenetic fallacy: Systematic Zoology, v. 16, pp. 289–95.

———. 1970. Type-specimens: their status and use: Systematic Zoology, v. 19, pp. 251–53.

———. 1972. "Basic taxa" and the "unit of classification": Systematic Zoology, v. 21, pp. 65–8.

———. 1985. On "character" and related terms: Systematic Zoology, v. 34, pp. 229–33.

Colwell, R. K. 1992. Niche: a bifurcation in the conceptual lineage of the term, *in* Keller, E. F., and Lloyd, E. A., eds., Keywords in Evolutionary Biology: Cambridge, Harvard University Press, pp. 241–48.

Cooter, R., and Kornhauser, L. 1980. Can litigation improve the law without the help of judges: The Journal of Legal Studies, v. 9, pp. 139–63.

Cowan, S. T. 1955. Introduction: the philosophy of classification: Journal of General Microbiology, v. 12, pp. 314–21.

Coyne, J. A. 1992. Genetics and speciation: Nature, v. 355, pp. 511–15.

Coyne, J. A., Orr, H. A., and Futuyma, D. J. 1988. Do we need a new species concept?: Systematic Zoology, v. 37, pp. 190–200.

Cracraft, J. 1986. Origin and evolution of continental biotas: speciation and historical congruence within the Australian avifauna: Evolution, v. 40, pp. 977–96.

———. 1987. Species concepts and the ontology of evolution: Biology and Philosophy, v. 2, pp. 329–46.

———. 1988. Deep-history biogeography: retrieving the historical pattern of evolving continental biotas: Systematic Zoology, v. 37, pp. 221–36.

————. 1992. The species of the birds-of-paradise (Paradisaeidae): applying the phylogenetic species concept to a complex pattern of diversification: Cladistics, v. 8, pp. 1–43.

Crease, T. J., Stanton, D. J., and Hebert, P. D. N. 1989. Polyphyletic origins of asexuality in *Daphnia pulex*. II. Mitochondrial DNA variation: Evolution, v. 43, pp. 1016–26.

Cronquist, A. 1978. Once again, what is a species?, *in* ch. 1 *of* Romberger, J. A., ed., Biosystematics in Agriculture, 2 *of* Beltsville Symposia in Agricultural Research: New York, John Wiley & Sons, pp. 3–20.

Cruse, D. A. 1979. On the transitivity of the part-whole relation: Journal of Linguistics, v. 15, pp. 29–38.

Cuellar, O., 1986. The evolution of parthenogenesis: a historical perspective, *in* Moens, P. B., ed., Meiosis: New York, Academic Press, pp. 43–104.

Cuénot, L. 1951. L'Évolution Biologique: les Faits, les Inertitudes: Paris, Masson & Cie, ix + 592 pp.

Callsen-Cencic, P., and Flügel, H. J. 1995. Larval development and the formation of the gut of *Siboglinum poseidoni* Flügel & Langhof (Pogonophora, Perviata). Evidence of protostomian affinity: Sarsia, v. 80, pp. 73–89.

Cutler, E. B., and Gibbs, P. E. 1985. A phylogenetic analysis of higher taxa in the phylum Sipuncula: Systematic Zoology, v. 34, pp. 162–73.

Damuth, J. 1985. Selection among "species": a formulation in terms of natural functional units: Evolution, v. 39, pp. 1132–46.

Damuth, J., and Heisler, I. L. 1988. Alternative formulations of multilevel selection: Biology and Philosophy, v. 3, pp. 407–30.

D'Andrade, R. 1995. The Development of Cognitive Anthropology: Cambridge, Cambridge University Press, xiv + 272 pp.

Danto, A. C. 1985. Narration and Knowledge (Including the Integral Text of Analytical Philosophy of History) [2nd ed.]: New York, Columbia University Press, xvii + 399 pp.

Darwin, C. 1842. The Structure and Distribution of Coral Reefs, Being the First Part of the Geology of the Voyage of the Beagle [1st ed.]: London, Smith and Elder, xii + 214 pp.; Second Editon, 1874.

————. 1846. Geological Observations on South America. Being the Third Part of the Geology of the Voyage of the Beagle, under the Command of Capt. Fitz-Roy, R.N., during the Years 1832–36: London, Smith, Elder, vii + 279 pp.; Second Edition 1874.

————. 1851. A Monograph on the Sub-Class Cirripedia, with Figures of all the Species. The Lepadidae; or, Pedunculated Cirripedes: London, Ray Society, xii + 400 pp.; Pl. I–X.

————. 1854. A Monograph on the Sub-Class Cirripedia, with Figures of all the Species. The Balanidae (or Sessile Cirripedes); the Verrucidae, Etc., Etc., Etc: London, Ray Society, viii + 684 pp.; Pls. I–XXX.

————. 1859. On the origin of species by means of natural selection, or the preservation of favoured races in the struggle for life [1st ed.]: London, John Murray, ix + 502 pp.; Sixth Edition 1872.

————. 1862. On the Various Contrivances by which British and Foreign Orchids are Fertilised by Insects, and on the Good Effects of Intercrossing [1st ed.]: London, John Murray, vi + 365 pp.; Second edition, 1877.

————. 1868. The Variation of Animals and Plants under Domestication [1st ed.]: London, John Murray; 2nd ed., 1875.

————. 1871. The Descent of Man, and Selection in Relation to Sex [1st ed.]: London, John Murray; Second Edtion, 1874.

————. 1873. On the males and complemental males of certain cirripedes, and on rudimentary structures: Nature, v. 8, pp. 431–32.

————. 1875. The Variation of Animals and Plants Under Domestication [2nd ed.]: London, John Murray.

Darwin, E. 1800. Phytologia; or the Philosophy of Agriculture and Gardening. With the Theory of Draining Morasses, and with an Improved Construction of the Drill Plough: London, J. Johnson, vii + 612 pp.; 12 Pls.

Davis, B. H. 1987. A legal point, *in* ch. 5 *of* Hoenigswald, H. M., and Wiener, L. F., eds., Biological Metaphor and Cladistic Classification: an Interdisciplinary Perspective: Philadelphia, University of Pennsylvania Press, pp. 115–21.

Davis, D. D. 1964. The giant panda: a morphological study of evolutionary mechanisms: Fieldiana, Zoology, v. 3, pp. 1–339.

Dawkins, R. 1976. The Selfish Gene: Oxford, Oxford University Press, xi + 224 pp.

————. 1982. The Extended Phenotype: the Gene as the Unit of Selection: Oxford, Oxford University Press, xii + 307 pp.

————. 1986. The Blind Watchmaker: New York and London, Norton, xiv + 332 pp.

De Beer, G. R. 1938. Embryology and evolution, *in* De Beer, G. R., ed., Evolution: Essays on Aspects of Evolutionary Biology Presented to E.S. Goodrich: London, Oxford University Press, pp. 57–78.

De Beer, G. 1956. The evolution of ratites: Bulletin of the British Museum (Natural History), Zoology, v. 4, pp. 59–70.

De Beer, G. R. 1958. Embryos and Ancestors [3rd ed.]: Oxford, Clarendon Press, xii + 197 pp.

Delsol, M. 1965. Relations entre phylogénèse et ontogénèse: Biologie Médicale, v. 54, pp. 41–64.

———. 1977. Embryogenesis, morphogenesis, genetics and evolution, *in* Hecht, M. K., Goody, P. C., and Hecht, B. M., eds., Major Patterns in Vertebrate Evolution: New York, Plenum Press, pp. 119–38.

Dennett, D. C. 1983. Intentional systems in cognitive ethology: the "Panglossian paradigm" defended: The Behavioral and Brain Sciences, v. 6, pp. 343–90; with commentary.

———. 1995. Darwin's Dangerous Idea: Evolution and the Meanings of Life: New York, Simon & Schuster, 507 pp.

De Queiroz, K. 1985. The ontogenetic method for determining character polarity and its relevance to phylogenetic systematics: Systematic Zoology, v. 34, pp. 280–99.

———. 1992. Phylogenetic definitions and taxonomic philosophy: Biology and Philosophy, v. 7, pp. 295–313.

———. 1994. Replacement of an essentialistic perspective on taxonomic definitions as exemplified by the definition of "Mammalia": Systematic Biology, v. 43, pp. 497–510.

———. 1995. The definitions of species and clade names: a reply to Ghiselin: Biology and Philosophy, v. 10, pp. 223–28.

De Queiroz, K., and Gauthier, J. A. 1990. Phylogeny as a central principle in taxonomy: phylogenetic definitions of taxon names: Systematic Zoology, v. 39, pp. 307–22.

———. 1992. Phylogenetic taxonomy: Annual Review of Ecology and Systematics, v. 23, pp. 449–80.

———. 1994. Toward a phylogenetic system of biological nomenclature: Trends in Ecology and Evolution, v. 9, pp. 27–31.

De Saussure, F. 1959. Course in General Linguistics: New York, Philosophical Library, xvi + 240 pp.

Desmond, A. 1982. Archetypes and Ancestors: Paleontology in Victorian London 1850–1875: Chicago, University of Chicago Press, 287 pp.

De Vries, H. 1912. Species and Varieties: their Origin by Mutation [3rd ed.]: Chicago, Open Court, xviii + 847 pp.; 1st ed., 1905: this one corrected only.

Dewey, J. 1910. The Influence of Darwin on Philosophy, and Other Essays in Contemporary Thought: New York, Henry Holt & Company, vi + 309 pp.

Diamond, J. M. 1965. Zoological classification system of a primitive people: Science, v. 151, pp. 1102–04.

Dobell, C. C. 1911. The principles of protistology: Archiv für Protistenkunde, v. 23, pp. 269–310.

Dobzhansky, T. 1935. A critique of the species concept in biology: Philosophy of Science, v. 2, pp. 344–55.

————. 1937. Genetics and the Origin of Species [1st ed.]: New York, Columbia University Press, xvi + 364 pp.

————. 1939. *Drosophila miranda*, a new species: Genetics, v. 20, pp. 377–91.

————. 1940. Speciation as a stage in evolutionary divergence: American Naturalist, v. 74, pp. 312–21.

————. 1941. Genetics and the Origin of Species [2nd ed.]: New York, Columbia University Press, xviii + 446 pp.

————. 1975. Review: Darwinian or "oriented" evolution?: Evolution, v. 29, pp. 376–78.

Dobzhansky, T., and Sturtevant, A. H. 1938. Inversions in the chromosomes of *Drosophila pseudoobscura*: Genetics, v. 23, pp. 28–64.

Dohrn, A. 1870. Untersuchungen über Bau und Entwicklung der Arthropoden: Leipzig, Wilhelm Engelmann, v + vi + 193 pp.; XVII Taf.

————. 1871. Geschichte des Krebsstammes, nach embryologischen, anatomischen und palaeontologischen Quellen. Ein Versuch: Jenaische Zeitschrift für Naturwissenschaft, v. 6, pp. 96–156.

————. 1875. Der Ursprung der Wirbelthiere und das Princip des Functionswechsels.— Genealogische Skizzen von Anton Dohrn: Leipzig, Wilhelm Engelmann, xv + 87 pp.

Dollo, L. 1893. Les lois de l'évolution: Bulletin de la Société Belge de Géologie, de Paléontologie et d'Hydrologie, v. 7, pp. 164–66.

Donellan, K. S. 1977. Reference and definite descriptions, *in* ch. 1 *of* Schwartz, S. P., ed., Naming, Necessity, and Natural Kinds: Ithaca, Cornell University Press, pp. 42–65.

Donoghue, M. J. 1992. Homology, *in* Keller, E. F., and Lloyd, E. A., eds., Keywords in Evolutionary Biology: Cambridge, Harvard University Press, pp. 170–79.

Donoghue, M. J., and Sanderson, M. J. 1994. Complexity and homology in plants, *in* ch. 12 *of* Hall, B. K., ed., Homology: the Hierarchical Basis of Comparative Biology: San Diego, Academic Press, pp. 393–421.

Donovan, S. K. 1989. Palaeontological criteria for the recognition of mass extinctions, *in* ch. 2 *of* Donovan, S. K., ed., Mass Extinction: Processes and Evidence: New York, Columbia University Press, pp. 19–36.

Dougherty, E. C. 1955. Comparative evolution and the origin of sexuality: Systematic Zoology, v. 4, pp. 147–90.

Douglas, M. 1986. How Institutions Think: Syracuse, Syracuse University Press, xi + 146 pp.

Dray, W. 1957. Laws and Explanation in History: Oxford, Oxford University Press.

Dretske, F. I. 1967. Can events move?: Mind, v. 76, pp. 479–92.

Driesch, H. 1893. Entwicklungsmechanische Studien. VII. Exogastrula and Anenteria (über die Wirkung von Wärmezufuhr auf die Larvenentwicklung der Echiniden). VIII. Über Variation der Mikromerenbildung (Wirking von Verdünnung des Meerwassers). IX. Über die Vertretbarkeit der »Anlagen« von Ektoderm und Entoderm. X. Über einige allgemeine entwicklungsmechanische Ergebnisse: Mittheilungen aus der Zoologischen Station zu Neapel, v. 11, pp. 221–54; Taf. 11.

Dunbar, R. I. M. 1982. Adaptation, fitness and the evolutionary tautology, *in* ch. 1 *of* Wrangham, R. W., Rubenstein, D. I., Dunbar, R. I. M., Bertram, B. C. R., and Clutton-Brock, T. H., eds., Current Problems in Sociobiology: Cambridge, Cambridge University Press, pp. 9–28.

Dupré, J. 1993. The Disorder of Things: Metaphysical Foundations of the Disunity of Science: Cambridge, Harvard University Press, xi + 308 pp.

Dupuis, C. 1984. Willi Hennig's impact on taxonomic thought: Annual Review of Ecology and Systematics, v. 15, pp. 1–24.

Durham, W. H. 1991. Coevolution: Genes, Culture, and Human Diversity: Stanford, Stanford University Press, xxii + 629 pp.

Durkheim, É., and Mauss, M. 1963. Primitive Classification [2nd ed.]: Chicago, University of Chicago Press, xlviii + 96 pp.; Translation and Introduction by Rodney Needham.

Dybdahl, M. F., and Lively, C. M. 1995. Diverse, endemic and polyphyletic clones in mixed populations of a freshwater snail (*Potamopyrgus antipodarum*): Journal of Evolutionary Biology, v. 8, pp. 385–98.

Eberhard, W. G. 1992. Species-isolation, genital mechanics, and the evolution of species-specific genitalia in three species of *Macrodactylus* beetles (Coleoptera, Scarabeidae, Melonthinidae): Evolution, v. 44, pp. 1774–83.

Eernisse, D. J., and Kluge, A. G. 1993. Taxonomic congruence versus total evidence, and amniote phylogeny inferred from fossils, molecules, and morphology: Molecular Biology and Evolution, v. 10, pp. 1170–95.

Ehrlich, P. R. 1961. Has the biological species concept outlived its usefulness?: Systematic Zoology, v. 10, pp. 167–76.

Ehrlich, P. R., and Holm, R. W. 1962. Patterns and populations: Science, v. 137, pp. 652–57.

Ehrlich, P. R., and Raven, P. H. 1969. Differentiation of populations: Science, v. 165, pp. 1228–32.

Eldredge, N. 1971. The allopatric model and phylogeny in Paleozoic invertebrates: Evolution, v. 25, pp. 156–67.

———. 1985. Unfinished Synthesis: Biological Hierarchies and Modern Evolutionary Thought: New York, Oxford University Press, ix + 327 pp.

———. 1986. Information, economics, and evolution: Annual Review of Ecology and Systematics, v. 17, pp. 351–69.

———. 1989. Macroevolutionary Dynamics: Species, Niches, and Adaptive Peaks: New York, McGraw-Hill Publishing Company, xii + 226 pp.

———. 1993. What, if anything, is a species?, in ch. 1 of Kimbel, W. H., and Martin, L. B., eds., Species, Species Concepts, and Primate Evolution: New York, Plenum Press, pp. 3–20.

Eldredge, N., and Cracraft, J. 1980. Phylogenetic Patterns and the Evolutionary Process: Method and Theory in Comparative Biology: New York, Columbia University Press, viii + 349 pp.

Eldredge, N., and Gould, S. J. 1972. Punctuated equilibria: an alternative to phylogenetic gradualism, in ch. 5 of Schopf, T. J. M., ed., Models in Paleobiology: San Francisco, Freeman and Cooper, pp. 82–115.

Eldredge, N., and Grene, M. 1992. Interactions: the Biological Context of Social Theory: New York, Columbia University Press, vii + 242 pp.

Eldredge, N., and Salthe, S. N. 1984. Hierarchy and evolution: Oxford Surveys in Evolutionary Biology., v. 1, pp. 82–206.

Ellegård, A. 1990. Darwin and the General Reader: the Reception of Darwin's Theory of Evolution in the British Periodical Press, 1859–72; with a new Foreword by David L. Hull [2nd ed.]: Chicago, University of Chicago Press, Originally published in 1958 as Volume 8 of the series Gothenberg Studies in English., 394 pp.

Elton, C. 1927. Animal Ecology: Oxford, Clarendon Press, viii + 207 pp.

Emig, C. C. 1975. Phylogenèse des Phoronida. Les lophophorates et le concept Archimerata: Zeitschrift für zoologische Systematik und Evolutions-Forschung, v. 14, pp. 10–24.

Ereshefsky, M. 1988. Axiomatics and individuality: a reply to Williams's "Species are individuals": Philosophy of Science, v. 55, pp. 427–34.

————. 1991a. The semantic approach to evolutionary theory: Biology and Philosophy, v. 6, pp. 59–80.

————. 1991b. Species, higher taxa, and the units of evolution: Philosophy of Science, v. 58, pp. 84–101.

————. 1992. The historical nature of evolutionary theory, *in* Nitecki, M. H., and Nitecki, D. V., eds., History and Evolution: Albany, State University of New York Press, pp. 81–99.

————. 1994. Some problems with the Linnaean hierarchy: Philosophy of Science, v. 61, pp. 186–205.

Falconer, H. 1856. On Prof. Huxley's attempted refutation of Cuvier's laws of correlation, in the reconstruction of extinct vertebrate forms: Annals and Magazine of Natural History, v. (2)17, pp. 476–93.

Farris, J. S. 1967. Comment on psychologism: Systematic Zoology, v. 16, pp. 345–47.

————. 1977. Phylogenetic analysis under Dollo's law: Systematic Zoology, v. 26, pp. 77–88.

Faulkner, D. J., and Ghiselin, M. T. 1983. Chemical defense and the evolutionary ecology of dorid nudibranchs and some other opisthobranch gastropods: Marine Ecology Progress Series, v. 13, pp. 295–301.

Fedonkin, M. A. 1996. The Precambrian fossil record: new insight of life: Memorie della Società Italiana di Scienze Naturali e del Museo Civico di Storia Naturale di Milano, v. 27, pp. 41–8.

Field, H. 1973. Theory change and the indeterminacy of reference: The Journal of Philosophy, v. 70, pp. 462–81.

Field, K. G., Olsen, Lane, D. J., Giovannoni, S., Ghiselin, M. T., Raff, E. C., Pace, N. R., and Raff, R. A. 1988. Molecular phylogeny of the animal kingdom: Science, v. 239, pp. 748–53.

Fink, W. L. 1982. The conceptual relationship between ontogeny and phylogeny: Paleobiology, v. 8, pp. 254–64.

Fisher, R. A. 1930. The Genetical Theory of Natural Selection: London, Oxford University Press.

Flessa, K. W., and Jablonski, D. 1983. Extinction is here to stay: Paleobiology, v. 9, pp. 315–21.

Foote, M. 1996. Evolutionary patterns in the fossil record: Evolution, v. 50, pp. 1–11.

Fortey, R. A., and Jefferies, R. P. S. 1982. Fossils and phylogeny—a compromise approach, *in* ch. 6 *of* Joysey, K. A., and Friday, A. K., eds., Problems of Phylogenetic Reconstruction: London, Academic Press, pp. 197–234.

Fortey, R. A. 1985. Pelagic trilobites as an example for deducing life habits of extinct arthropods: Transactions of the Royal Society of Edinburgh (Earth Sciences), v. 76, pp. 219–30.

———. 1989. There are extinctions and extinctions: examples from the Lower Palaeozoic: Philosophical Transactions of the Royal Society of London, v. B325, pp. 327–55.

Fortey, R. A., Briggs, D. E. G., and Wills, M. A. 1996. The Cambrian evolutionary 'explosion': decoupling cladogenesis from morphological disparity: Biological Journal of the Linnaean Society, v. 57, pp. 13–33.

Fristrup, K. 1992. Character: current usages, *in* Keller, E. F., and Lloyd, E. A., eds., Keywords in Evolutionary Biology: Cambridge, Harvard University Press, pp. 45–51.

Frost, D. R., and Kluge, A. G. 1994. A consideration of epistemology in systematic biology, with special reference to species: Cladistics, v. 10, pp. 259–94.

Futuyma, D. J. 1988. *Sturm und Drang* and the evolutionary synthesis: Evolution, v. 42, pp. 217–26.

———. 1989. Macroevolutionary consequences of speciation: inferences from phytophagous insects, *in* ch. 22 *of* Otte, D., and Endler, J. A., eds., Speciation and its Consequences: Sunderland, Sinauer Associates, pp. 557–78.

Garcia-Bellido, A. 1977. Homeotic and atavistic mutations in insects: American Zoologist, v. 17, pp. 613–29.

Gardiner, B. 1982. Tetrapod classification: Zoological Journal of the Linnean Society, v. 74, pp. 207–32.

Gardiner, S. L., and Jones, M. L. 1994. On the significance of larval and juvenile morphology for suggesting phylogenetic relationships of the Vestimentifera: American Zoologist, v. 34, pp. 513–22.

Gause, G. F. 1934. The Struggle for Existence: Baltimore, William & Wilkins, ix + 163 pp.

Gauthier, J., Kluge, A. G., and Rowe, T. 1988. Amniote phylogeny and the importance of fossils: Cladistics, v. 4, pp. 105–209.

Gavrilets, S., and Hastings, A. 1996. Founder effect speciation: a theoretical assessment: American Naturalist, v. 147, pp. 466–91.

Geoffroy Saint-Hilaire, É. 1818. Philosophie Anatomique. Des Organes Respiratoires sus le Rapport de la Détermination et l'Identité de leurs Pièces Osseuses: Paris, Méquignon-Marvis, xxxix + 518 pp.; Pl. 1–10.

Geoffroy Saint-Hilaire, I. 1832–1837. Histoire Générale et Particulière des Anomalies de L'Organisation chez L'Homme et les Animaux, Ouvrage comprenant des Recherches sur les Caractères, la Classification, L'Influence Physiologique et Pathologique, les Rapports Généraux, les Lois et les Causes des Monstruosités, des Variétes et Vices de Confirmation, ou Traité de Tératologie: Paris, J.-B. Baillière, viii + 746 + 571 + 618 pp.; Atlas 8 + xx pp., XX Pl.

―――. 1847. Vie, Travaux et Doctrine Scientifique D'Étienne Geoffroy Saint-Hilaire: P. Bertrand, Paris, vi + 479 pp.

Ghiselin, M. T. 1966a, An application of the theory of definitions to systematic principles: Systematic Zoology, v. 15, pp. 127–30.

―――. 1966b. On psychologism in the logic of taxonomic controversies: Systematic Zoology, v. 15, pp. 207–15.

―――. 1966c. On semantic pitfalls of biological adaptation: Philosophy of Science, v. 33, pp. 147–53.

―――. 1967. Further remarks on logical errors in systematic theory: Systematic Zoology, v. 16, pp. 347–48.

―――. 1969a. The evolution of hermaphroditism among animals: Quarterly Review of Biology, v. 44, pp. 189–208.

―――. 1969b. Non-phenetic evidence in phylogeny: Systematic Zoology, v. 18, pp. 460–62.

―――. 1969c. The principles and concepts of systematic biology, *in* Sibley, C. G., ed., Systematic Biology: Proceedings of an International Conference: Washington, National Academy of Sciences, pp. 45–55.

―――. 1969d. The Triumph of the Darwinian Method [1st ed.]: Berkeley, University of California Press, x + 287 pp.

―――. 1974a. The Economy of Nature and the Evolution of Sex: Berkeley, University of California Press, xii + 364 pp.

―――. 1974b. Embryology and Phylogeny in Annelids and Arthropods by D. T. Anderson: Systematic Zoology, v. 23, pp. 150–51.

―――. 1974c. A radical solution to the species problem: Systematic Zoology, v. 23, pp. 536–44; 1975.

―――. 1976. The nomenclature of correspondence: a new look at "homology" and "analogy", *in* Masterton, R. B., Hodos, W., and Jerrison, H., eds., Evolution, Brain and Behavior: Persistent Problems: Hillsdale, New Jersey, Lawrence Erlbaum Associates, pp. 129–32.

―――. 1978. The economy of the body: American Economic Review, v. 68, pp. 233–37.

―――. 1980a. Evolutionary anatomy and language: The Behavioral and Brain Sciences, v. 3, pp. 20–21.

―――. 1980b. The failure of morphology to assimilate Darwinism, *in* Mayr, E., and Provine, W. B., eds., The Evolutionary Synthesis: Cambridge, Harvard University Press, pp. 180–93.

―――. 1981a. Categories, life, and thinking: The Behavioral and Brain Sciences, v. 4, pp. 269–313; with commentary.

―――. 1981b. The metaphysics of phylogeny: Paleobiology, v. 7, pp. 139–45.

―――. 1983. Lloyd Morgan's canon in evolutionary context: The Behavioral and Brain Sciences, v. 6, pp. 362–63.

―――. 1984a. B.F. Skinner and Dr. Pangloss: The Behavioral and Brain Sciences, v. 7, pp. 687–88.

―――. 1984b. "Definition," "character," and other equivocal terms: Systematic Zoology, v. 33, pp. 104–10.

―――. 1984. Narrow approaches to phylogeny: a review of nine books on cladism: Oxford Surveys in Evolutionary Biology., v. 1, pp. 209–22.

―――. 1985. Can Aristotle be reconciled with Darwin?: Systematic Zoology, v. 34, pp. 457–60.

―――. 1986a. The assimilation of Darwinism in developmental psychology: Human Development, v. 29, pp. 12–21.

―――. 1986b. Principles and prospects for general economy, *in* Radnitzky, G., and Bernholz, P., eds., Economic Imperialism: the Economic Approach Applied outside the Field of Economics: New York, PWPA Press, pp. 21–31.

―――. 1986c. We are all contraptions: New York Times Book Review, v. 1986, pp. 18–9.

―――. 1987a. Bioeconomics and the metaphysics of selection: Journal of Social and Biological Structures, v. 10, pp. 361–69.

―――. 1987b. Classification as an evolutionary problem, *in* ch. 6 *of* Costall, A., and Still, A., eds., Cognitive psychology in question: Brighton, Harvester Press, pp. 70–86.

———. 1987c. Evolutionary aspects of marine invertebrate reproduction, *in* ch. 9 *of* Giese, A. C., Pearse, J. S., and Pearse, V. B., eds., Reproduction of Marine Invertebrates: Palo Alto, Blackwell Scientific Publications, v. 9, pp. 609–65.

———. 1987d. Hierarchies and their components: Paleobiology, v. 13, pp. 108–11.

———. 1987e. Response to commentary on the individuality of species: Biology and Philosophy, v. 2, pp. 207–12.

———. 1987f. Species concepts, individuality, and objectivity: Biology and Philosophy, v. 2, pp. 127–43.

———. 1988. The origin of molluscs in the light of molecular evidence: Oxford Surveys in Evolutionary Biology., v. 5, pp. 66–95.

———. 1989a. Individuality, history and laws of nature in biology, *in* Ruse, M., ed., What the philosophy of biology is: Dordrecht, Kluwer Academic Publishers, pp. 53–66.

———. 1989b. Summary of our current knowledge of metazoan phylogeny, *in* Fernholm, B., Bremer, K., and Jörnvall, H., eds., The hierarchy of life: molecules and morphology in phylogenetic analysis: Amsterdam, Exerpta Medica, pp. 261–72.

———. 1991. Classical and molecular phylogenetics: Bollettino di Zoologia, v. 58, pp. 289–94.

———. 1992. Biology, economics, and bioeconomics, *in* ch. 1 *of* Radnitzky, G., ed., Universal Economics: Assessing the Achievements of the Economic Approach: New York, Paragon House, pp. 71–118.

———. 1994a. Darwin's language may seem teleological, but his thinking is another matter: Biology and Philosophy, v. 9, pp. 489–92.

———. 1994b. The Origin of Vertebrates and the Principle of Succession of Functions. Genealogical Sketches by Anton Dohrn. 1875. An English translation from the German, introduction and bibliography: History and Philosophy of Life Sciences, v. 16, pp. 5–98.

———. 1995a. Darwin, progress, and economic principles: Evolution, v. 49, pp. 1029–37.

———. 1995b. Ostensive definitions of the names of species and clades: Biology and Philosophy, v. 10, pp. 219–22.

Gibson, G. D., and Chia, Fu-S. 1989. Developmental variability (Pelagic and Benthic) in *Haminoea callidegenita* (Opisthobranchia: Cephalaspidea) is influenced by egg mass jelly: Biological Bulletin, v. 176, pp. 103–10.

Gibson, J. J. 1966. The Senses Considered as Perceptual Systems: Boston, Houghton Mifflin, xiv + 335 pp.

Gilinsky, N. L. 1981. Stabilizing species selection in the Archaeogastropoda: Paleobiology, v. 7, pp. 316–33.

Gilmour, J. S. L. 1937. A taxonomic problem: Nature, v. 139, pp. 1040–42.

———. 1940. Taxonomy and philosophy, *in* Huxley, J., ed., The New Systematics: Oxford, Oxford University Press, pp. 461–74.

Gilson, É. 1984. From Aristotle to Darwin and Back Again: a Journey in Final Causality, Species, and Evolution: Notre Dame, University of Notre Dame Press, xx + 209 pp.

Glen, W. 1982. The Road to Jaramillo: Critical Years of the Revolution in Earth Sciences: Stanford, Stanford University Press, xix + 459 pp.

Gochfeld, M. 1974. Terms for highly similar species: Systematic Zoology, v. 23, pp. 445–46.

Gombrich, E. H. 1977. Art and Illusion: a Study in the Psychology of Pictorial Representation [5th ed.]: London, Phaidon Press, xiv + 386 pp.

———. 1989. Evolution in the arts: the altar painting, its ancestry and progeny, *in* ch. 6 *of* Grafen, A., ed., Evolution and its Influence: Oxford, Clarendon Press, pp. 107–25.

Goodman, J. C. 1978. An economic theory of the evolution of common law: The Journal of Legal Studies, v. 7, pp. 393–406.

Goodman, N. 1972. Problems and Projects: Indianapolis, Bobbs-Merrill, xii + 463 pp.

Goodrich, E. S. 1924. Living Organisms: An Account of their Origin & Evolution: Oxford, Clarendon Press, ii + 200 pp.

Goosens, W. K. 1977. Underlying trait terms, *in* ch. 5 *of* Schwartz, S. P., ed., Naming, Necessity, and Natural Kinds: Ithaca, Cornell University Press, pp. 133–54.

Gosliner, T. M., and Ghiselin, M. T. 1984. Parallel evolution in opisthobranch gastropods and its implications for phylogenetic methodology: Systematic Zoology, v. 33, pp. 255–74.

Gould, S. J. 1965. Is uniformitarianism necessary?: American Journal of Science, v. 263, pp. 223–28.

———. 1970. Dollo on Dollo's law: irreversability and the status of evolutionary laws: Journal of the History of Biology, v. 3, pp. 189–212.

———. 1973. Systematic pluralism and the uses of history: Systematic Zoology, v. 22, pp. 322–24.

———. 1977. Ontogeny and Phylogeny: Cambridge, Harvard University Press, xiv + 501 pp.

————. 1983. Irrelevance, submission, and partnership: the changing role of palaeon-tology in Darwin's three centennials, and a modest proposal for macroevolution, *in* ch. 17 *of* Bendall, D. S., ed., Evolution from Molecules to Men: Cambridge, Cambridge University Press, pp. 347–66.

————. 1989. Wonderful Life: the Burgess Shale and the Nature of History: New York, Norton, 347 pp.

————. 1994. Common pathways of illumination: Natural History, v. 103, pp. 10–20.

Gould, S. J., and Calloway, C. B. 1980. Clams and brachiopods—ships that pass in the night: Paleobiology, v. 6, pp. 383–96.

Gould, S. J., and Lewontin, R. C. 1979. The spandrels of San Marco and the Pangloss-ian paradigm: a critique of the adaptationist programme: Proceedings of the Royal Society of London, B, v. 205, pp. 581–98.

Gracia, J. J. E. 1988. Individuality: an Essay on the Foundations of Metaphysics: Albany, State University of New York Press, xx + 315 pp.

Granger, H. 1987. Deformed kids and the fixity of species: Classical Quarterly, v. 37, pp. 110–16.

Grantham, T. A. 1995. Hierarchical approaches to macroevolution: recent work on species selection and the "effect hypothesis": Annual Review of Ecology and Sys-tematics, v. 26, pp. 301–21.

Gray, R. D. 1952. Goethe the Alchemist: a Study of the Alchemical Symbolism in Goethe's Literary and Scientific Works: Cambridge, Cambridge University Press, x + 312 pp.

Graybeal, A. 1995. Naming species: Systematic Biology, v. 44, pp. 237–50.

Greene, J. C. 1977. Darwin as a social evolutionist: Journal of the History of Biology, v. 10, pp. 1–27.

Gregg, J. R. 1950. Taxonomy, language and reality: American Naturalist, v. 84, pp. 419–35.

————. 1954. The Language of Taxonomy: an Application of Symbolic Logic to the Study of Classificatory Systems: New York, Columbia University Press, xi + 71 pp.

Gregory, W. K. 1935. Reduplication in evolution: Quarterly Review of Biology, v. 10, pp. 272–90.

————. 1936. The transformation of organic designs: a review of the origin and deployment of the earliest vertebrates: Biological Reviews of the Cambridge Philo-sophical Society, v. 11, pp. 311–44.

Grehan, J. R., and Ainsworth, R. 1985. Orthogenesis and evolution: Systematic Zool-ogy, v. 34, pp. 174–92.

Grene, M. 1987. Hierarchies in biology: American Scientist, v. 75, pp. 504–10.

———. 1989. Interaction and evolution, *in* Ruse, M., ed., What the Philosophy of Biology Is: Essays Dedicated to David Hull: Dordrecht, Kluwer Academic Publishers, pp. 67–73.

Griesemer, J. R. 1992. Niche: historical perspectives, *in* Keller, E. F., and Lloyd, E. A., eds., Keywords in Evolutionary Biology: Cambridge, Harvard University Press, pp. 231–40.

Griffiths, G. C. D. 1974. On the foundations of biological systematics: Acta Biotheoretica, v. 23, pp. 85–131.

Grinnell, J. 1904. The origin and distribution of the chestnut-backed Chickadee: The Auk, v. 21, pp. 364–82.

Grossmann, R. 1983. The Categorical Structure of the World: Bloomington, Indiana University Press, xvi + 431 pp.

Gustus, R. M., and Cloney, R. A. 1972. Ultrastructural similarities between setae of brachiopods and polychaetes: Acta Zoologica, Stockholm, v. 53, pp. 229–33.

Haas, O., and Simpson, G. G. 1946. Analysis of some phylogenetic terms, with attempts at redefinition: Proceedings of the American Philosophical Society, v. 90, pp. 319–49.

Haeckel, E. 1866. Generelle Morphologie der Organismen. Allgemeine Grundzüge der organischen Formen-Wissenschaft, mecanisch begründet durch die von Charles Darwin reformirte Descendenz-Theorie: Berlin, Verlag von Georg Reimer, xxxii + 574 + clx + 462 pp.; Taf. I–II, I–VIII.

———. 1872. Die Kalkschwämme: Berlin, Georg Reimer, xvi + 484 + viii + 418 pp.; Taf. 1–60.

———. 1873. Die Gastraeatheorie, die phylogenetische Klassifikation des Thierreichs und die Homologie der Keimblätter: Jenaische Zeitschrift für Naturwissenschaft, v. 8, pp. 1–55; Taf. I.

———. 1875. Die Gastrula und die Eifurchung der Thiere: Jenaische Zeitschrift für Naturwissenschaft, v. 9, pp. 402–508; Taf. XIX–XXV.

———. 1877. Nachträge zur Gastraea-Theorie: Jenaische Zeitschrift für Naturwissenschaft, v. 11, pp. 55–98.

———. 1878. Ueber die Individualität des Thierkörpers: Jenaische Zeitschrift für Naturwissenschaft, NF, v. 12, pp. 1–20.

———. 1912. Zur Geschichte der Entwicklungslehre: Das Freie Wort, v. 12.

Häuser, C. L. 1987. The debate about the biological species concept—a review: Zeitschrift für zoologische Systematik und Evolutions-Forschung, v. 25, pp. 241–57.

Halanych, K. M. 1995. The phylogenetic position of the pterobranch hemichordates based on 18S rRNA sequence data: Molecular Phylogenetics and Evolution, v. 4, pp. 72–6.

Halanych, K. M., Bacheller, J. D., Aguinaldo, A. M. A., Liva, S. M., Hillis, D. M., and Lake, J. A. 1995. Evidence from 18S ribosomal DNA that the lophophorates are protostome animals: Science, v. 267, pp. 1641–43.

Haldane, J. B. S. 1932. The Causes of Evolution: London, Longmans, Green, vi + 235 pp.

Hall, B. K. 1984. Developmental mechanisms underlying the formation of atavisms: Biological Reviews of the Cambridge Philosophical Society, v. 59, pp. 89–124.

———. 1994. Homology: the Hierarchical Basis of Comparative Biology: San Diego, Academic Press, xvi + 483 pp.

———. 1994. Introduction, *in* Hall, B. K., ed., Homology: the Hierarchical Basis of Comparative Biology: San Diego, Academic Press, pp. 1–19.

Hallam, A. 1973. A Revolution in the Earth Sciences: from Continental Drift to Plate Tectonics: Oxford, Clarendon Press, xiii + 127 pp.

Hamilton, W. D. 1967. Extraordinary sex ratios: Science, v. 156, pp. 477–88.

Hampé, A. 1959. Contribution à l'étude du développement et de la regulation des déficiences et des excédents dans la patte de l'embryon du poulet: Archives de Anatomie Microscopique et de Morphologie Expérimentale, v. 48, pp. 345–78.

Hampton, J. A. 1982. A demonstration of intransitivity in natural categories: Cognition, v. 12, pp. 151–64.

Hanken, J., and Wake, D. B. 1993. Miniaturization of body size: organismal consequences and evolutionary significance: Annual Review of Ecology and Systematics, v. 24, pp. 501–19.

Hansen, T. A. 1982. Modes of larval development in early Tertiary neogastropods: Paleobiology, v. 8, pp. 367–77.

———. 1983. Modes of larval development and rates of speciation in early Tertiary neogastropods: Science, v. 220, pp. 501–02.

Hardy, A. C. 1954. Escape from specialization, *in* Huxley, J., Hardy, A. C., and Ford, E. B., eds., Evolution as a Process: London, George Allen & Unwin, pp. 122–42.

Harland, W. B., Armstrong, R. L., Cox, A. V., Craig, L. E., Smith, A. G., and Smith, D. G. 1990. A Geologic Time Scale 1989: Cambridge, Cambridge University Press, xvi + 263 pp.

Harper, C. W., Jr. 1976. Phylogenetic inference in paleontology: Journal of Paleontology, v. 50, pp. 180–93.

Harvey, W. 1847. The Works of William Harvey, M.D. Translated from the Latin with a Life of the Author by Robert Willis, M.D: London, Sydenham Society, xcvi + 624 pp.

Haszprunar, G. 1992. The types of homology and their significance for evolutionary biology and phylogenetics: Journal of Evolutionary Biology, v. 5, pp. 13–24.

Hattiangadi, J. N. 1987. How is Language Possible? Philosophical Reflections on the Evolution of Language and Knowledge: La Salle, Open Court, xxi + 224 pp.

Havel, J. A., Hebert, P. D. V., and Delorme, L. D. 1990. Genetic diversity of asexual Ostracoda from a low arctic site: Journal of Evolutionary Biology, v. 3, pp. 391–410.

Heath, D. J. 1977. Simultaneous hermaphroditism: cost and benefit: Journal of Theoretical Biology, v. 64, pp. 363–73.

Heberer, G. 1959. Theorie der additiven Typogenese, in Heberer, G., ed., Die Evolution der Organismen: Ergebnisse und Probleme der Astammungslehre [2nd ed.]: Stuttgart, Gustav Fischer Verlag, v. 2, pp. 857–914.

Hebert, P. D. N., Beaton, M. J., Schwartz, S. S., and Stanton, D. J. 1989. Polyphyletic origins of asexuality in Daphnia pulex. I. breeding-system variation and levels of clonal diversity: Evolution, v. 43, pp. 1004–15.

Hebert, P. D. N., and Crease, T. J. 1983. Clonal diversity in populations of Daphnia pulex reproducing by obligate parthenogenesis: Heredity, v. 51, pp. 353–69.

Hebert, P. D. N., Ward, R. D., and Weider, L. J. 1988. Clonal-diversity patterns and breeding system variation in Daphnia pulex, an asexual-sexual complex: Evolution, v. 42, pp. 1024–35.

Heise, H., and Starr, M. P. 1968. Nomenifers: are they christened or classified?: Systematic Zoology, v. 17, pp. 458–67.

———. 1969. A reply to Buchannan's critique: Systematic Zoology, v. 18, pp. 345–47.

Hempel, C. G. 1959. The logic of functional analysis, in Gross, L., ed., Symposium on Sociological Theory: New York, Harper & Row, pp. 271–307; Reprinted in Hempel, 1965.

———. 1965. Aspects of Scientific Explanation and other Essays in the Philosophy of Science: New York, The Free Press, vii + 504 pp.

Hengeveld, R. 1988. Mayr's ecological species criterion: Systematic Zoology, v. 37, pp. 47–55.

Hennig, W. 1950. Grundzüge einer Theorie der phylogenetischen Systematik: Berlin, Deutscher Zentralverlag.

———. 1966. Phylogenetic Systematics [2nd ed.]: Urbana, University of Illinois Press, Translated by D. Dwight Davis and Rainer Zangerl., vi + 263 pp.; 1979 printing has additional foreword.

Hertwig, O. 1922. Das Werden der Organismen: zur Widerlegung von Darwins Zufallstheorie durch das Gesetz in der Entwicklung [3rd ed.]: Jena, Verlag von Gustav Fischer, xx + 686 pp.

Hickman, C. S. 1988. Analysis of form and function in fossils: American Zoologist, v. 28, pp. 775–93.

Hillis, D. M. 1994. Homology in molecular biology, *in* ch. 10 *of* Hall, B. K., ed., Homology: the Hierarchical Basis of Comparative Biology: San Diego, Academic Press, pp. 339–68.

Hirsch, E. 1982. The Concept of Identity: New York, Oxford University Press, x + 318 pp.

Hirshleifer, J. 1982. Evolutionary models in economics and law: Cooperation versus conflict strategies: Research in Law and Economics, v. 4, pp. 1–60.

Hoagland, K. E., and Robertson, R. 1988. An assessment of poecilogony in marine invertebrates: phenomenon or fantasy?: Biological Bulletin, v. 174, pp. 109–25.

Hodge, M. J. S. 1991. Darwin, Whewell, and natural selection: Biology and Philosophy, v. 6, pp. 457–70.

Hodos, W. 1976. The concept of homology and the evolution of behavior, *in* Masterton, R. B. W., Hodos, W., and Jerison, H., eds., Evolution, Brain, and Behavior: Persistent Problems: Hillsdale, Lawrence Erlbaum Associates, pp. 153–67.

Hoenigswald, H. M., and Wiener, L. F. 1987. Biological Metaphor and Cladistic Classification: an Interdisciplinary Perspective: Philadelphia, University of Pennsylvania Press, xiii + 286 pp.

Hogan, M. 1992. Natural kinds and ecological niches—response to Johnson's paper: Biology and Philosophy, v. 7, pp. 203–08.

Holland, N. D. 1996. Homology, homeobox genes, and the early evolution of the vertebrates: Memoirs of the California Academy of Sciences, n. 20, pp. 63–70.

Holmes, O. W., Jr. 1881. The Common Law: Boston, Little, Brown, and Co., xvi + 422 pp.

House, M. R. 1988. Extinction and survival in the Cephalopoda, *in* ch. 7 *of* Larwood, G. P., ed., Extinction and Survival in the Fossil Record: Oxford, Clarendon Press, pp. 139–54; Systematics Association Special Volume No. 34.

———. 1989. Ammonoid extinction events: Philosophical Transactions of the Royal Society of London, v. B325, pp. 307–26.

Howard, D. J., and Gregory, P. G. 1993. Post-insemination signalling systems and rein-forcement: Philosophical Transactions of the Royal Society of London, B, v. 340, pp. 231–36.

Hu, Wei-S., and Temin, H. M. 1990. Retroviral recombination and reverse transcription: Science, v. 250, pp. 1227–33.

Hull, D. L. 1964. Consistency and monophyly: Systematic Zoology, v. 13, pp. 1–11.

———. 1965. The effect of essentialism on taxonomy—two thousand years of stasis: British Journal for the Philosophy of Science, v. 15, pp. 314–26; Concluded Vol. 16, pp. 1–18.

———. 1967. Certainty and circularity in evolutionary taxonomy: Evolution, v. 21, pp. 174–89.

———. 1968. The operational imperative: sense and nonsense in operationism: Systematic Zoology, v. 17, pp. 438–57.

———. 1969. What philosophy of biology is not: Journal of the History of Biology, v. 2, pp. 241–68.

———. 1984. Can Kripke alone save essentialism? A reply to Kitts: Systematic Zoology, v. 33, pp. 110–12.

———. 1987. Genealogical actors in ecological plays: Biology and Philosophy, v. 1, pp. 168–84.

———. 1988a. Progress in ideas of progress, in Nitecki, M. H., ed., Evolutionary Progress: Chicago, University of Chicago Press, pp. 27–48.

———. 1988b. Science as a Process: an Evolutionary Account of the Social and Conceptual Development of Science: Chicago, University of Chicago Press, xiii + 583 pp.

———. 1992. Biological species: an inductivist's nightmare, in ch. 4 of Douglas, M., and Hull, D., eds., How Classification Works: Nelson Goodman among the Social Sciences: Edinburgh, Edinburgh University Press, pp. 42–68.

Humphries, C. J. 1988. Ontogeny and Systematics: New York, Columbia University Press, xiii + 236 pp.

Hutchinson, G. E. 1965. The Ecological Theater and the Evolutionary Play: New Haven, Yale University Press, xiii + 139 pp.

Huxley, J. S. 1940. Towards the new systematics, in Huxley, J., ed., The New Systematics: Oxford, Oxford University Press, pp. 1–46.

Huxley, J. 1942. Evolution: the Modern Synthesis: London, George Allen & Unwin, 0 + 645 pp.

Huxley, J. S. 1958. Evolutionary processes and taxonomy with special reference to grades: Uppsala Universitets Årsskrift, v. 1958, pp. 21–39.

Huxley, T. H. 1852. Upon animal individuality: Proceedings of the Royal Institution, v. 1, pp. 184–89.

———. 1856. On the method of palaeontology: Annals of Natural History, v. 18, pp. 43–54.

Hyman, L. H. 1940. The Invertebrates: New York, McGraw-Hill, v. 1: Protozoa through Ctenophera, x + 726 pp.

Inglis, W. G. 1966. The observational basis of homology: Systematic Zoology, v. 15, pp. 219–28.

———. 1991. Characters: the central mystery of taxonomy and systematics: Biological Journal of the Linnaean Society, v. 44, pp. 121–39.

Jablonski, D. 1986. Larval ecology and macroevolution in marine invertebrates: Bulletin of Marine Science, v. 39, pp. 565–87.

Jackson, J. F., and Pounds, J. A. 1979. Comments on assessing the dedifferentiating effect of gene flow: Systematic Zoology, v. 28, pp. 78–85.

Jacobshagen, E. 1925. Allgemeine vergleichende Formenlehre der Tiere: Leipzig, Werner Klinkhardt Verlag, viii + 258 pp.

Janzen, D. H. 1977. What are dandelions and aphids?: American Naturalist, v. 111. pp. 586–89.

Jardine, N. 1967. The concept of homology in biology: British Journal for the Philosophy of Science, v. 18, pp. 125–39.

———. 1969. The observational and theoretical components of homology: a study based on the morphology fo the dermal skull-roofs of rhiphidistian fishes: Biological Journal of the Linnaean Society, v. 1. pp. 327–61.

Jefferies, R. P. S. 1979. The origin of chordates—a methodological essay, *in* ch. 17 *of* House, M. R., ed., The Origin of Major Invertebrate Groups, No. 12 *of* The Systematics Association Special Volume: London, Academic Press, pp. 443–78.

Jepsen, G. L. 1949. Selection, "orthogenesis," and the fossil record: Proceedings of the American Philosophical Society, v. 93, pp. 479–500.

Johannes, R. E. 1981. Words of the Lagoon: Fishing and Marine Lore in the Palau District of Micronesia: Berkeley, University of California Press, xiv + 245 pp.

Johnson, D. M. 1990. Can abstractions be causes?: Biology and Philosophy, v. 5, pp. 63–77.

Johnson, L. E. 1995. Species: on their nature and moral standing: Journal of Natural History, v. 29, pp. 843–49.

Johnson, M. S., Murray, J., and Clarke, B. 1986. Allyzymic similarities among species of *Partula* on Moorea: Heredity, v. 56, pp. 319–27.

———. 1993a. The ecological genetics and adaptive radiation of *Partula* on Moorea: Oxford Surveys in Evolutionary Biology., v. 9, pp. 167–238.

———. 1993b. Evolutionary relationships and extreme genital variation in a closely related group of *Partula*: Malacologia, v. 35, pp. 43–61.

Jones, G. V. 1982. Stacks not fuzzy sets: an ordinal basis for prototype theory of concepts: Cognition, v. 12, pp. 281–90.

Jones, M. L. 1985. On the Vestimentifera, a new phylum: six new species, and other taxa, from hydrothermal vents and elsewhere: Bulletin of the Biological Society of Washington, v. 6, pp. 117–58.

Jung, K. G. 1971. Psychological Types: Princeton, Princeton University Press, xv + 608 pp.

Kälin, J. 1935. Über einige Grundbegriffe in der vergleichenden Anatomie und ihre Bedeutung für die Erforschung der Baupläne im Tierreich, *in* Anonymous, ed., Comptes Rendus du XIIe Congrès International de Zoologie, pp. 647–64.

Katz, J. J. 1977. A proper theory of names: Philosophical Studies, v. 31. pp. 1–80.

Keil, F. C. 1979. Semantic and Conceptual Development: an Ontological Perspective: Cambridge, Harvard University Press, xv + 214 pp.

———. 1986a. The acquisition of natural kind and artifact terms, *in* ch. 7 *of* Demopoulos, W., and Marras, A., eds., Language Learning and Concept Acquisition: Norwood, Ablex, pp. 133–53.

———. 1986. On the structure-dependent nature of stages of cognitive development, *in* ch. 6 *of* Levin, I., ed., Stage and Structure: Norwood, Ablex, pp. 145–63.

———. 1989. Concepts, Kinds, and Cognitive Development: Cambridge, MIT Press, xv + 328 pp.

Keil, F. C., and Batterman, N. 1984. A characteristic-to-defining shift in the development of word meaning: Journal of Verbal Learning and Verbal Behavior, v. 23, pp. 221–36.

Kellogg, D. E. 1988. "And then a miracle occurs"—weak links in the chain of argument from punctuation to hierarchy: Biology and Philosophy, v. 3, pp. 3–28.

Kimbel, W. H., and Rak, Y. 1993. The importance of species taxa in paleoanthropology and an argument for the phylogenetic concept of the species category, *in* ch. 18 *of*

Kimbel, W. H., and Martin, L. B., eds., Species, Species Concepts, and Primate Evolution: New York, Plenum Press, pp. 461–84.

Kitcher, P. 1984a. Against the monism of the moment: a reply to Elliott Sober: Philosophy of Science, v. 51, pp. 613–30.

———. 1984b. Species: Philosophy of Science, v. 51, pp. 308–33.

———. 1986. Bewitchment of the biologist: Nature, v. 320, pp. 649–50.

———. 1987. Ghostly whispers: Mayr, Ghiselin, and the "Philosophers" on the ontological status of species: Biology and Philosophy, v. 2, pp. 184–92.

———. 1989. Some puzzles about species, *in* Ruse, M., ed., What the Philosophy of Biology Is: Essays Dedicated to David Hull: Dordrecht, Kluwer Academic Publishers, pp. 183–208.

———. 1992. Gene: current usages, *in* Keller, E. F., and Lloyd, E. A., eds., Keywords in Evolutionary Biology: Cambridge, Harvard University Press, pp. 128–31.

Kitts, D. B. 1984. The names of species: a reply to Hull: Systematic Zoology, v. 33, pp. 112–15.

Kitts, D. B., and Kitts, D. J. 1979. Biological species as natural kinds: Philosophy of Science, v. 46, pp. 613–22.

Kluge, A. G. 1985. Ontogeny and phylogenetic systematics: Cladistics, v. 1. pp. 13–27.

———. 1988. The characterization of ontogeny, *in* ch. 3 *of* Humphries, C. J., ed., Ontogeny and Systematics: New York, Columbia University Press, pp. 57–81.

———. 1990. Species as historical individuals: Biology and Philosophy, v. 5, pp. 417–31.

Kluge, A. G., and Strauss, R. E. 1985. Ontogeny and systematics: Annual Review of Ecology and Systematics, v. 16, pp. 247–68.

Knowlton, N. 1993. Sibling species in the sea: Annual Review of Ecology and Systematics, v. 24, pp. 189–216.

Kollar, E. J., and Fisher, C. 1980. Tooth induction in chick epithelium: expression of quiescent genes for enamel synthesis: Science, v. 207, pp. 993–95.

Kornet, D. J. 1993. Permanent splits as speciation events: a formal reconstruction of the internodal species concept: Journal of Theoretical Biology, v. 164, pp. 407–35.

Kowalevsky, A. O. 1866. Entwickelungsgeschichte der einfachen Ascidien: Mémoires de l'Académie Impériale des Sciences de Saint-Pétersbourg, 7, v. 10, pp. 1–19; Taf. I–III.

———. 1871. Weitere Studien über die Entwicklung der einfachen Ascidien: Archiv für Mikroskopische Anatomie, v. 7, pp. 101–30; Taf. X–XIII.

————. 1874. Studies on the development of the Brachiopoda: Izvestia Imperatorskago Obshchestva Lubitelei Estestvoznaniya Antropologii i Etnografii, v. 14, pp. 1–40; Pl. I–V. In Russian.

Kraus, F. 1988. An empirical evaluation of the use of the ontogeny polarization criterion in phylogenetic inference: Systematic Zoology, v. 37, pp. 106–41.

Kripke, S. A. 1980. Naming and Necessity [2nd ed.]: Oxford, Basil Blackwell, viii + 172 pp.; 1st. ed., 1972.

Krishtalka, L. 1993. Anagenetic angst: species boundaries in eocene primates, *in* ch. 13 *of* Kimbel, W. H., and Martin, L. B., eds., Species, Species Concepts, and Primate Evolution: New York, Plenum Press, pp. 331–44.

Kristensen, R. M. 1983. Loricifera, a new phylum with aschelminthes characters from the meiobenthos: Zeitschrift für zoologische Systematik und Evolutions-Forschung, v. 21, pp. 163–80.

Kroeber, A. L., and Kluckhohn, C. 1952. Culture: a Critical Review of Concepts and Definitions, Volume 47, No. 1, *of* Papers of the Peabody Museum of American Archaeology and Ethnology, Harvard University: Cambridge, Peabody Museum of American Archaeology and Ethnology, Harvard University, viii + 22i 3 pp.

Kryzanowsky, S. G. 1939. Das Rekapitulationsprinzip: Acta Zoologica, Stockholm, v. 20, pp. 1–87.

Kusch, M. 1995. Psychologism: a Case Study in the Sociology of Phiosophical Knowledge: London, Routledge, xviii + 327 pp.

Lambert, D. M., Michaux, B., and White, C. S. 1987. Are species self-defining?: Systematic Zoology, v. 36, pp. 196–205.

Lan, R., and Reeves, P. R. 1996. Gene transfer is a major factor in bacterial evolution: Molecular Biology and Evolution, v. 13, pp. 47–55.

Landa, J. T. 1994. Trust, Ethnicity, and Identity: Ann Arbor, University of Michiga Press, xvi + 225 pp.

Laudan, R. 1992. What's so special about the past?, *in* Nitecki, M. H., and Nitecki, D. V., eds., History and Evolution: Albany, State University of New York Press, pp. 55–67.

Lauder, G. V. 1988. Atavisms and homology of hyprobranchial elements in lower vertebrates: Journal of Morphology, v. 195, pp. 237–45.

Laycock, H. 1978. Some questions of ontology: Philosophical Review, v. 81, pp. 3–42.

Lemche, H. 1957. A new living deep-sea mollusc of the Cambro-Devonian class Monoplacophora: Nature, v. 179, pp. 413–16.

Lemche, H., and Wingstrand, K. G. 1959. The anatomy of *Neopilina galatheae* Lemche, 1957 (Mollusca Triblidiacea): Galathea Reports, v. 3, pp. 9–71.

Lennox, J. G. 1992. Teleology, *in* Keller, E. F., and Lloyd, E. A., eds., Keywords in Evolutionary Biology: Cambridge, Harvard University Press, pp. 324–33.

———. 1993. Darwin *was* a teleologist: Biology and Philosophy, v. 8, pp. 409–21.

———. 1994. Teleology by another name: a reply to Ghiselin: Biology and Philosophy, v. 9, pp. 493–95.

Lenoir, T. 1982. The Strategy of Life: Teleology and Mechanics in Nineteenth-Century German Biology: Utrecht, D. Reidel, xii + 314 pp.

Leonard, H. S., and Goodman, N. 1940. The calculus of individuals and its uses: Journal of Symbolic Logic, v. 5, pp. 45–55.

Levinton, J. 1988. Genetics, Paleontology, and Macroevolution: Cambridge, Cambridge University Press, xv + 637 pp.

Levinton, J. S., and Simon, C. M. 1980. A critique of the punctuated equilibria model and implications for the detection of speciation in the fossil record: Systematic Zoology, v. 29, pp. 130–42.

Linnaeus, C. 1758. Systema Naturae per Regna Tria Naturae, secundum Classes, Ordines, Genera, Species, cum Characteribus, Differentiis, Synonymis, Locis [10th ed.]: Holmiae, Laurentii Salvii, v. 1, iv + 824 pp.

Liou, L. W., and Price, T. D. 1994. Speciation by reinforcement of premating isolation: Evolution, v. 48, pp. 1451–59.

Little, F. J., Jr. 1966. Bispecific chimerid sponges from the Antarctic: Nature, v. 211, pp. 436–38.

Littlejohn, M. J. 1981. Reproductive isolation: a critical review, *in* ch. 15 *of* Atchley, W. S., and Woodruff, D. S., eds., Evolution and Speciation: Essays in Honour of M.J.D. White: Cambridge, Cambridge University Press, pp. 298–334.

Lloyd, E. A. 1988. The Structure and Confirmation of Evolutionary Theory, No. 37, *of* Contributions in Philosophy: New York, Greenwood Press, viii + 235 pp.

Lloyd, E. A., and Gould, S. J. 1993. Species selection on variability: Proceedings of the National Academy of Sciences USA, v. 90, pp. 595–99.

Löther, R. 1972. Die Beherrschung der Mannigfaltigkeit: Philosophische Grundlagen der Taxonomie: Jena, VEB Gustav Fischer Verlag, 285 pp.

———. 1991. Über die Natur der Arten und monophyletischen Taxa: Mitteilungen aus dem Zoologischen Museum in Berlin, v. 67, pp. 17–23.

Lotsy, J. P. 1931. On the species of the taxonomist in its relation to evolution: Genetica, v. 13, pp. 1–16.

Lovejoy, A. O. 1936. The Great Chain of Being: Cambridge, Harvard University Press.

————. 1959. Recent criticism of the Darwinian theory of recapitulation: its grounds and its initiator, *in* ch. 15 *of* Glass, B., Temkin, O., and Straus, W. L., eds., Forerunners of Darwin: 1745–1859: Baltimore, Johns Hopkins University Press, pp. 438–58.

Lyell, C. 1830. Principles of Geology, being an Attempt to Explain the Former Changes of the Earth's Surface, by Reference to Causes Now in Operation [1st ed.]: London, John Murray, v. 1, xv + 511 pp.

————. 1832. Principles of Geology, being an Attempt to Explain the Former Changes of the Earth's Surface, by Reference to Causes Now in Operation [1st ed.]: London, John Murray, v. 2, xii + 330 pp.

Mabee, P. M. 1989a. Assumptions underlying the use of ontogenetic sequences for determining character state order: Transactions of the American Fisheries Society, v. 118, pp. 159–66.

————. 1989b. An empirical rejection of the ontogenetic polarity criterion: Cladistics, v. 5, pp. 409–16.

————. 1993. Phylogenetic interpretation of ontogenetic change: sorting out the actual and artefactual in an empirical case study of centrarchid fishes: Zoological Journal of the Linnaean Society, v. 107, pp. 175–291.

Mahner, M. 1993. What is a species?: Journal for General Philosophy of Science, v. 24, pp. 103–26.

————. 1994. Phenomenalistische Erblast in der Biologie: Biologisches Zentralblatt, v. 113, pp. 435–48.

Mandler, J. M., Bauer, P. J., and McDonough, L. 1991. Separating the sheep from the goats: differentiating global categories: Cognitive Psychology, v. 23, pp. 263–98.

Mangold-Wirz, K., and Fioroni, P. 1970. Die Sonderstellung der Cephalopoden: Zoologischer Jahrbücher Abteilung für Systematik, v. 97, pp. 522–631.

Manton, S. M. 1977. The Arthropoda: Habits, Functional Morphology and Evolution: Oxford, Clarendon Press, xxii + 527 pp.

Marcus, R. B. 1974. Classes, collections, and individuals: American Philosophical Quarterly, v. 11, pp. 227–32.

————. 1993. Modalities: Philosophical Essays: New York, Oxford University Press, xiv + 266 pp.

Markman, E. 1973. Facilitation of part-whole comparisons by use of the collective noun "family": Child Development, v. 44, pp. 837–40.

————. 1979. Classes and collections: conceptual organization and numerical abilities: Cognitive Psychology, v. 11, pp. 395–411.

————. 1989. Categorization and Naming in Children: Problems of Induction: Cambridge, MIT Press, xii + 250 pp.

Markman, E., and Seibert, J. 1976. Classes and collections: internal organization and resulting holistic properties: Cognitive Psychology, v. 8, pp. 561–77.

Masters, J. C. 1993. Primates and paradigms: problems with the identification of genetic species, *in* ch. 3 *of* Kimbel, W. H., and Martin, L. B., eds., Species, Species Concepts, and Primate Evolution: New York, Plenum Press, pp. 43–64.

Masters, J. C., and Spencer, H. G. 1989. Why we need a new genetic species concept: Systematic Zoology, v. 38, pp. 270–79.

Maulitz, R. C. 1971. Schwann's way: cells and crystals: Journal of the History of Medicine and Allied Sciences, v. 26, pp. 422–37.

Maurer, B. A., Brown, J. H., and Rusler, R. 1992. The micro and the macro in body size evolution: Evolution, v. 46, pp. 939–53.

Maxwell, W. D. 1989. The end Permian mass extinction, *in* ch. 8 *of* Donovan, S. K., ed., Mass Extinction: Processes and Evidence: New York, Columbia University Press, pp. 152–73.

Maynard Smith, J. 1977. Why the genome does not congeal: Nature, v. 268, pp. 693–96.

Maynard Smith, J. 1978. The Evolution of Sex: Cambridge, Cambridge University Press, x + 222 pp.

Mayr, E. 1940. Speciation phenomena in birds: American Naturalist, v. 74, pp. 249–78.

————. 1942. Systematics and the Origin of Species: New York, Columbia University Press, ? + 334 pp.

————. 1959. Darwin and the evolutionary theory in biology, *in* Meggers, B. J., ed., Evolution and Anthropology: a Centennial Appraisal: Washington, The Anthropological Society of Washington, pp. 1–10.

————. 1963a. Animal Species and Evolution: Cambridge, Harvard University Press, xiv + 797 pp.

————. 1963b. The taxonomic evaluation of fossil hominids, *in* Washburn, S. L., ed., Classification and Human Evolution: Chicago, Aldine Publishing Company, pp. 332–46.

————. 1965. Numerical phenetics and taxonomic theory: Systematic Zoology, v. 14, pp. 73–97.

————. 1969. Principles of Systematic Zoology [1st ed.]: New York, McGraw-Hill, xi + 428 pp.

————. 1970. Populations, Species, and Evolution. An Abridgement of *Animal Species and Evolution*: Cambridge, Harvard University Press, xv + 453 pp.

————. 1976. Is the species a class or an individual?: Systematic Zoology, v. 25, pp. 192.

————. 1980. The role of systematics in the evolutionary synthesis, *in* Mayr, E., and Provine, W. B., eds., The Evolutionary Synthesis: Perspectives on the Unification of Biology: Cambridge, Harvard University Press, pp. 123–36.

————. 1981. Biological classification: toward a synthesis of opposing methodologies: Science, v. 214, pp. 510–16.

————. 1982a. The growth of biological thought: Diversity, evolution, and inheritance: Cambridge, Harvard University Press, xi + 974 pp.

————. 1982b. Speciation and macroevolution: Evolution, v. 36, pp. 1119–32.

————. 1983. The concept of finality in Darwin and after Darwin: Scientia, v. 118, pp. 97–117.

————. 1983. How to carry out the adaptationist program?: American Naturalist, v. 121, pp. 324–34.

————. 1987a. Answers to these comments: Biology and Philosophy, v. 2, pp. 212–20.

————. 1987b. The ontological status of species: Scientific progress and philosophical terminology: Biology and Philosophy, v. 2, pp. 145–66.

————. 1988a. Die Darwinische Revolution und die Widerstände gegen die Selektionstheorie, *in* Meier, H., ed., Die Herausforderung der Evolutionsbiologie: München, Piper, pp. 221–49.

————. 1988b. The why and how of species: Biology and Philosophy, v. 3, pp. 431–41.

————. 1989. Attaching names to objects, *in* Ruse, M., ed., What the Philosophy of Biology Is: Essays Dedicated to David Hull: Dordrecht, Kluwer Academic Publishers, pp. 235–43.

————. 1991a. The ideological resistance to Darwin's theory of natural selection: Proceedings of the American Philosophical Society, v. 135, pp. 123–39.

————. 1991b. One Long Argument: Charles Darwin and the Genesis of Modern Evolutionary Thought: Cambridge, Harvard University Press, xiv . +. 195 pp.

————. 1992a. Controversies in retrospect: Oxford Surveys in Evolutionary Biology., v. 8, pp. 1–34.

————. 1992b. Darwin's principle of divergence: Journal of the History of Biology, v. 25, pp. 343–59.

————. 1992c. The idea of teleology: Journal of the History of Ideas, v. 53, pp. 117–35.

————. 1994a. Reasons for the failure of theories: Philosophy of Science, v. 61, pp. 529–33.

————. 1994b. Recapitulation reinterpreted: the somatic program: Quarterly Review of Biology, v. 69, pp. 223–32.

————. 1995. Systems of ordering data: Biology and Philosophy, v. 10, pp. 419–34.

Mayr, E., and Ashlock, P. D. 1991. Principles of Systematic Zoology [2nd ed.]: New York, McGraw-Hill, xx + 475 pp.

Mayr, E., and Bock, W. J. 1994. Provisional classification versus standard avian sequences. Heuristics and communication in ornithology: Ibis, v. 136, pp. 12–18.

McCafferty, W. P., and Chandler, L. 1974. Denotations of some comparative systematic terminology: Systematic Zoology, v. 23, pp. 139–40.

McGinnis, W. 1994. A century of homeosis, a decade of homeoboxes: Genetics, v. 137, pp. 607–11.

McKitrick, M. C. 1994. On homology and the ontological relationships of parts: Systematic Biology, v. 43, pp. 1–10.

Medin, D. L., and Schaffer, M. M. 1978. Context theory of classification learning: Psychological Review, v. 85, pp. 207–38.

Medin, D. L., and Smith, E. E. 1984. Concepts and concept formation: Annual Review of Psychology, v. 35, pp. 113–39.

Medin, D. L., and Wattenmaker, W. D. 1987. Family resemblance, conceptual cohesiveness, and category construction: Cognitive Psychology, v. 19, pp. 242–79.

Menard, H. W. 1986. The Ocean of Truth: a Personal History of Global Tectonics: Princeton, Princeton University Press, xiv + 353 pp.

Metschnikoff, E. 1876. Beiträge zur Morphologie der Spongien: Zeitschrift für wissenschaftliche Zoologie, v. 27, pp. 275–86.

————. 1879. Spongiologische Studien: Zeitschrift für wissenschaftliche Zoologie, v. 32, pp. 349–87; Taf. XX–XXIII.

————. 1880. Über die intracelluläre Verdauung bei Coelenteraten: Zoologischer Anzeiger, v. 3, pp. 261–63.

————. 1882. Vergleichend-embryologische Studien. 3. Über die Gastrula einiger Metazoen: Zeitschrift für wissenschaftliche Zoologie, v. 37, pp. 286–313.

————. 1883. Untersuchungen über die intracelluläre Verdauung bei Wirbellosen Thieren: Arbeiten aus dem Zoologischen Institute der Universität Wien, v. 5, pp. 141–68; Taf. I–II.

————. 1886. Embryologische Studien an Medusen. Ein Beitrag zur Genealogie der Primitiv-Organe: Wien, Alfred Holder, vi + 159 pp.; 12 Taf.

Michaux, B. 1989. Homology: a question of form or a product of genealogy?: Rivista di Biologia, v. 82, pp. 217–34.

Michod, R. E., and Levin, B. R. 1988. The Evolution of Sex: an Examination of Current Ideas: Sunderland, Sinauer Associates, ix + 342 pp.

Mill, J. S. 1872. A System of Logic Ratiocinative and Inductive Being a Connected View of the Principles of Evidence and the Methods of Scientific Investigation [8th ed.]: London, Longmans, Green, xvi + 622 pp.

Miller, G. A. 1956. The magical number seven, plus or minus two: some limits on our capacity for processing information: Psychological Review, v. 63, pp. 81–97.

Millikan, R. G. 1989. In defense of proper functions: Philosophy of Science, v. 56, pp. 288–302.

Mills, S. K., and Beatty, J. H. 1979. The propensity interpretation of fitness: Philosophy of Science, v. 46, pp. 263–86.

Milne-Edwards, H. 1851. Introduction à la Zoologie Générale ou Considérations sur les Tendances de la Nature: Paris, Victor Masson, iv + 180 pp.

Minelli, A. 1993. Biological Systematics: the State of the Art: London, Chapman & Hall, xvii + 387 pp.

Mishler, B. D. 1988. Relationships between Ontogeny and Phylogeny, with Reference to Bryophytes, *in* ch. 5 *of* Humphries, C. J., ed., Ontogeny and Systematics: New York, Columbia University Press, pp. 117–36.

Mishler, B. D., and Brandon, R. N. 1987. Individuality, pluralism, and the phylogenetic species concept: Biology and Philosophy, v. 2, pp. 397–414.

Mitton, J. B., Berg, C. J. J., and Orr, K. S. 1989. Population structure, larval dispersal, and gene flow in the queen conch, *Strombus gigas*, of the Caribbean: Biological Bulletin, v. 177, pp. 356–62.

Mokyr, J. 1990. The Lever of Riches: Technological Creativity and Economic Progress: New York, Oxford University Press, xi + 349 pp.

Moore, F. B.-G., and Tonsor, S. J. 1994. A simulation of Wright's shifting-balance process: migration and the three phases: Evolution, v. 48, pp. 69–80.

Morse, E. S. 1873. The systematic position of the Brachiopoda: Proceedings of the Boston Society of Natural History, v. 15, pp. 315–72.

Morss, J. R. 1991. Against Ontogeny, *in* Griffiths, P., ed., Trees of Life: Essays in Philosophy of Biology: New York, Kluwer, pp. 1–29.

Müller, F. 1864. Für Darwin: Leipzig, Verlag von Wilhelm Engelmann, iii + 91 pp.

————. 1869. Facts and Arguments for Darwin: with Additions by the Author. Translated by W.S. Dallas: London, John Murray, vii + 144 pp.

Müller, F. M. 1871. Lectures on the Science of Language: London, Longmans, Green & Co., xx + 481 + viii + 668 pp.

Müller, G. B., and Wagner, G. P. 1996. Homology, *Hox* genes, and developmental integration: American Zoologist, v. 36, pp. 4–13.

Muller, H. J. 1932. Some genetic aspects of sex: American Naturalist, v. 66, pp. 118–38.

————. 1939. Reversibility in evolution considered from the standpoint of genetics: Biological Reviews of the Cambridge Philosophical Society, v. 14, pp. 261–80.

Munson, R. 1975. Is biology a provincial science?: Philosophy of Science, v. 42, pp. 428–47.

Murdock, G. P. 1953. The processing of anthropological materials, *in* Kroeber, A. L., ed., Anthropology Today: an Encyclopedic Inventory: Chicago, University of Chicago Press, pp. 476–87.

Murray, J., and Clarke, B. 1968. Partial reproductive isolation in the genus *Partula* (Gastropoda) on Moorea: Evolution, v. 22, pp. 684–98.

————. 1980. The genus *Partula* on Moorea: speciation in progress: Proceedings of the Royal Society of London, v. B211, pp. 83–117.

————. 1984. Movement and gene flow in *Partula taeniata*: Malacologia, v. 25, pp. 343–48.

Naef, A. 1917. Die individuelle Entwicklung organischer Formen als Urkunde ihrer Stammesgeschichte: Jena, Verlag von Gustav Fischer, iii + 77 pp.

Nagel, E. 1977. Teleology revisited: The Journal of Philosophy, v. 74, pp. 261–301.

Neff, N. A. 1986. A rational basis for a priori character weighting: Systematic Zoology, v. 35, pp. 118–23.

Nelson, G. J. 1971. Paraphyly and polyphyly: redefinitions: Systematic Zoology, v. 20, pp. 471–72.

————. 1978. Ontogeny, phylogeny, paleontology, and the biogenetic law: Systematic Zoology, v. 27, pp. 324–45.

————. 1985. Class and individual: a reply to M. Ghiselin: Cladistics, v. 1, pp. 386–89.

Nelson, R. R., and Winter, S. B. 1982. An Evolutionary Theory of Economic Change: Cambridge, Harvard University Press.

Neuhaus, B. 1994. Ultrastructure of alimentary canal and body cavity, ground pattern, and phylogenetic relationships of the Kinorhyncha: Microfauna Marina, v. 9, pp. 61–156.

Neurath, H., Walsh, K. A., and Winter, W. P. 1967. Evolution of structure and function of proteases: Science, v. 158, pp. 1638–44.

Newell, N. D. 1949. Phyletic size increase, an important trend illustrated by fossils: Evolution, v. 3, pp. 103–24.

Newman, W. A. 1980. A review of extant *Scillaelepas* (Cirripedia: Scalpellidae) including recognition of new species from the North Atlantic, Western Indian Ocean and New Zealand: Tethys, v. 9, pp. 379–98.

———. 1992. Origin of Maxillopoda: Acta Zoologica, Stockholm, v. 73, pp. 319–22.

Nielsen, C. 1991. The development of the brachiopod *Crania (Neocrania) anomala* (O.F. Müller) and its phylogenetic significance: Acta Zoologica, Stockholm, v. 72, pp. 7–28.

Niven, B. S. 1989. Formalization of the Paterson concept of an animal species: Rivista di Biologia, v. 82, pp. 191–207.

Nixon, K. C., and Wheeler, Q. D. 1990. An amplification of the phylogenetic species concept: Cladistics, v. 6, pp. 211–23.

———. 1992. Extinction and the origin of species, *in* Novacek, M. J., and Wheeler, Q. D., eds., Extinction and Phylogeny: New York, Columbia University Press, pp. 119–43.

O Foighil, D., and Smith, M. J. 1995. Evolution of asexuality in the cosmopolitan marine clam *Lasaea*: Evolution, v. 49, pp. 140–50.

O'Hara, R. J. 1993. Systematic generalization, historical fate, and the species problem: Systematic Biology, v. 42, pp. 231–46.

———. 1996a. Trees of history in systematics and phylogeny: Memorie della Società Italiana di Scienze Naturali e del Museo Civico di Storia Naturale di Milano, v. 27, pp. 81–88.

———. 1996b. Mapping the space of time: temporal representation in the historical sciences: Memoirs of the California Academy of Sciences, n. 20, pp. 7–17.

Oken, L. 1805. Die Zeugung: Bamberg, Joseph Anton Goebhardt, viii + 216 pp.

———. 1807. Über die Bedeutung der Schädelknochen: Bamberg & Würzburg, J. a. Göbhardt, 18 pp.

———. 1847. Elements of Physiophilosophy: London, Ray Society, Translated by Alfred Tulk., xxx + 665 pp.; Translation of 1831 ed.

Oppenheimer, J. M. 1967. Essays in the History of Embryology and Biology: Cambridge, M.I.T. Press, ix + 374 pp.

Osborn, H. F. 1902. Homoplasy as a law of latent or potential homology: American Naturalist, v. 36, pp. 259–71.

———. 1917. The Origin and Evolution of Life: on the Theory of Action Reaction and Interaction of Energy: New York, Charles Scribner's Sons, xxi + 322 pp.

Osche, G. 1982. Rekapitulatonsentwicklung und ihre Bedeutung für die Phylogenetik— Wann gilt die "Biogenetische Grundregel?": Verhandlungen des Naturwissenschaftlichen Vereins in Hamburg, NF, v. 25, pp. 5–31.

Oudemans, A. C. 1920. Das Dollosche Gesetz der Irreversibilität: Archiv für Naturgeschichte, v. 86, pp. 1–10.

Owen, R. 1843. Lectures on the Comparative Anatomy and Physiology of the Invertebrate Animals, Delivered at the Royal College of Surgeons, in 1843: London, Longman, Brown, Green, and Longmans., i + 392 pp.

———. 1848. On the Archetype and Homologies of the Vertebrate Skeleton: London, Richard and John E. Taylor, vi + 203 pp.

———. 1849. On the Nature of Limbs. A Discourse Delivered on Friday, February 9, at an Evening Meeting of the Royal Institution of Great Britain: London, John van Voorst, ii + 119 pp.; 2 Plates.

———. 1866. On the Anatomy of Vertebrates: London, Longmans, Green, v. 2. Birds and Mammals., viii + 592 pp.

Padian, K., and Clemens, W. A. 1985. Terrestrial vertebrate diversity: episodes and insights, *in* ch. 2 *of* Valentine, J. W., ed., Phanerozoic Diversity Trends: Profiles in Macroevolution: Princeton, Princeton University Press, pp. 41–96.

Paley, W. 1802. Natural Theology: or, Evidences of the Existence and Attributes of the Deity, Collected from the Appearances of Nature: London, R. Faulder, xii + 586 pp.

Palmer, A. R., and Strathmann, R. R. 1981. Scale of dispersal in varying environments and its implications for life histories of marine invertebrates: Oecologia, v. 48, pp. 308–18.

Panchen, A. L. 1992. Classification, Evolution, and the Nature of Biology: Cambridge, Cambridge University Press, xi + 403 pp.

Paneth, F. 1963. The epistemological status of the concept of an element: British Journal for the Philosophy of Science, v. 13, pp. 1–14, 144–92.

Parker, E. D., Jr. 1979. Ecological implications of clonal diversity in parthenogenetic morphospecies: American Zoologist, v. 19, pp. 753–62.

Paterson, H. E. H. 1982. Perspectives on speciation by reinforcement: South African Journal of Science, v. 78, pp. 53–7.

———. 1985. The recognition concept of species, *in* Vrba, E. S., ed., Species and Speciation: Pretoria, Transvaal Museum, pp. 21–9.

————. 1988. On defining species in terms of sterility: problems and alternatives: Pacific Science, v. 42, pp. 65–71.

Patterson, C. 1978. Arthropods and ancestors: Antenna, v. 2, pp. 99–103.

————. 1982. Morphological characters and homology, *in* ch. 2 *of* Joysey, K. A., and Friday, A. E., eds., Problems of Phylogenetic Reconstruction: London, Academic Press, pp. 21–74.

————. 1983. How does phylogeny differ from ontogeny, *in* Goodwin, B. C., Holder, N., and Wylie, C. C., eds., Development and Evolution: Cambridge, Cambridge University Press, pp. 1–31.

Pederson, H. 1931. The Discovery of Language: Linguistic Science in the Nineteenth Century: Cambridge, Harvard University Press, x + 360 pp.

Pellegrin, P. 1986. Aristotle's Classification of Animals: Biology and the Conceptual Unity of the Aristotelian Corpus [2nd ed.]: Berkeley, University of California Press, Translated by Anthony Preus., xiv + 235 pp.

Pelseneer, P. 1895. Hermaphroditism in Mollusca: Quarterly Journal of Microscopical Science, NS, v. 37, pp. 19–46; Pl. 4–6.

Perrier, E. 1898. Les Colonies Animales et la Formation des Organismes [2nd ed.]: Paris, Masson et Cie, xxxii + 797 pp.

Peterson, C. R., and Beach, L. R. 1967. Man as an intuitive statistician: Psychological Bulletin, v. 68, pp. 29–46.

Petren, K., Bolger, D. T., and Case, T. J. 1993. Mechanisms in the competitive success of an invading sexual gecko over an asexual native: Science, v. 259, pp. 354–58.

Piaget, J. 1971. Insights and Illusions of Philosophy: New York, World Publishing Company, xvii + 232 pp.

Pinna, G. 1995. La Natura Paleontologica dell'Evoluzione: Torino, Giulio Einaudi, iii + 500 pp.

Pittendrigh, C. S. 1958. Adaptation, natural selection, and behavior, *in* ch. 18 *of* Roe, A., and Simpson, G. G., eds., Behavior and Evolution: New Haven, Yale University Press, pp. 390–416.

Platnick, N. I. 1979. Philosophy and the transformation of cladistics: Systematic Zoology, v. 28, pp. 537–46.

————. 1985. Philosophy and the transformation of cladistics revisited: Cladistics, v. 1, pp. 87–94.

Popper, K. R. 1945. The Open Society and its Enemies: London, Routledge & Sons, vii + 268 + v + 352 pp.

————. 1962. Conjectures and Refutations: the Growth of Scientific Knowledge: London, Basic Books, xi + 412 pp.

————. 1966. The Open Society and its Enemies [5th ed.]: London, Routledge & Kegan Paul, xi + 361 + v + 420 pp.

————. 1972. Objective Knowledge: an Evolutionary Approach: Oxford, Clarendon Press, x + 380 pp.

————. 1978. Natural selection and the emergence of mind: Dialectica, v. 32, pp. 339–55.

Priest, G. L. 1977. The common law process and the selection of efficient rules: The Journal of Legal Studies, v. 6, pp. 65–82.

Prior, E. W. 1985. What is wrong with etiological accounts of biological function?: Pacific Philosophical Quarterly, v. 66, pp. 310–38.

Provine, W. B. 1986. Sewall Wright and Evolutionary Biology: Chicago, University of Chicago Press, xvi + 545 pp.

Putnam, H. 1977. Is semantics possible?, *in* ch. 3 *of* Schwartz, S. P., ed., Naming, Necessity, and Natural Kinds: Ithaca, Cornell University Press, pp. 102–32.

Quattro, J. M., Avise, J. C., and Vrijenhoek, R. C. 1991. Molecular evidence for multiple origins of hybridogenetic fish clones (Poeciliidae: *Poeciliopsis*): Genetics, v. 127, pp. 391–98.

Quine, W. V. 1977. Natural kinds, *in* ch. 6 *of* Schwartz, S. P., ed., Naming, Necessity, and Natural Kinds: Ithaca, Cornell University Press, pp. 155–75.

Quiring, R., Walldorf, U., Kloter, U., and Gehring, W. J. 1994. Homology of the *eyeless* gene of *Drosophila* to the *small eye* gene in mice and *aniridia* in humans: Science, v. 265, pp. 785–89.

Radnitzky, G. 1992. Universal Economics: Assessing the Achievements of the Economic Approach: New York, Paragon House, xii + 446 pp.

Radnitzky, G., and Bernholz, P. 1986. Economic Imperialism: the Economic Approach Applied outside the Field of Economics: New York, Paragon House.

Raff, R. A., Field, K. G., Olsen, G. J., Giovannoni, S. J., Lane, D. J., Ghiselin, M. T., Pace, N. R., and Raff, E. C. 1989. Metazoan phylogeny based on analysis of 18S ribosomal RNA, *in* ch. 18 *of* Fernholm, B., Bremer, K., and Jörnvall, H., eds., The Hierarchy of Life: Molecules and Morphology in Phylogenetic Analysis: Amsterdam, Exerpta Medica, pp. 247–60.

Raikow, R. J. 1975. The evolutionary reappearance of ancestral muscles as developmental anomalies in two species of birds: Condor, v. 77, pp. 514–17.

Rainger, R. 1989. What's the use: William King Gregory and the functional morphology of fossil vertebrates: Journal of the History of Biology, v. 22, pp. 103–39.

————. 1991. An Agenda for Antiquity: Henry Fairfield Osborn & Vertebrate Paleontology at the American Museum of Natural History: Tuscaloosa, University of Alabama Press, xiii + 360 pp.

Raubenheimer, D., and Crowe, T. M. 1987. The recognition species concept: is it really an alternative?: South African Journal of Science, v. 83, pp. 530–34.

Reed, E. S. 1978. Darwin's evolutionary philosophy: the laws of change: Acta Biotheoretica, v. 27, pp. 201–35.

————. 1979. The role of symmetry in Ghiselin's "radical solution to the species problem": Systematic Zoology, v. 28, pp. 71–8.

————. 1981. The lawfulness of natural selection: American Naturalist, v. 118, pp. 61–71.

Regan, C. T. 1926. Organic evolution: Report of the British Association for the Advancement of Science, v. 1925, pp. 75–86.

Reid, T. 1785. Essays on the Intellectual Powers of Man: Edinburgh.

Remane, A. 1952. Die Grundlagen des natürlichen Systems, der vergleichenden Anatomie und der Phylogenetik [1st ed.]: Leipzig, Akademische Verlagsgesellschaft Geest & Portig; Second edition 1956.

————. 1959. Die Geschichte der Tiere, in Heberer, G., ed., Die Evolution der Organismen: Ergebnisse und Probleme der Astammungslehre [2nd ed.]: Stuttgart, Gustav Fischer Verlag, v. 1, pp. 340–422.

————. 1960. Die Beziehungen Zwischen Phylogenie und und Ontogenie: Zoologischer Anzeiger, v. 164, pp. 306–37.

————. 1963. The evolution of the Metazoa from colonial flagellates vs. plasmodial ciliates, in ch. 2 of Dougherty, E. C., ed., The Lower Metazoa: Comparative Biology and Phylogeny: Berkeley, University of California Press, pp. 23–32.

Rensch, B. 1939. Typen der Artbildung: Biological Reviews of the Cambridge Philosophical Society, v. 14, pp. 186–222.

————. 1959. Evolution above the Species Level: New York, Columbia University Press, xvii + 419 pp.

Ricci, C. 1992. Rotifera: parthenogenesis and heterogony, in Dallai, R., ed., Sex Origin and Evolution: Modena, Mucchi, pp. 329–41.

Richardson, R. C., and Kane, T. C. 1988. Orthogenesis and evolution in the 19th century, in Nitecki, M. H., ed., Evolutionary Progress: Chicago, University of Chicago Press, pp. 149–68.

Ridley, M. 1989. The cladistic solution to the species problem: Biology and Philosophy, v. 4, pp. 1–16.

———. 1990. Comments on Wilkinson's commentary: Biology and Philosophy, v. 5, pp. 447–50.

Riedl, R. 1979. Order in Living Organisms: a Systems Analysis of Evolution: New York, John Wiley & Sons, xx + 313 pp.

Rieger, R. M., Haszprunar, G., and Schuchert, P. 1991. On the origin of the Bilateria: traditional views and recent alternative concepts, *in* Simonetta, A. M., and Conway Morris, S., eds., The Early Evolution of Metazoa and the Significance of Problematic Taxa: Cambridge, Cambridge University Press, pp. 107–12.

Rieppel, O. 1979. Ontogeny and the recognition of primitive character states: Zeitschrift für zoologische Systematik und Evolutions-Forschung, v. 17, pp. 57–61.

———. 1986. Species as individuals: a review and critique of the argument: Evolutionary Biology, v. 20, pp. 283–317.

———. 1992. Homology and logical fallacy: Journal of Evolutionary Biology, v. 5, pp. 701–15.

———. 1993. The conceptual relationship of ontogeny, phylogeny, and classification: the taxic approach: Evolutionary Biology, v. 27, pp. 1–32.

———. 1994. Homology, topology, and typology, *in* ch. 2 *of* Hall, B. K., ed., Homology: the Hierarchical Basis of Comparative Biology: San Diego, Academic Press, pp. 62–100.

Robinson, R. 1954. Definition: Oxford, Clarendon Press, viii + 207 pp.

Robson, G. C. 1928. The Species Problem: an Introduction to the Study of Evolutionary Divergence in Natural Populations: Edinburgh, Oliver and Boyd, vii + 283 pp.

Rodrigues, P. D. 1986. On the term character: Systematic Zoology, v. 35, pp. 140–41.

Romer, A. S. 1945. Vertebrate Paleontology [2nd ed.]: Chicago, University of Chicago Press, ix + 687 pp.

———. 1949. Time series and trends in animal evolution, *in* ch. 7 *of* Jepsen, G. L., Mayr, E., and Simpson, G. G., eds., Genetics, Paleontology, and Evolution: Princeton, Princeton University Press, pp. 103–20.

Rorty, R. 1961. Pragmatism, categories, and language: Philosophical Review, v. 70, pp. 197–233.

Rosch, E. 1978. Principles of categorization, *in* ch. 2 *of* Rosch, E., and Lloyd, B. B., eds., Cognition and Categorization: Hillsdale, Lawrence Erlbaum Associates, pp. 99–133.

Rosen, D. E. 1978. Vicariant patterns and historical explanation in biogeography: Systematic Zoology, v. 27, pp. 159–88.

Rosenberg, A. 1987. Why does the nature of species matter? Comments on Ghiselin and Mayr: Biology and Philosophy, v. 2, pp. 192–97.

————. 1989. From reductionism to instrumentalism?, *in* Ruse, M., ed., What the Philosophy of Biology Is: Essays Dedicated to David Hull: Dordrecht, Kluwer Academic Publishers, pp. 245–62.

Rosenberg, A., and Williams, M. 1986. Fitness as primitive and propensity: Philosophy of Science, v. 53, pp. 412–18.

Rosenblatt, R. H., and Wapples, R. S. 1986. A genetic comparision of allopatric populations of shore fish species from the eastern and central Pacific Ocean: dispersal and vicariance: Copeia, v. 1982, pp. 275–84.

Roth, E. M., and Shoben, E. J. 1983. The effect of context on the structure of categories: Cognitive Psychology, v. 15, pp. 346–78.

Roth, V. L. 1984. On homology: Biological Journal of the Linnaean Society, v. 22, pp. 13–29.

————. 1988. The biological basis of homology, *in* ch. 1 *of* Humphries, C. J., ed., Ontogeny and Systematics: New York, Columbia University Press, pp. 1–26.

————. 1994. Within and between organisms: replicators, lineages, and homologues, *in* ch. 9 *of* Hall, B. K., ed., Homology: the Hierarchical Basis of Comparative Biology: San Diego, Academic Press, pp. 301–37.

Rowe, F. W. E., Baker, A. N., and Clark, H. E. S. 1988. The morphology, development and taxonomic status of *Xyloplax* Baker, Rowe and Clark (1986) (Echinodermata: Concentricycloidea), with the description of a new species: Proceedings of the Royal Society of London, v. B233, pp. 431–59; Plates 1–7.

Rowe, T. 1987. Definition and diagnosis in the phylogenetic system: Systematic Zoology, v. 36, pp. 208–11.

————. 1988. Definition, diagnosis, and origin of Mammalia: Journal of Vertebrate Paleontology, v. 8, pp. 241–64.

————. 1996. Brain heterochrony and origin of the mammalian ear: Memoirs of the California Academy of Sciences, n. 20, pp. 71–95.

Rowe, T., and Gauthier, J. 1992. Ancestry, paleontology, and the definition of the name Mammalia: Systematic Biology, v. 41, pp. 372–78.

Rudwick, M. J. S. 1964. The inference of function from structure in fossils: British Journal for the Philosophy of Science, v. 15, pp. 27–40.

Ruhlen, M. 1987. A Guide to the World's Languages: Stanford, Stanford University Press, v. 1: Classification., xxv + 433 pp.

Rupke, N. A. 1994. Richard Owen: Victorian Naturalist: New Haven, Yale University Press, xviii + 462 pp.

Ruse, M. 1970. Are there laws in biology?: Australasian Journal of Philosophy, v. 48, pp. 234–46.

———. 1973. The Philosophy of Biology: London, Hutchinson University Library, 231 pp.

———. 1987. Species: natural kinds, individuals, or what?: British Journal for the Philosophy of Science, v. 38, pp. 225–42.

———. 1988. Philosophy of Biology Today: Albany, State University of New York Press, x + 155 pp.

Russell, B. 1959. My Philosophical Development: New York, Simon and Schuster, 279 pp.

Ryan, M. J. 1990. Signals, species, and sexual selection: American Scientist, v. 78, pp. 46–78.

Ryle, G. 1949. The Concept of Mind: London, Hutchinson, 334 pp.

———. 1953. Categories, in Flew, A. G. N., ed., Logic and Language: Oxford, Oxford University Press, pp. 65–81.

Salmon, N. U. 1981. Reference and Essence: Princeton, Princeton University Press, xvi + 293 pp.

Salthe, S. N. 1985. Evolving Hierarchical Systems: Their Structure and Representation: New York, Columbia University Press, x + 343 pp.

Salvini-Plawen, L. von. 1980. Phylogenetischer Status und Bedeutung der mesenchymaten Bilateria: Zoologischer Jahrbücher Abteilung für Anatomie, v. 103, pp. 354–73.

Salvini-Plawen, L. von, and Mayr, E. 1977. On the evolution of photoreceptors and eyes: Evolutionary Biology, v. 10, pp. 207–63.

Sander, K. 1982. Rekapitulation aus der Sicht eines Entwicklungsphysiologien: Die konservierende Rolle funktioneller Verknüpfung in der Ontogenese: Verhandlungen des Naturwissenschaftlichen Vereins in Hamburg, NF, v. 25, pp. 35–50.

Sanderson, M. J. 1993. Reversibility in evolution: a maximum likelihood approach to character gain loss bias in phylogenies: Evolution, v. 47, pp. 236–52.

Schaffer, B. 1952. Rates of evolution in the coelacanth and dipnoan fishes: Evolution, v. 6, pp. 101–11.

Schindewolf, O. H. 1950. Grundfragen der Paläontologie: Stuttgart, E. Schweizerbart'sche Verlagsbuchhandlung, 506 pp.

Schleicher, A. 1869. Darwinism Tested by the Science of Language: London, John Camden Hotten, Transl. Alex. V.W. Bikkers., 69 pp.

Schoch, R. M. 1986. Phylogeny Reconstruction in Paleontology: New York, Van Nostrand Reinhold, xii + 353 pp.

Schopf, J. M. 1960. Emphasis on holotype (?): Science, v. 131, pp. 1043.

Schopf, T. J. M. 1982. A critical assessment of punctuated equilibria I. Duration of taxa: Evolution, v. 36, pp. 1144–57.

Schwann, T. 1839. Mikroskopische Untersuchungen über die Uebereinstimmung in der Struktur und dem Wachstum der Thiere und Pflanzen: Berlin, Sanders, xviii + 270 pp.; Pl. I–IV.

Searle, J. R. 1958. Proper names: Mind, v. 67, pp. 166–73.

————. 1960. Speech Acts: an Essay in the Philosophy of Language: Cambridge, Cambridge University Press, viii + 203 pp.

Seger, J., and Hamilton, W. D. 1988. Parasites and sex, *in* ch. 11 *of* Michod, R. E., and Levin, B. R., eds., The Evolution of Sex: an Examination of Current Ideas: Sunderland, Sinauer Associates, pp. 176–93.

Settle, T. 1993. "Fitness" and "altruism": traps for the unwary, bystander and biologist alike: Biology and Philosophy, v. 8, pp. 61–83.

Sewertzoff, A. N. 1924. Die Factoren der progressiven Entwickelung der niederen Wirbeltiere: Zoologicheskii Zhurnal, v. 4, pp. 12–60.

————. 1927. Über die Beziehungen zwischen der Ontogenese und der Phylogenese der Tiere: Jenaische Zeitschrift für Naturwissenschaft, v. 63, pp. 51–180.

————. 1931. Morphologische Gesetzmässigkeiten der Evolution: Jena, Verlag von Gustav Fischer, xiv + 371 pp.

Shapere, D. 1982. The concept of observation in science and philosophy: Philosophy of Science, v. 49, pp. 485–526.

Shapiro, K. J., and Alexander, I. E. 1975. The Experience of Introversion. An Integration of Phenomenological, Empirical, and Jungian Approaches: Durham, Duke University Press, xii + 180 pp.

Sharvy, R. 1983. Mixtures: Philosophy and Phenomenological Research, v. 44, pp. 227–39.

Sheldon, P. R. 1987. Parallel gradualistic evolution of Ordovician trilobites: Nature, v. 330, pp. 561–63.

Shields, W. M. 1982. Philopatry, Inbreeding, and the Evolution of Sex: Albany, State University of New York Press, ix + 245 pp.

Short, L. L. 1969. "Suture-zones," secondary contacts, and hybridization: Systematic Zoology, v. 18, pp. 458–60.

———. 1970. A reply to Uzzell and Ashmole: Systematic Zoology, v. 19, pp. 199–202.

Shrock, R. R. 1948. Sequence in Layered Rocks: a Study of Features and Structures Useful for Determining Top and Bottom or Order of Succession in Bedded and Tabular Rock Bodies: New York, McGraw-Hill, xiii + 507 pp.

Shulman, M. J., and Bermingham, E. 1995. Early life histories, ocean currents, and the population genetics of Caribbean reef fishes: Evolution, v. 49, pp. 897–910.

Siewing, R. 1972. Zur Deszendenz der Chordaten—Erwiderung und Versuch einer Geschichte der Archicoelomaten: Zeitschrift für zoologische Systematik und Evolutions-Forschung, v. 10, pp. 269–71.

———. 1979. Homology of cleavage types?, *in* Siewing, R., ed., Ontogenese und Phylogenese: Hamburg, Verlag Paul Parey, pp. 7–18.

———. 1980. Körpergliederung und phylogenetisches System: Zoologischer Jahrbücher Abteilung für Anatomie, v. 103, pp. 196–210.

Signor, P. W., III. 1985. Real and apparent trends in species richness through time, *in* ch. 4 *of* Valentine, J. W., ed., Phanerozoic Diversity Trends: Profiles in Macroevolution: Princeton, Princeton University Press, pp. 129–50.

Signor, P. W., III, and Lipps, J. H. 1982. Sampling bias, gradual extinction patterns and catastrophes in the fossil record, *in* Silver, L. T., and Schultz, P. H., eds., Geological Implications of Impacts of Large Asteroids and Coments on the Earth, 190 *of* Geological Society of America Special Paper: pp. 291–96.

Simon, H. A. 1962. The architecture of complexity: Proceedings of the American Philosophical Society, v. 106, pp. 467–82.

Simons, P. 1987. Parts: a Study in Ontology: Oxford, Oxford University Press, xiii + 390 pp.

Simpson, G. G. 1944. Tempo and Mode in Evolution: New York, Columbia University Press, xi + 237 pp.

———. 1945. The principles of classification and a classification of mammals: Bulletin of the American Museum of Natural History, v. 85, pp. xvi + 1–350.

———. 1951. The species concept: Evolution, v. 5, pp. 285–98.

———. 1953. The Major Features of Evolution: New York, Columbia University Press, xx + 434 pp.

———. 1961. Principles of Animal Taxonomy: New York, Columbia University Press.

Slatkin, M. 1985. Gene flow in natural populations: Annual Review of Ecology and Systematics, v. 16, pp. 393–430.

Sloan, P. R. 1986. From logical universals to historical individuals: Buffon's idea of biological species, *in* ch. 6 *of* Anonymous, ed., Histoire du Concept d'Espèce dans les Sciences de la Vie: Paris, Fondation Singer-Polignac, pp. 101–40.

Smart, J. 1940. Entomological systematics examined as a practical problem, *in* Huxley, J., ed., The New Systematics: Oxford, Oxford University Press, pp. 475–92.

Smart, J. J. C. 1963. Philosophy and Scientific Realism: London, Routledge & Kegan Paul, viii + 160 pp.

Smit, P. 1961. Ontogenesis and Phylogenesis: their Interrelation and their Interpretation: Leiden, E. J. Brill, vi + 103 pp.

Smith, A. 1776. An Inquiry into the Nature and Causes of the Wealth of Nations: London, W. Strahan and T. Cadell, xii + 510 + iv + 587 pp.

Smith, A. B., and Patterson, C. 1988. The influence of taxonomic method on the perception of patterns of evolution: Evolutionary Biology, v. 23, pp. 127–216.

Smith, E. E., and Medin, D. L. 1981. Categories and Concepts: Cambridge, Harvard University Press, viii + 203 pp.

Smith, L. B. 1979. Perceptual development and category generalization: Child Development, v. 50, pp. 705–15.

Sneath, P. H. A. 1995. Thirty years of numerical taxonomy: Systematic Biology, v. 44, pp. 281–98.

Sneath, P. H. A., and Sokal, R. R. 1973. Numerical Taxonomy: the Principles and Practice of Numerical Classification: San Francisco, W. H. Freeman and Company, xv + 573 pp.

Sober, E. 1980. Evolution, population thinking, and essentialism: Philosophy of Science, v. 47, pp. 350–83.

———. 1984. Sets, species, and evolution: comments on Philip Kitcher's "Species": Philosophy of Science, v. 51, pp. 334–41.

———. 1988. Reconstructing the Past: Parsimony, Evolution, and Inference: Cambridge, MIT Press, xv + 265 pp.

———. 1992. Monophyly, *in* Keller, E. F., and Lloyd, E. A., eds., Keywords in Evolutionary Biology: Cambridge, Harvard University Press, pp. 202–08.

Sokal, R. R., and Sneath, P. H. 1963. Principles of Numerical Taxonomy [1st ed.]: San Francisco, W. H. Freeman, 359 pp.

Spencer, H. 1850. Social Statics; or, the Conditions Essential to Human Happiness Specified, and the First of them Developed: London.

Spencer, H. G., McArdle, B., and Lambert, D. M. 1986. A theoretical investigation of speciation by reinforcement: American Zoologist, v. 128, pp. 241–62.

Stanley, S. M. 1973. An explanation for Cope's rule: Evolution, v. 27, pp. 1–26.

———. 1975a. Clades versus clones in evolution: why we have sex: Science, v. 190, pp. 382–83.

———. 1975b. A theory of evolution above the species level: Proceedings of the National Academy of Sciences USA, v. 72, pp. 646–50.

———. 1979. Macroevolution: Pattern and Process: San Francisco, W.H. Freeman and Company, xi + 332 pp.

Starr, P. 1992. Social categories and claims in the liberal state, *in* ch. 8 *of* Douglas, M., and Hull, D., eds., How Classification Works: Nelson Goodman among the Social Sciences: Edingurgh, Edinburgh University Press, pp. 154–79.

Steadman, P. 1979. The Evolution of Designs. Biological Analogy in Architecture and the Applied Arts: Cambridge, Cambridge University Press, xi + 276 pp.

Stebbins, G. L. 1987. Species concepts: semantics and actual situations: Biology and Philosophy, v. 2, pp. 198–203.

Steiner, H. 1966. Atavismen bei Artbastarden und ihre Bedeutung zur Feststellung von Verwandtschaftsbeziehungen. Kreuzungsergebnisse innerhalb der Sinfvogelfamilie der *Spermestidae*: Revue Suisse de Zoologie, v. 73, pp. 321–37.

Stern, C. 1949. Gene and character, *in* ch. 2 *of* Jepsen, G. L., Mayr, E., and Simpson, G. G., eds., Genetics, Paleontology, and Evolution: Princeton, Princeton University Press, pp. 13–22.

Stewart, J. E. 1993. The maintenance of sex: Evolutionary Theory, v. 10, pp. 195–202.

Storch, V., and Welsch, U. 1972. Über Bau und Entstehung der Mantelrandstacheln von *Lingula unguis* L. (Brachiopoda): Zeitschrift für wissenschaftliche Zoologie, v. 183, pp. 181–89.

Strathmann, R. 1974. The spread of sibling larvae of sedentary marine invertebrates: American Naturalist, v. 108, pp. 29–44.

———. 1978. The evolution and loss of feeding larval stages of marine invertebrates: Evolution, v. 32, pp. 894–906.

———. 1993. Hypotheses on the origins of marine larvae: Annual Review of Ecology and Systematics, v. 24, pp. 89–117.

Strauss, E. G., Strauss, J. H., and Levine, A. J. 1990. Virus evolution, *in* ch. 9 *of* Fields, B. N., and Knipe, D. M., eds., Virology [2nd ed.]: New York, Raven Press, pp. 167–97.

Strawson, P. F. 1959. Individuals: an Essay in Descriptive Metaphysics: London, Methuen, 255 pp.

Strickland, H. E., Phillips, J., Richardson, J., Owen, R., Jenyns, L., Broderip, W. J., Henslow, J. S., Shuckard, W. E., Waterhouse, G. R., Yarrell, W., Darwin, C., and Westwood, J. O. 1843. Report of a committee appointed to consider of the rules by which the nomenclature of zoology may be established on a uniform and permanent basis: Report of the British Association for the Advancement of Science, v. 12, pp. 104–21; Manchester, June, 1842.

Sturtevant, A. H., and Dobzhansky, T. 1936. Inversions in the third chromosome of wild races of *Drosophila pseudo-obscura*, and their use in the study of the history of the species: Proceedings of the National Academy of Sciences USA, v. 22, pp. 448–50.

Swedmark, B. 1964. The interstitial fauna of marine sand: Biological Reviews of the Cambridge Philosophical Society, v. 39, pp. 1–42.

Szalay, F. S. 1993. Species concepts: the tested, the untestable, and the redundant, *in* ch. 2 *of* Kimbel, W. H., and Martin, L. B., eds., Species, Species Concepts, and Primate Evolution: New York, Plenum Press, pp. 21–41.

Szalay, F. S., and Bock, W. J. 1991. Evolutionary theory and systematics: relationships between process and pattern: Zeitschrift für zoologische Systematik und Evolutions-Forschung, v. 29, pp. 1–39.

Szidat, L. 1956. Geschichte, Anwendung und einige Folgerungen aus den parasitogenetischen Regeln: Zeitschrift für Parasitenkunde, v. 17, pp. 237–68.

Taub, L. 1993. Evolutionary ideas and "empirical" methods: the analogy between language and species in works by Lyell and Schleicher: British Journal for the History of Science, v. 26, pp. 171–93.

Templeton, A. R. 1980. Modes of speciation and inferences based on genetic distances: Evolution, v. 34, pp. 719–29.

———. 1989. The meaning of species and speciation: a genetic perspective, *in* ch. 1 *of* Otte, D., and Endler, J. A., eds., Speciation and its Consequences: Sunderland, Sinauer Associates, pp. 3–27.

Thompson, J. V. 1828–1834. Zoological Researches, and Illustrations; or, Natural History of Nondescript or Imperfectly Known Amimals, in a Series of Memoirs, Illustrated by numerous Figures: Cork, King and Ridings.

———. 1835. Discovery of the metamorphosis in the second type of the Cirripedes, viz. the Lepades, completing the natural history of these singular animals, and confirming their affinity with the Crustacea: Philosophical Transactions of the Royal Society of London, v. 125, pp. 355–58.

Thompson, M. 1967. Categories, *in* Edwards, P., ed., The Encyclopedia of Philosophy: New York, Macmillan, v. 2, pp. 46–55.

Thompson, P. 1989. The Structure of Biological Theories: Albany, State University of New York Press, x + 148 pp.

Thorpe, W. H. 1940. Ecology and the future of systematics, *in* Huxley, J., ed., The New Systematics: Oxford, Oxford University Press, pp. 341–64.

Tomlinson, J. T. 1966. The advantages of hermaphroditism and parthenogenesis: Journal of Theoretical Biology, v. 11, pp. 54–8.

Turner, J. R. G. 1967. Why does the genotype not congeal?: Evolution, v. 21, pp. 645–56.

Tylor, E. B. 1871. Primitive Culture: Researches into the Development of Mythology, Philosophy, Religion, Art, and Custom: London, John Murray, ix + 453 + viii + 426 pp.

Urbanek, A. 1996. The origin and maintenance of diversity: a case study of Upper Silurian graptoloids: Memorie della Società Italiana di Scienze Naturali e del Museo Civico di Storia Naturale di Milano, v. 27, pp. 119–27.

Uzell, T., and Ashmole, N. P. 1970. Suture zones: an alternative view: Systematic Zoology, v. 19, pp. 197–99.

Valentine, J. W., and Jablonski, D. 1983. Larval adaptations and patterns of brachiopod diversity in space and time: Evolution, v. 37, pp. 1052–61.

———. 1986. Mass extinctions: sensitivity of marine larval types: Proceedings of the National Academy of Sciences USA, v. 83, pp. 6912–14.

Valentine, J. W., and May, C. M. 1996. Hierarchies in biology and paleontology: Paleobiology, v. 22, pp. 23–33.

Van der Hammen, L. 1989. Structuralism in evolutionary biology and systematics, *in* Goodwin, B., Sibatani, A., and Webster, G., eds., Dynamic Structures in Biology: Edinburgh, Edinburgh University Press, pp. 131–42.

Van der Steen, W. 1991. Natural selection as natural history: Biology and Philosophy, v. 6, pp. 41–4.

———. 1994. New ways to look at fitness: History and Philosophy of Life Sciences, v. 16, pp. 479–92.

Van Fraassen, B. C. 1989. Laws and Symmetry: Inventory, Clarendon Press, xv + 395 pp.

Van Holde, K. E., Miller, K. I., and Lang, W. H. 1992. Molluscan hemocyanins: structure and function: Advances in Comparative and Environmental Physiology, v. 13, pp. 257–300.

Van Valen, L. 1973. A new evolutionary law: Evolutionary Theory, v. 1, pp. 1–30.

————. 1976. Individualistic classes: Philosophy of Science, v. 43, pp. 539–48; Actual year of publication 1977.

————. 1982. Homology and causes: Journal of Morphology, v. 173, pp. 305–12.

Van Valen, L. M., and Maiorana, V. C. 1991. HeLa, a new microbial species: Evolutionary Theory, v. 10, pp. 71–4.

Vavilov, N. I. 1922. The law of homologous series in variation: Journal of Genetics, v. 12, pp. 47–89.

Vermeij, G. J. 1978. Biogeography and Adaptation: Patterns of Marine Life: Cambridge, Harvard University Press, xiii + 332 pp.

————. 1987. Evolution and Escalation: an Ecological History of Life: Princeton, Princeton University Press, xv + 527 pp.

————. 1995. Economics, volcanoes, and Phanerozoic revolutions: Paleobiology, v. 21, pp. 125–52.

Verrell, P. A. 1988. Stabilizing selection, sexual selection, and speciation: a view of specific mate recognition systems: Systematic Zoology, v. 37, pp. 209–15.

Virchow, R. 1860. Cellular Pathology as Based upon Physiological and Pathological Histology: London, John Churchill, Translated from the Second Edition by Frank Chace., xviii + 511 pp.

Vrba, E. S., and Eldredge, N. 1984. Individuals, hierarchies, and processes: towards a more complete evolutionary theory: Paleobiology, v. 10, pp. 146–71.

Vrijenhoek, R. C. 1989a, Genetic and ecological constraints on the origins and establishment of unisexual vertebrates, in Dawley, R. M., and Bogart, J. P., eds., Evolution and Ecology of Unisexual Vertebrates: Albany, The New York State Museum, pp. 24–31.

————. 1989b, Genotypic diversity and coexistence among sexual and clonal lineages of *Poeciliopsis*, in ch. 16 of Otte, D., and Endler, J. A., eds., Speciation and its Consequences: Sunderland, Sinauer Associates, pp. 386–400.

Vrijenhoek, R. C., Dawley, R. M., Cole, C. J., and Bogart, J. C. 1989. A list of the known unisexual vertebrates, in Dawley, R. M., and Bogart, J. P., eds., Evolution and Ecology of Unisexual Vertebrates: Albany, The New York State Museum, pp. 19–23.

Wagner, G. P. 1982. The logical structure of irreversible systems transformations: a theorem concerning Dollo's law and chaotic movement: Journal of Theoretical Biology, v. 96, pp. 337–46.

————. 1989. The biological homology concept: Annual Review of Ecology and Systematics, v. 20, pp. 51–69.

————. 1994. Homology and the mechanisms of development, *in* ch. 8 *of* Hall, B. K., ed., Homology: the Hierarchical Basis of Comparative Biology: San Diego, Academic Press, pp. 273–99.

————. 1996. Homologues, natural kinds and the evolution of modularity: American Zoologist, v. 36, pp. 36–43.

Waisbren, S. J. 1988. The importance of morphology in the evolutionary synthesis as demonstrated by the contributions of the Oxford group: Goodrich, Huxley, and de Beer: Journal of the History of Biology, v. 21, pp. 291–330.

Wake, D. B. 1996. Evolutionary developmental biology—prospects for an evolutionary synthesis at the developmental level: Memoirs of the California Academy of Sciences, n. 20, pp. 97–107.

Wake, M. H. 1992. Morphology, the study of form and function, in modern evolutionary biology: Oxford Surveys in Evolutionary Biology., v. 8, pp. 288–346.

Wallace, A. F. C. 1961. On being just complicated enough: Proceedings of the National Academy of Sciences USA, v. 47, pp. 458–64.

Walsh, W. H. 1967. An Introduction to the Philosophy of History [3rd ed.]: London, Hutchinson University Library, 215 pp.

Wang, W. S.-Y. 1987. Representing language relationsips, *in* ch. 13 *of* Hoenigswald, H. M., and Wiener, L. F., eds., Biological Metaphor and Cladistic Classification: an Interdisciplinary Perspective: Philadelphia, University of Pennsylvania Press, pp. 243–56.

Webster, G. 1989. Structuralism and Darwinism: concepts for the study of form, *in* Goodwin, B., Sibatani, A., and Webster, G., eds., Dynamic Structures in Biology: Edinburgh, Edinburgh University Press, pp. 1–15.

Weismann, A. 1904. Vorträge über Deszendenztheorie gehalten an der Universität Freiburg im Breisgau [2nd ed.]: Jena, Verlag von Gustav Fischer, xii + 340 + vi + 344 pp.; Taf I–III.

Weston, P. H. 1988. Indirect and direct methods in systematics, *in* ch. 2 *of* Humphries, C. J., ed., Ontogeny and Systematics: New York, Columbia University Press, pp. 27–56.

Wheeler, Q. D. 1986. Character weighting and cladistic analysis: Systematic Zoology, v. 35, pp. 102–09.

White, C. S., Michaux, B., and Lambert, D. M. 1990. Species and Neo-Darwinism: Systematic Zoology, v. 39, pp. 399–413.

White, M. 1965. Foundations of Historical Knowledge: New York, Harper & Row, ix + 299 pp.

Whitney, W. D. 1868. Language and the Study of Language: Twelve Lectures on the Principles of Linguistic Science: New York, Charles Scribner & Company, xi + 505 pp.

———. 1883. The Life and Growth of Language [4th ed.]: London, Kegan Paul, Trench & Co., vii + 526 pp.

Whitten, J. M. 1959. The tracheal system as a systematic character in larval Diptera: Systematic Zoology, v. 8, pp. 130–39.

Wiener, P. P. 1949. Evolution and the Founders of Pragmatism: Harvard, Harvard University Press, xiv + 288 pp.

Wiggins, D. 1980. Sameness and Substance: Cambridge, Harvard University Press, xi + 238 pp.

Wiley, E. O. 1975. Karl R. Popper, systematics, and classification—a reply to Walter Bock and other evolutionary systematists: Systematic Zoology, v. 24, pp. 233–43.

———. 1978. The evolutionary species concept reconsidered: Systematic Zoology, v. 27, pp. 17–26.

———. 1980. Is the evolutionary species fiction? A consideration of classes, individuals, and historical entities: Systematic Zoology, v. 29, pp. 76–80.

———. 1981. Phylogenetics: the Theory and Practice of Phylogenetic Systematics: New York, John Wiley and Sons, xv + 439 pp.

Wilkinson, M. 1990. A commentary on Ridley's cladistic solution to the species problem: Biology and Philosophy, v. 5, pp. 433–46.

Williams, G. C. 1966. Adaptation and Natural Selection: a Critique of Some Current Evolutionary Thought: Princeton, Princeton University Press, x + 307 pp.

———. 1975. Sex and Evolution, 8 *of* Monographs in Population Biology: Princeton, Princeton University Press, x + 200 pp.

———. 1992. Natural Selection: Domains, Levels, and Challenges: New York, Oxford University Press, x + 208 pp.

Williams, M. B. 1970. Deducing the consequences of evolution: a mathematical model: Journal of Theoretical Biology, v. 29, pp. 343–85.

———. 1973. Falsifiable predictions of evolutionary theory: Philosophy of Science, v. 40, pp. 518–37.

———. 1981. Is biology a different type of science?, *in* Sumner, Slater, and Wilson, eds., Pragmatism and Purpose: Essays Presented to Thomas A. Goudge: Toronto, University of Toronto Press, pp. 279–88.

———. 1985. Species are individuals: theoretical foundations for the claim: Philosophy of Science, v. 52, pp. 578–89.

———. 1989. Evolvers are individuals: extension of the species as individuals claim, *in* Ruse, M., ed., What the Philosophy of Biology Is: Essays Dedicated to David Hull: Dordrecht, Kluwer Academic Publishers, pp. 301–08.

———. 1992. Species: current usages, *in* Keller, E. F., and Lloyd, E. A., eds., Keywords in Evolutionary Biology: Cambridge, Harvard University Press, pp. 318–23.

Willmann, R. 1989. Evolutionary or biological species?: Abhandlungen des Naturwissenschaftlichen Vereins in Hamburg, v. NF28, pp. 95–110.

Wills, M. A., Briggs, D. E. G., and Fortey, R. A. 1994. Disparity as an evolutionary index: a comparison of Cambrian and Recent arthropods: Paleobiology, v. 20, pp. 93–130.

Wilson, B. E. 1995. A (not-so-radical) solution to the species problem: Biology and Philosophy, v. 10, pp. 339–56.

Wilson, D. S. 1983. The group selection controversy: history and present status: Annual Review of Ecology and Systematics, v. 14, pp. 159–87.

———. 1992. Group selection, *in* Keller, E. F., and Lloyd, E. A., eds., Keywords in Evolutionary Biology: Cambridge, Harvard University Press, pp. 137–44.

Wilson, D. S., and Sober, E. 1989. Reviving the superorganism: Journal of Theoretical Biology, v. 136, pp. 337–56.

Winnepenninckx, B., Backeljau, T., and De Wachter, R. 1995. Phylogeny of protostome worms derived from 18S rRNA sequences: Molecular Biology and Evolution, v. 12, pp. 641–49.

Winter, S. G. 1964. Economic "natural selection" and the theory of the firm: Yale Economic Essays, v. 4, pp. 225–72.

Wittgenstein, L. 1961. Tractatus Logico-Philosophicus: London, Routledge & Kegan Paul, xii + 166 pp.

Wolff, C. F. 1896. Theoria Generationis: Leipzig, Verlag von Wilhelm Engelmann, Translated by Paul Samassa, 95 + 97 pp.; Taf. I–II.

Wood, R. 1990. Reef-building sponges: American Scientist, v. 78, pp. 224–35.

Woodger, J. H. 1929. Biological Principles: a Critical Study: London, Routledge & Kegan Paul, xii + 496 pp.

———. 1945. On biological transformations, *in* Le Gros Clark, W. E., and Medawar, P. B., eds., Essays on Growth and Form Presented to D'A. W. Thompson: London, Oxford University Press, pp. 95–120.

Wright, L. 1973. Functions: Philosophical Review, v. 82, pp. 139–68.

Wright, S. 1931. Evolution in Mendelian populations: Genetics, v. 16, pp. 97–159.

———. 1932. The roles of mutation, inbreeding, crossbreeding and selection in evolution. Proceedings of the Sixth International Congress of Genetics 1, pp. 356–66.

———. 1977. Evolution and the Genetics of Populations: Chicago, University of Chicago Press, v. 3, v + 613 pp.

———. 1982. Character change, speciation, and the higher taxa: Evolution, v. 36, pp. 427–43.

Wynne-Edwards, V. C. 1962. Animal Dispersion in Relation to Social Nehaviour: New York, Hafner Publishing Company, xi + 653 pp.

Young, B. A. 1993. On the necessity of an archetypal concept in morphology: with special reference to the concepts of "structure" and "homology": Biology and Philosophy, v. 8, pp. 225–48.

Young, J. C. 1978. Illness categories and action strategies in a Tarascan town: American Ethnologist, v. 5, pp. 81–97.

Zadeh, L. A. 1982. A note on prototype theory and fuzzy sets: Cognition, v. 12, pp. 291–97.

Zanzi, L. 1991. Dalla Storia all'Epistemologia: lo Storicismo Scientifico. Principi di una Teoria della Storicizzazione: Milano, Jaca Book, vii + 480 pp.

Ziff, P. 1977. About proper names: Mind, v. 76, pp. 319–32.

Zink, S. 1963. The meaning of proper names: Mind, v. 72, pp. 401–99.

INDEX